RFID
HANDBOOK

RFID HANDBOOK

Radio-Frequency Identification Fundamentals and Applications

KLAUS FINKENZELLER

Giesecke & Devrient GmbH, Munich, Germany

Translated by
Rachel Waddington
Swadlincote, UK

JOHN WILEY & SON, LTD
Chichester • New York • Weinheim • Brisbane • Singapore •Toronto

First published under the title *RFID-Handbuch* by Carl Hanser Verlag

National 01243 779777
International (+44) 1243 779777

e-mail (for orders and customer service enquiries): cs-books@wiley.co.uk

Visit our Home Page on http://www.wiley.co.uk or http://www.wiley.com

Other Wiley Editorial Offices

John Wiley & Sons, Inc., 605 Third Avenue,
New York, NY 10158-0012, USA

Weinheim • Brisbane • Singapore • Toronto

Library of Congress Cataloging-in-Publication Data

Finkenzeller, Klaus.
[RFID Handbuch. English]
RFID handbook : radio-frequency identification fundamentals and applications / Klaus Finkenzeller : translated by Rachel Waddington
p. cm.
Includes bibliographical references and index.
ISBN 0-471-98851-0 (alk. paper)
1. Inventory control—Automation. 2. Radio frequency identification systems. I. Title.
TS160.F5513 1999 99-16221
658.7—dc21 CIP

British Library Cataloguing in Publication Data

A catalogue record for this book is available from the British Library

ISBN 0 471 98851 0

Produced from PostScript files supplied by the translator
Printed and bound in Great Britain by Antony Rowe Ltd, Chippenham
This book is printed on acid-free paper responsibly manufactured from sustainable forestry in which at least two trees are planted for each one used for paper production.

Contents

Preface

This book is aimed at an extremely wide range of readers. First and foremost it is intended for students and engineers who find themselves confronted with RFID technology for the first time. A few basic chapters are provided for this audience describing the functionality of RFID technology and the physical and IT-related principles underlying this field. The book is also intended for practitioners who, as users, wish to or need to obtain as comprehensive and detailed an overview of the various technologies, the legal framework or the possible applications of RFID as possible.

Although a wide range of individual articles are now available on this subject, the task of gathering all this scattered information together when it is needed is a tiresome and time-consuming one – as researching this book has proved. This book therefore aims to fill a gap in the range of literature on the subject of RFID.

This book uses numerous pictures and diagrams to attempt to give a graphic representation of RFID technology in the truest sense of the word. Particular emphasis is placed on practical considerations. For this reason the chapter entitled "Example Applications" is particularly comprehensive.

Technological developments in the field of RFID technology are proceeding at such a pace that although a book like this can explain the general scientific principles it is not dynamic enough to be able to explore the latest trends regarding the most recent products on the market. I am therefore grateful for any suggestions and advice – particularly from the field of industry. The basic concepts and underlying physical principles remain, however, and provide a good background for understanding the latest developments.

At this point I would also like to express my thanks to those companies who were kind enough to contribute to the success of this project by providing numerous technical data sheets, lecture manuscripts and photographs.

Munich, January 1998 Klaus Finkenzeller

List of Abbreviations

μP	Microprocessor
μs	Microsecond (10^{-6} seconds)
ABS	Acrylnitrilbutadienstyrol
AFC	Automatic Fare Collection
AM	Amplitude Modulation
ASIC	Application Specific Integrated Circuit
ASCII	American Standard Code for Information Interchange
ASK	Amplitude Shift Keying
BAPT	Bundesamt für Post und Telekommunikation
Bd	Baud, transmission speed in bit/s
BMBF	Bundesministerium für Bildung und Forschung (Ministry for Education and Research, was BMFT)
BP	Bandpass filter
C	Capacitance (of a capacitor)
CEN	Comité Européen de Normalisation
CEPT	Conférence Européene des Postes et Télécommunications
CICC	Close Coupling Integrated Circuit Chip Card
CLK	Clock (timing signal)
CRC	Cyclic Redundancy Checksum
CCITT	Comité Consultatif International Télégraphique et Téléphonique
dBm	Logarithmic measure of power, related to 1 mW HF-power (0 dBm = 1 mW, 30 dBm = 1W)
DBP	Differential Bi-Phase encoding
DIN	Deutsche Industrienorm (German industrial standard)
EAN	European Article Number (barcode on groceries and goods)
EAS	Electronic Article Surveillance
EC	Eurocheque or electronic cash

EEPROM	Electric Erasable and Programmable Read Only Memory
ERP	Equivalent Radiated Power
ETCS	European Train Control System
ETS	European Telecommunication Standard
ETSI	European Telecommunication Standards Institute
EMC	Electro-Magnetic Compatibility
EVC	European Vital Computer (part of ETCS)
FDX	Full-Duplex
FM	Frequency modulation
FRAM	Ferroelectric Random Access Memory
FSK	Frequency Shift Keying
GSM	Global System for Mobile Communication (was Groupe Spécial Mobile)
HDX	Half-Duplex
HF	High Frequency (3 ... 30 MHz)
ICC	Integrated Chip Card
ID	Identification
ISM	Industrial Scientific Medical (frequency range)
ISO	International Standardization Organization
L	Loop (inductance of a coil)
LAN	Local Area Network
LF	Low Frequency (30 ... 300 kHz)
LPD	Low Power Device (low power radio system for the transmission of data or speech over a few hundred meters)
LRC	Longitudinal Redundancy Check
LSB	Least Significant Bit
MAD	MIFARE® Application Directory
MSB	Most Significant Bit
nomL	Non-public mobile land radio (industrial radio, transport companies, taxi radio, etc.)
NRZ	Non-Return to Zero Encoding
NTC	Negative Temperature Coefficient (thermal resistor)
OCR	Optical Character Recognition
OEM	Original Equipment Manufacturer
OTP	One Time Programmable
PC	Personal Computer
PICC	Proximity Integrated Circuit Chip Card

PP	Plastic Package
PPS	Polyphenylensulfide
PSK	Phase Shift Keying
PVC	Polyvinylchloride
RADAR	Radio Detecting and Ranging
RAM	Random Access Memory
RCS	Radar Cross Section
RFID	Radio Frequency Identification
RTI	Road Transport Information System
RWD	Read Write Device
SAM	Security Authentication Module
SAW	Surface Acoustic Wave
SEQ	Sequential System
SMD	Surface Mounted Devices
SRAM	Static Random Access Memory
TR	Technische Richtlinie (Technical Guideline)
UHF	Ultra High Frequency (300 MHz ... 3 GHz)
UPC	Universal Product Code
VDE	Verein Deutscher Elektrotechniker (German Association of Electrical Engineers)
VICC	Vicinity Integrated Circuit Chip Card
XOR	eXclusive-OR
ZV	Zulassungsvorschrift (Licensing Regulation)

1

Introduction

In recent years automatic identification procedures (Auto ID) have become very popular in many service industries, purchasing and distribution logistics, industry, manufacturing companies and material flow systems. Automatic identification procedures exist to provide information about people, animals, goods and products in transit.

The omnipresent barcode labels that triggered a revolution in identification systems some considerable time ago, are being found to be inadequate in an increasing number of cases. Barcodes may be extremely cheap, but their stumbling block is their low storage capacity and the fact that they cannot be reprogrammed.

The technically optimal solution would be the storage of data in a silicon chip. The most common form of electronic data carrying device in use in everyday life is the smart card based upon a contact field (telephone smart card, bank cards). However, the mechanical contact used in the smart card is often impractical. A contactless transfer of data between the data carrying device and its reader is far more flexible. In the ideal case, the power required to operate the electronic data carrying device would also be transferred from the reader using contactless technology. Because of the procedures used for the transfer of power and data, contactless ID systems are called *RFID systems* (Radio Frequency Identification).

The number of companies that are actively involved in the development and sale of RFID systems indicates that this is a market that should be taken seriously. Total worldwide sales of RFID systems for the year 2000 are estimated at above 2 billion US$. The *RFID market* therefore belongs to the fastest growing sector of the radio technology industry, including mobile phones and cordless telephones.

Furthermore, in recent years contactless identification has been developing into an independent interdisciplinary field, which no longer fits into any of the conventional pigeon holes. It brings together elements from extremely varied fields: HF technology and EMC, semiconductor technology, data protection and cryptography, telecommunications, manufacturing technology and many related areas.

As an introduction, the following chapter gives a brief overview of different auto ID systems, that perform similar functions to RFID.

1.1 Automatic Identification Systems

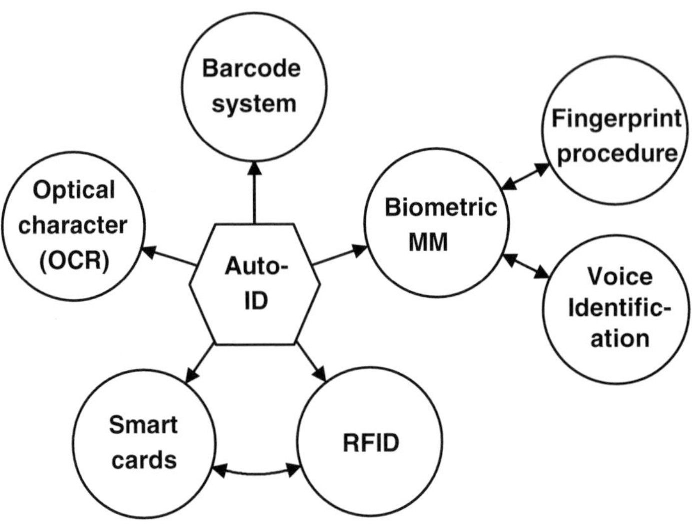

Figure 1.1: Overview of the most important Auto ID procedures

1.1.1 Barcode systems

Barcodes have successfully held their own against other identification systems over the past 20 years. According to experts, the turnover volume for barcode systems totalled around 3 billion DM in Western Europe at the beginning of the 1990s [virnich].

The barcode is a binary code comprising a field of bars and gaps arranged in a parallel configuration. They are arranged according to a predetermined pattern and represent data elements that refer to an associated symbol. The sequence, made up of wide and narrow bars and gaps can be interpreted numerically and alphanumerically. It is read by optical laser scanning, i.e. by the different reflection of a laser beam from the black bars and white gaps [ident 1]. However, despite being identical in their physical design, there are considerable differences between the code layouts in the approximately ten different barcode types currently in use.

The most popular barcode by some margin is the *EAN code* (European Article Number), which was designed specifically to fulfil the requirements of the grocery industry in 1976. The EAN code represents a development of the UPC (Universal Product Code) from the USA, which was introduced in the USA as early as 1973. Today, the UPC represents a subset of the EAN code, and is therefore compatible with it [virnich].

The EAN code is made up of 13 digits: the country identifier, the company identifier, the manufacturer's item number and a check digit.

Country Identifer		Company Identifier					Manufacturer's Item Number					CD
4	0	1	2	3	4	5	0	8	1	5	0	9
FRG		Company Name 1 Road Name 80001 München					Chocolate Rabbit 100g					

Figure 1.2: Example of the structure of a barcode in EAN coding (EAN = European Article Number)

In addition to the EAN code, the following barcodes are popular in other industrial fields:

- Code Codabar: Medical / clinical applications, fields with high safety requirements.
- Code 2/5 interleaved: Automotive industry, goods storage, pallets, shipping containers and heavy industry.
- Code 39: Processing industry, logistics, universities and libraries.

Figure 1.3: This barcode is printed on the back of this book and contains the ISBN number of the book.

1.1.2 Optical character recognition

Optical character recognition (OCR) was first used in the 1960s. Special fonts were developed for this application that stylised characters so that they could be read both in the normal way by people and automatically by machines. The most important advantage of OCR systems is the high density of information and the possibility of reading data visually in an emergency (or simply for checking) [virnich].

Today, OCR is used in production, service and administrative fields, and also in banks for the registration of cheques (personal data (name, account number) is printed on the bottom line of a cheque in OCR type).

However, OCR systems have failed to become universally applicable because of their high price and the complicated readers that they require in comparison with other ID procedures.

1.1.3 Biometric procedure

Biometrics is defined as the science of counting and (body) measurement procedures involving living beings. In the context of identification systems, biometry is the general term for all procedures that identify people by comparing unmistakable and individual physical characteristics. In practice, these are finger printing and hand printing procedures, voice identification and, less commonly, retina (or iris) identification.

1.1.3.1 Voice identification

Recently, specialised systems have become available to identify individuals using speaker verification (speaker recognition). In such systems, the user talks into a microphone linked to a computer. This equipment converts the spoken words into digital signals, which are evaluated by the identification software.

The objective of speaker verification is to check the supposed identity of the person based upon their voice. This is achieved by checking the speech characteristics of the speaker against an existing reference pattern. If they correspond, then a reaction can be initiated (e.g. "open door").

1.1.3.2 Finger printing procedures (dactyloscopy)

Criminology has been using finger printing procedures for the identification of criminals since the turn of the century. This process is based upon the comparison of papillae and dermal ridges of the finger tips, which can be obtained not only from the finger itself, but also from objects that the individual in question has touched.

When finger printing procedures are used for personal identification, usually for entrance procedures, the finger tip is placed upon a special reader. The system calculates a data record from the pattern it has read and compares this with a stored reference pattern. Modern finger print ID systems require less than half a second to recognise and check a finger print. In order to prevent violent frauds, fingerprint ID systems have even been developed that can detect whether the finger placed on the reader is that of a living person [schmidhäusler].

1.1.4 Smart cards

A *smart card* is an electronic data storage system, possibly with additional computing capacity (microprocessor card), which – for convenience – is incorporated into a plastic card the size of a credit card. The first smart cards in the form of prepaid telephone smart cards were launched in 1984. Smart cards are placed in a reader, that makes a galvanic connection to the contact surfaces of the smart card using contact springs. The smart card is supplied with energy and a clock pulse from the reader via the contact surfaces. Data transfer between the reader and the card takes place using a bidirectional serial interface (I/O port). It is possible to differentiate between two basic types of smart card based upon their internal functionality: memory card and microprocessor card.

One of the primary advantages of the smart card is the fact that the data stored on it can be protected against undesired (read) access and manipulation. Smart cards make all services that relate to information or financial transactions simpler, safer and cheaper. For this reason, 200 million smart cards were issued worldwide in 1992. In 1995 this figure had risen to 600 million, of which 500 million were memory cards and 100 million were microprocessor cards. The *smart card market* therefore represents one of the fastest growing submarkets of the microelectronics industry.

One disadvantage of contact-based smart cards is the vulnerability of the contacts to wear, corrosion and dirt. Readers that are used frequently are expensive to maintain due to their tendency to malfunction. In addition, readers that are accessible to the public (telephone boxes) cannot be protected against vandalism.

1.1.4.1 Memory cards

In *memory cards* the memory – usually an EEPROM – is accessed using a sequential logic (state machine). It is also possible to incorporate simple security algorithms, e.g. stream ciphering using this system. The functionality of the memory card in question is usually optimised for a specific application. Flexibility of application is highly limited but, on the positive side, memory cards are very cost effective. For this reason, memory cards are predominantly used in price sensitive, large-scale applications [rankl]. One example of this is the national insurance card used by the state pension system in Germany [lemme].

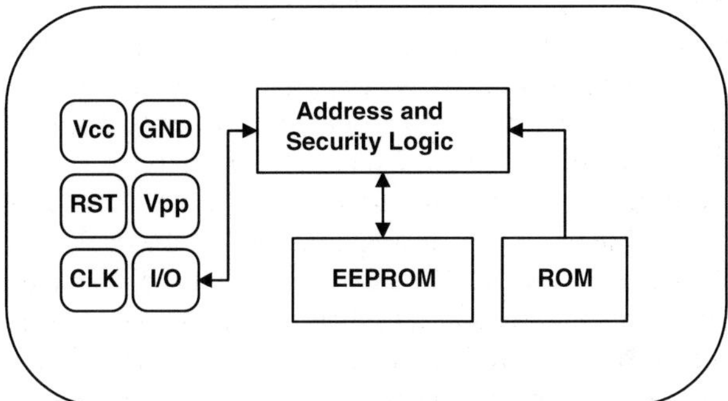

Figure 1.4: Typical architecture of a memory card with security logic

1.1.4.2 Microprocessor cards

As the name suggests, *microprocessor cards* contain a microprocessor, which is connected to a segmented memory (ROM-, RAM- and EEPROM-segment).

The mask programmed ROM incorporates an *operating system* (higher programme code) for the microprocessor and is inserted during chip manufacture. The contents of the ROM are determined during manufacturing, are identical for all microchips from the same production batch, and cannot be overwritten.

The chip's EEPROM contains application data and application-related programme code. Reading from or writing to this memory area is controlled by the operating system.

The RAM is the microprocessor's temporary working memory. Data stored in the RAM are lost when the supply voltage is disconnected.

Microprocessor cards are very flexible. In modern smart card systems it is also possible to integrate different applications in a single card (multi-application). The application specific parts of the programme are not loaded into the EEPROM until after manufacture and can be initiated via the operating system.

Microprocessor cards are primarily used in security sensitive applications. Examples are smart cards for GSM mobile phones and the new EC (electronic cash) cards. The option of programming the microprocessor cards also facilitates rapid adaptation to new applications [rankl].

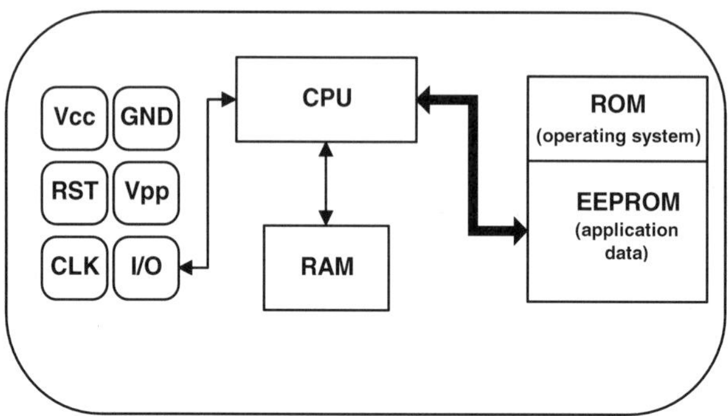

Figure 1.5: Typical architecture of a microprocessor card

1.1.5 RFID systems

RFID systems are closely related to the smart cards described above. Like smart card systems, data is stored on an electronic data carrying device – the transponder. However, unlike the smart card, the power supply to the data carrying device and the data exchange between data carrying device and reader are achieved without the use of galvanic contacts, using instead magnetic or electromagnetic fields. The underlying technical procedure is drawn from the fields of radio and radar engineering. The abbreviation RFID stands for radio frequency identification, i.e. information carried by radio waves. Due to the numerous advantages of RFID systems compared with other identification systems, RFID systems are now beginning to conquer new mass markets. One example is the use of contactless smart cards as tickets for short-distance public transport.

1.2 A Comparison of Different ID Systems

A comparison between the identification systems described above highlights the strengths and weakness of RFID in relation to other systems. Here too, there is a close relationship between contact based smart cards and RFID systems, however the latter circumvents all the disadvantages related to faulty contacting (sabotage, dirt, unidirectional insertion, time consuming insertion, etc.).

Table 1.1: Comparison of different RFID systems showing their advantages and disadvantages

System parameters:	Barcode	OCR	Voice recog.	Biometry	Smart card	RFID systems
Typical data quantity / byte:	1 ~ 100	1 ~ 100	-	-	16 ~ 64k	16 ~ 64k
Data density	low	low	high	high	very high	very high
Machine readability	good	good	expensive	expensive	good	good
Readability by people	limited	simple	simple	difficult	imposs.	imposs.
Influence of dirt / damp	v. high	v. high	-	-	possible (contacts)	no influence
Influence of (opt.) covering	total failure	total failure	-	possible	-	no influence
Influence of direction and position	low	low	-	-	unidirectional	no influence
Degradation / wear	limited	limited	-	-	contacts	no influence
Purchase cost / reading electronics	very low	medium	very high	very high	low	medium
Operating costs (e.g. printer)	low	low	none	none	medium (contacts)	none
Unauthorised copying / modification	slight	slight	possible* (audio tape)	imposs.	imposs.	imposs.
Reading speed (incl. handling of data carrier)	low ~ 4 s	low ~ 3 s	very low > 5 s	very low > 5...10 s	low ~ 4 s	very fast ~ 0,5 s
Maximum distance between data carrier and reader	0–50 cm	< 1 cm Scanner	0–50 cm	direct contact **	direct contact	0–5m, microwave

* The danger of "Replay" can be reduced by selecting the text to be spoken using a random generator, because the text that must be spoken is not known in advance.

** This only applies for fingerprint ID. In the case of retina or iris evaluation direct contact is not necessary or possible.

1.3 Components of an RFID System

An *RFID system* is always made up of two components:

- the *transponder*, which is located on the object to be identified,
- the interrogator or *reader*, which, depending upon design and the technology used, may be a read or write/read device (in this book – in accordance with normal colloquial usage – the data capture device is always referred to as the *reader*, regardless of whether it can only read data or is also capable of writing).

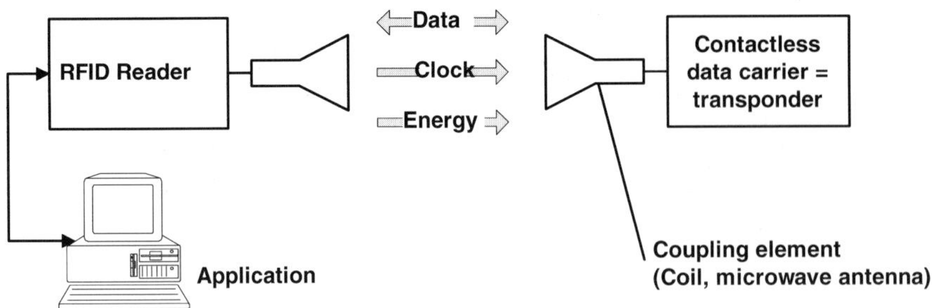

Figure 1.6: The reader and transponder are the main components of every RFID system

Figure 1.7: RFID reader and contactless smart card in practical use. (Reproduced by permission of Kaba Legic)

A reader typically contains a radio frequency module (transmitter and receiver), a control unit and a coupling element to the transponder. In addition, many readers are fitted with an additional interface (RS 232, RS 485, ...) to enable it to forward the data received to another system (PC, robot control system, ...).

The transponder, which represents the actual *data carrying device* of an RFID system, normally consists of a *coupling element* and an electronic *microchip*. When the transponder, which does not usually possess its own voltage supply (battery), is not within the interrogation zone of a reader it is totally passive. The transponder is only activated when it is within the interrogation zone of a reader. The power required to activate the transponder is supplied to the transponder through the coupling unit (contactless) as is the timing pulse and data.

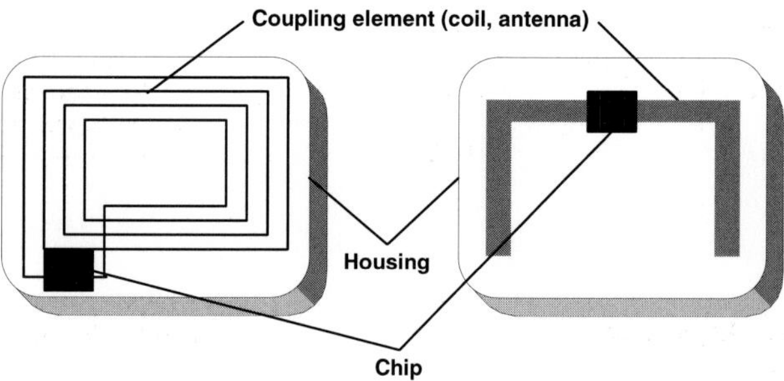

Figure 1.8: Basic layout of the RFID data carrying device, the transponder. Left: Inductively coupled transponder with antenna coil, right: Microwave transponder with dipolar antenna

2

Differentiation Features of RFID Systems

2.1 Fundamental Differentiation Features

RFID systems exist in countless variants, produced by an almost equally high number of manufacturers. If we are to maintain an overview of RFID systems we must seek out features that can be used to differentiate one RFID system from another.

RFID systems operate according to one of two basic procedures: full duplex (FDX) / half duplex (HDX) systems, and sequential systems (SEQ).

In *full* and *half duplex* systems the transponder's response is broadcast when the reader's RF field is switched on. Because the transponder's signal to the receiver antenna can be extremely weak in comparison with the signal from the reader itself, appropriate transmission procedures must be employed to differentiate the transponder's signal from that of the reader. In practice, data transfer from transponder to reader takes place using load modulation, load modulation using a subcarrier, but also (sub)harmonics of the reader's transmission frequency.

In contrast, *sequential procedures* employ a system whereby the field from the reader is switched off briefly at regular intervals. These gaps are recognised by the transponder and used for sending data from the transponder to the reader. The disadvantage of the sequential procedure is the loss of power to the transponder during the break in transmission, which must be smoothed out by the provision of sufficient auxiliary capacitors or batteries.

The data capacities of RFID transponders normally range from a few bytes to several kByte. So-called 1 bit transponders represent the exception to this rule: a data quantity of exactly 1 bit is just enough to signal two states to the reader: "transponder in the field" or "no transponder in the field". However, this is perfectly adequate to fulfil simple monitoring or signalling functions. Because a 1 bit transponder does not need an electronic chip, these transponders can be manufactured for a fraction of a penny. For this reason, vast numbers of 1 bit transponders are used in *Electronic Article Surveillance* (EAS) to protect goods in shops and businesses. If someone attempts to leave the shop with goods that have not been paid for the reader installed in the exit recognises the state "transponder in the field" and initiates the appropriate reaction. The 1 bit transponder is removed or deactivated at the till when the goods are paid for.

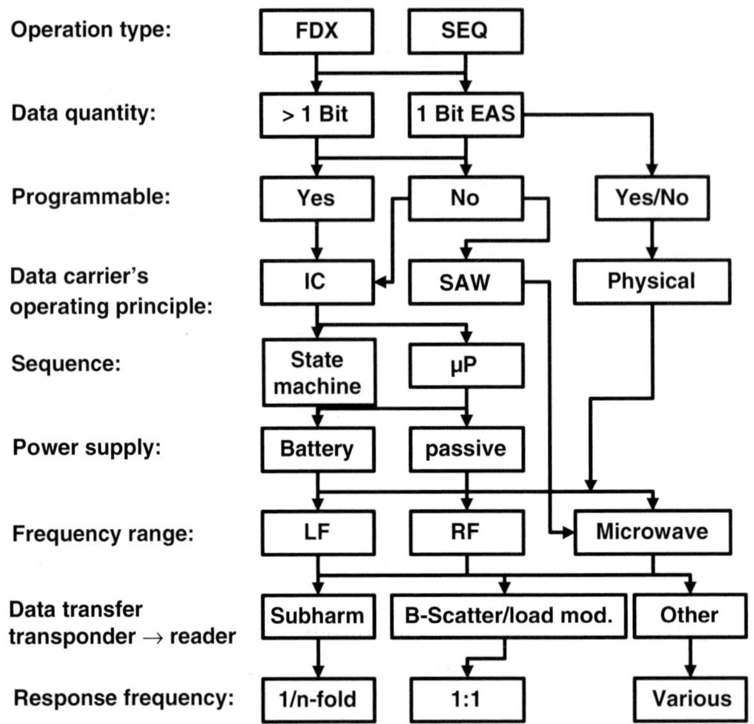

Figure 2.1: The various features of RFID systems [isd]

The possibility of writing data to the transponder provides us with another way of classifying RFID systems. In very simple systems the transponder's data record, usually a simple (serial) number, is incorporated when the chip is manufactured and cannot be altered thereafter. In writeable transponders, on the other hand, the reader can write data to the transponder. Three main procedures are used to store the data: in inductively coupled RFID systems EEPROMs (electrically erasable programmable read only memory) are dominant, however these have the disadvantage of a high power consumption during the writing operation and a limited number of write cycles (typically of the order of 100 000 to 1 000 000). FRAMs (ferromagnetic random access memory) have recently been used in isolated cases. The read power consumption of FRAMs is lower than that of EEPROMs by a factor of 100 and the writing time is 1000 times lower. Manufacturing problems have hindered its widespread introduction onto the market as yet.

Particularly common in microwave systems, static RAMs (static random access memory, SRAM) are also used for data storage, and facilitate very rapid write cycles. However, data retention requires an uninterruptible power supply from an auxiliary battery.

In programmable systems, write and read access to the memory and any requests for write and read authorisation must be controlled by the data carrier's "internal logic". In the simplest case these functions can be realised by a state machine. (See Chapter 10 "The Architecture of Electronic Data Carriers" for further information). Very complex sequences can be realised using *state machines*. However, the disadvantage of state machines is their inflexibility regarding changes to the programmed functions, because such changes necessitate changes to the circuitry of the silicon chip. In practice, this means redesigning the chip layout, with all the associated expense.

The use of a microprocessor improves upon this situation considerably. An operating system for the management of application data is incorporated into the processor during manufacture using a mask. Changes are thus cheaper to implement and, in addition, the software can be specifically adapted to perform very different applications.

In the context of contactless smart cards, writeable data carriers with a state machine are also known as "memory cards", to distinguish them from "processor cards".

In this context, we should also mention transponders that can store data by utilising physical effects. This includes the read only surface wave transponder and 1 bit transponders that can usually be deactivated (set to "0"), but can rarely be reactivated (set to "1").

One very important feature of RFID systems is the *power supply* to the transponder. *Passive transponders* do not have their own power supply, and therefore all power required for the operation of a passive transponder must be drawn from the (electrical / magnetic) field of the reader. Conversely, *active transponders* incorporate a battery, which supplies all or part of the power for the operation of a microchip.

One of the most important characteristics of RFID systems is the *operating frequency* and the resulting range of the system. The operating frequency of an RFID system is the frequency at which the reader transmits. The transmission frequency of the transponder is disregarded. In most cases it is the same as the *transmission frequency* of the reader (load modulation, backscatter). However, the transponder's "transmitting power" may always be set several powers of ten lower than that of the reader.

The different transmission frequencies are classified into the three basic ranges, LF (low frequency, 30 kHz – 300 kHz), HF (high frequency) / RF (radio frequency, 3 MHz – 30 MHz) and UHF (ultra high frequency, 300 MHz – 3 GHz)/microwave (> 3 GHz). A further subdivision of RFID systems according to range allows us to differentiate between close-coupling (0 – 1 cm), remote-coupling (0 – 1 m), and long-range systems (> 1 m).

The different procedures for sending data from the transponder back to the reader can be classified into three groups: the use of reflection or backscatter (the frequency of the reflected wave corresponds with the transmission frequency of the reader → frequency ratio 1:1) or load modulation (the reader's field is influenced by the transponder → frequency ratio 1:1), the use of subharmonics ($1/n$ fold) and the generation of harmonic waves (n-fold) in the transponder.

2.2 Transponder Construction Formats

2.2.1 Disks and coins

The most common construction format is the so-called *disk* (coin), a transponder in a round (ABS) injection moulded housing, with a diameter ranging from a few millimetres to 10 cm. There is usually a hole for a fastening screw in the centre. As an alternative to (ABS) injection moulding, polystyrol or even epoxy resin may be used to achieve a wider operating temperature range.

2.2.2 *Glass housing*

Glass transponders have been developed that can be injected under the skin of an animal for identification purposes (see Chapter 13 "Example Applications").

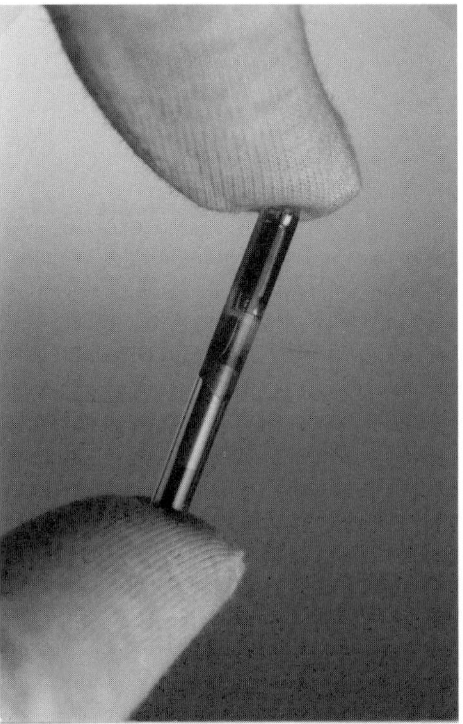

Figure 2.2: Close-up of a 32 mm glass transponder for the identification of animals or further processing into other construction formats (Reproduced by permission of Texas Instruments)

Glass tubes of just 12 to 32 mm contain a microchip mounted upon a carrier (PCB) and a chip capacitor to smooth the supply current obtained. The transponder coil incorporates wire of just 0.03 mm thickness wound onto a ferrite core. The internal components are embedded in a soft adhesive to achieve mechanical stability.

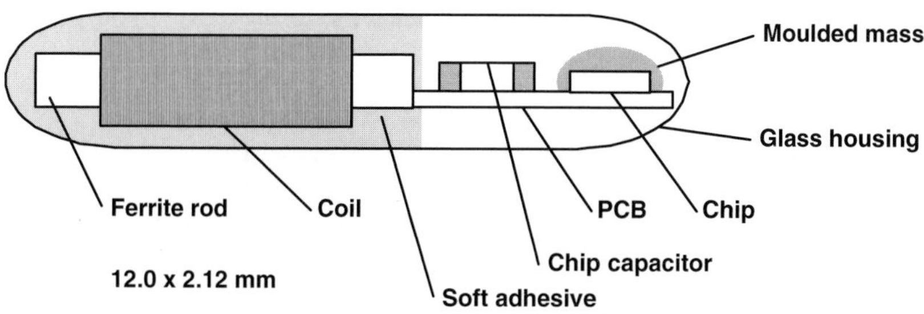

Figure 2.3: Mechanical layout of a glass transponder

2.2.3 Plastic housing

The *plastic housing* (*plastic package*, PP) was developed for applications involving particularly high mechanical demands. This housing can easily be integrated into other products, for example into *car keys* for *electronic immobilisation systems*.

Figure 2.4: Transponder in a plastic housing. (Reproduced by permission of Philips Electronics B.V.)

The wedge made of moulding substance (IC casting compound) contains almost the same components as the glass transponder, but its longer coil gives it a greater functional range. Further advantages are its ability to accept larger microchips and its greater tolerance to mechanical vibrations, which is required by the automotive industry, for example. The *PP Transponder* has proved completely satisfactory with regard to other quality requirements, such as temperature cycles or fall tests [bruhnke].

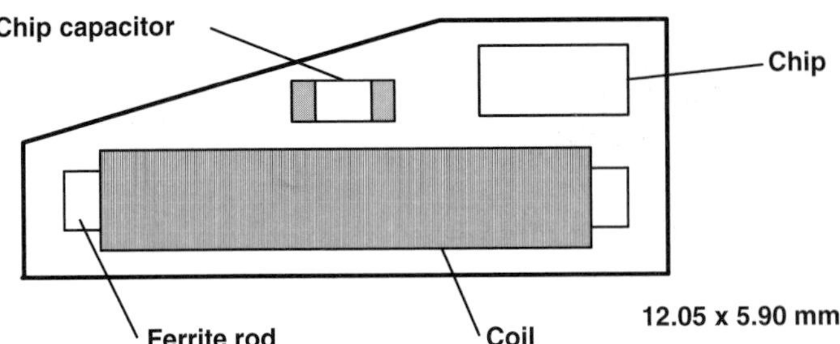

Figure 2.5: Mechanical layout of a transponder in a plastic housing. The housing is just 3 mm thick

2.2.4 *Tool and gas bottle identification*

Special construction formats have been developed to install inductively coupled transponders into *metal surfaces*. The transponder coil is wound in a ferrite pot core. The transponder chip is mounted on the reverse of the *ferrite pot core* and contacted with the transponder coil.

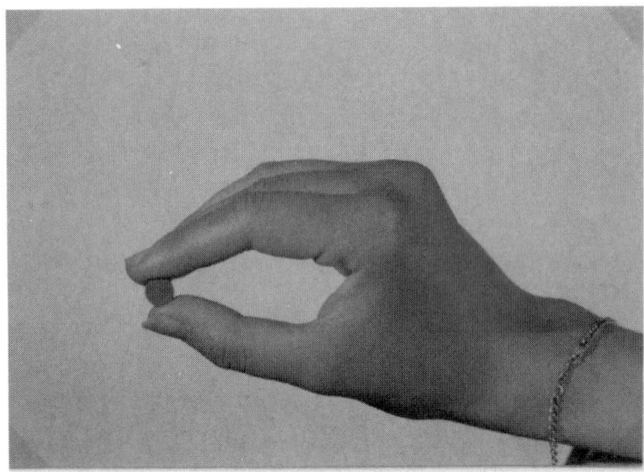

Figure 2.6: Transponder in a standardised construction format in accordance with ISO 69873, for fitting into one of the retention knobs of a CNC tool. (Reproduced by permission of Leitz GmbH & Co, Oberkochen)

In order to obtain sufficient mechanical stability, vibration and heat tolerance, transponder chip and ferrite pot core are cast into a PPS shell using epoxy resin [link]. The external dimensions of the transponder and their fitting area have been standardised in *ISO 69873* for incorporation into a retention knob or quick-release taper for tool identification. Different designs are used for the identification of gas bottles.

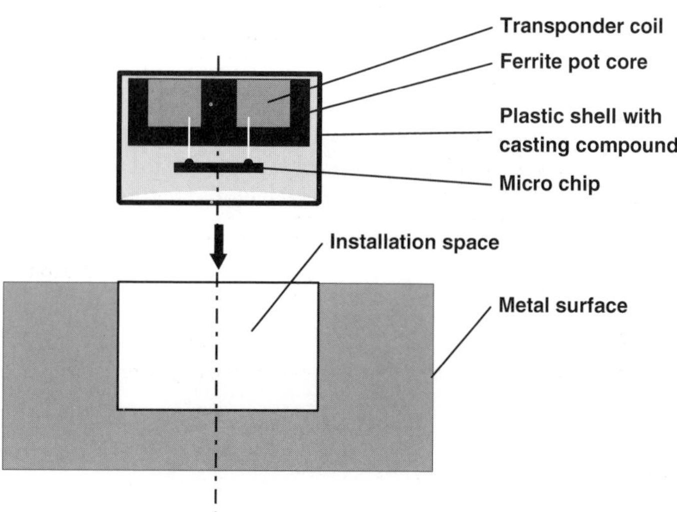

Figure 2.7: Mechanical layout of a transponder for fitting into metal surfaces. The transponder coil is wound around a U-shaped ferrite core and then cast into a plastic shell. It is installed with the opening of the U-shaped core uppermost

2.2.5 Keys and key fobs

Transponders are also integrated into mechanical keys for immobilisers or door locking applications with particularly high security requirements. These are generally based upon a transponder in a plastic housing, which is cast or injected into the key fob.

The keyring transponder design has proved very popular for systems providing access to office and work areas.

Figure 2.8: Keyring transponder for an access system (Reproduced by permission of Intermarketing)

2.2.6 Clocks

This construction format was developed at the beginning of the 1990s by the Austrian company Ski-Data and was first used in ski passes. These "*contactless clocks*" were also able to gain ground in access control systems. The clock contains a frame antenna with a small number of windings printed onto a thin printed circuit board, which follows the clock housing as closely as possible to maximise the area enclosed by the antenna coil – and thus the range.

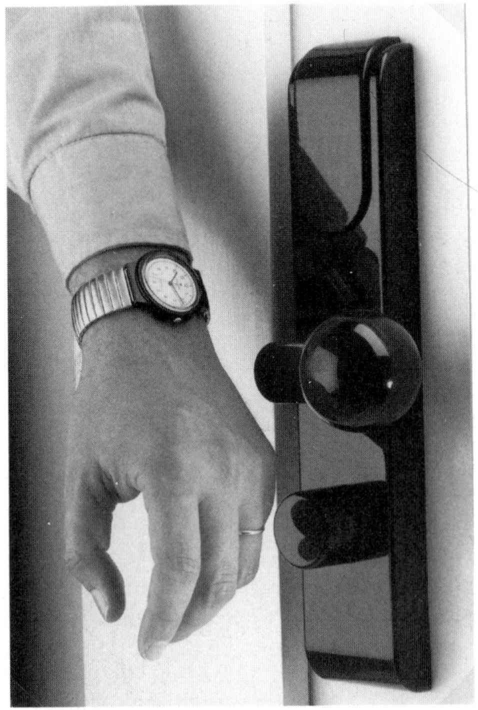

Figure 2.9: Watch with integral transponder in use in a contactless access authorisation system (Reproduced by permission of Junghans Uhren GmbH, Schramberg)

2.2.7 ID-1 format, contactless smart cards

The ID-1 format familiar from credit cards and telephone cards (85.72 × 54.03 mm × 0.76 mm ± tolerances) is becoming increasingly important for *contactless smart cards* in RFID systems. One advantage of this format for inductively coupled RFID systems is the large coil area, which increases the range of the smart cards.

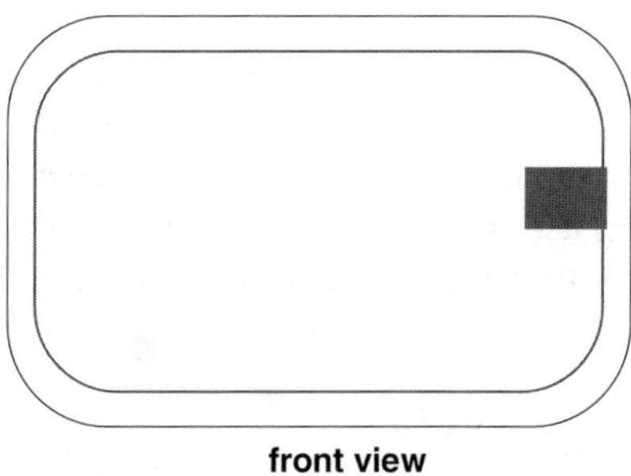

Figure 2.10: Layout of a contactless smart card: card body with transponder module and antenna

Contactless smart cards are produced by the lamination of a transponder between four PVC foils. The individual foils are baked at high pressure and temperatures above 100 °C to produce a permanent bond (the manufacture of contactless smart cards is described in detail in Chapter 12 "The Manufacture of Transponders and Contactless Smart Cards").

Contactless smart cards of the design ID-1 are excellently suited for carrying adverts and often have artistic overprints, like those on telephone cards, for example.

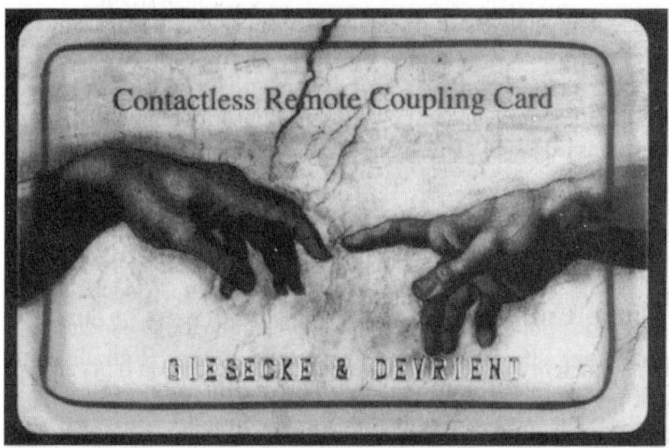

Figure 2.11: Semitransparent contactless smart card. The transponder antenna can be clearly seen along the edge of the card (Reproduced by permission of Giesecke & Devrient, Munich)

However, it is not always possible to adhere to the maximum thickness of 0.8 mm specified for ID-1 cards in ISO 7810. Microwave transponders in particular require a thicker design, because in this design the transponder is usually inserted between two PVC shells or packed using an (ABS) injection moulding procedure.

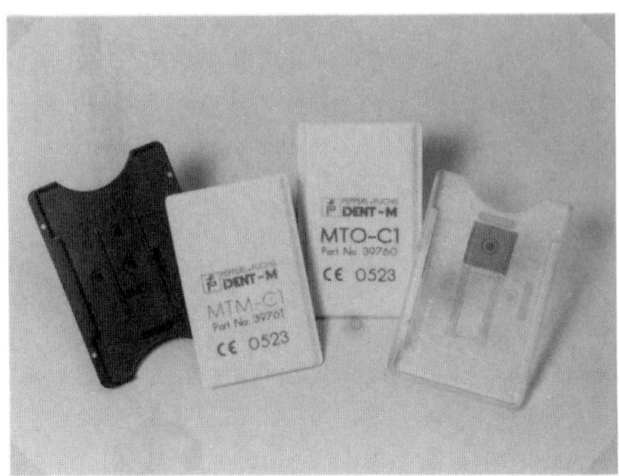

Figure 2.12: Microwave transponders in plastic shell housings (Reproduced by permission of Pepperl & Fuchs GmbH)

2.2.8 *Other formats*

In addition to these main designs, several application-specific special designs are also manufactured. Examples are the "racing pigeon transponder" or the "champion chip" for sports timing. Transponders can be incorporated into any design required by the customer. The preferred options are glass or PP transponders, which are then processed further to obtain the ultimate form.

2.3 Frequency, Range and Coupling

2.3.1 *Close coupling*

RFID systems with a very small range, typically in the region of 0 to 1 cm are known as close coupling systems The transponder must either be incorporated within a reader or positioned upon a surface provided for this purpose.

Close coupling systems can be operated at any desired frequency between DC and 30 MHz, because the operation of the transponder does not rely upon the radiation of electromagnetic waves. The close coupling between data carrier and reader also facilitates the provision of greater amounts of power, even for the operation of a microprocessor that does not have an optimal power consumption. Close-coupled systems are primarily used in applications that are subject to strict security requirements, but do not require a large range. Examples are electronic door locking systems or contactless smart card systems with payment functions. Close-coupled transponders are currently used exclusively as contactless smart cards in ID-1 format.

2.3.2 Remote coupling

Systems with write and read ranges of up to 1 m are known as remote-coupled. All *remote-coupled systems* are based upon an *inductive (magnetic) coupling* between reader and transponder. These systems are therefore also known as *inductive radio systems*. Around 90 – 95 % of all RFID systems purchased are inductively coupled systems.

Frequencies below 135 kHz or the frequencies 6.75 MHz, 13.56 MHz and 27.125 MHz are used as transmission frequencies. The power that can be transmitted by inductive coupling is very low and depends upon the distance between transponder and reader, so normally only read only data carriers with very low power consumption are used. However, high end systems with microprocessor transponders have also moved into the field of inductively coupled systems.

In order to avoid reference to a possibly erroneous ranges, this book uses only the term inductive (coupled) systems for classification.

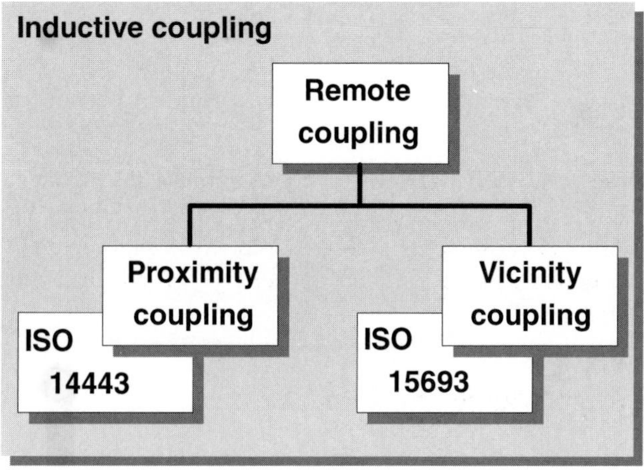

Figure 2.13: In contactless smart cards a differentiation is also made between proximity coupling (typically 15 cm) and vicinity coupling (approximately 1 m)

2.3.3 Long range

Long range systems are available for ranges typically between 1 m and 10 m. However, there are also individual systems that can cope with significantly greater ranges. All long range systems operate using electromagnetic waves in the *microwave range*. The transmission frequency is normally 2.45 GHz, however there are systems that operate at 915 MHz (not permitted in Europe), 5.8 GHz and 24.125 GHz.

The power supplied is never sufficient to provide the transponder with enough energy to operate a microchip. Long range systems (with the exception of surface wave transponders) therefore possess an auxiliary battery. This does not provide the power for data transfer between transponder and reader, but serves exclusively to supply the microchip and to retain stored data.

Communication between transponder and reader uses only the HF energy received from the reader. The backscatter procedure (modulated echo crosssection) is the standard procedure used for data transfer between transponder and reader.

In order to avoid reference to a possibly erroneous range figure, this book uses only the terms *microwave systems* or *backscatter system* for classification.

2.3.4 System performance

One option for classifying RFID systems is to divide them according to the range of functions offered by the system, the *system performance*. If we classify all the RFID systems available on the market by their range of functions, i.e. by the memory size of the data carrier, transaction speed, range and cryptographic functions, then low end and high end systems represent the two extremes of the total spectrum.

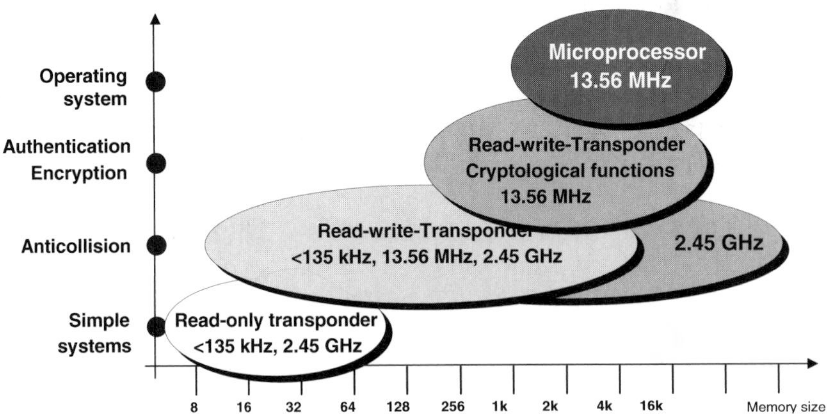

Figure 2.14: RFID systems can be classified into low end and high end systems according to their functionality

- *Read only systems* form the bottom end of *low end systems*. Read only means that data can be read from the data carrier, but cannot be written back to it. The data record of a read only chip normally consists only of an unique serial number made up of several bytes. If a read only transponder is placed in the HF field of a reader then the transponder begins to broadcast its own serial number. It is not possible for the reader to address a read only transponder – there is a unidirectional flow of data from the transponder to the reader. In practical operation of a read only system, it is also necessary to ensure that there is only ever one transponder in the reader's interrogation zone, otherwise the two or more transponders simultaneously transmitting would lead to a data collision. The reader would no longer be able to recognise any logical data.

  Despite this limitation, read only transponders are excellently suited for many applications in which the reading of one unique number is sufficient. Because of the simple function of a read only transponder, the chip area can be minimised, which gives low power consumption and a low manufacturing cost.

Read only systems are operated in the frequency range < 135 kHz or at 2.45 GHz. The achievable ranges are very high thanks to the low power consumption of the microchip.

- Read only systems are used where they can replace the functionality of barcode systems, for example in the control of product flows, in the identification of pallets, containers, gas bottles etc.
- The mid-range is occupied by a variety of systems with writeable data memory. The memory size ranges from 16 bytes to over 16 kByte EEPROM or SRAM. This is the field where diversity is greatest, making it difficult to make generalisations.

  These systems are operated at all frequencies available for RFID systems, in particular 135 kHz, 13.56 MHz, 27.125 MHz and 2.45 GHz.
- The *high end* segment is made up of systems with cryptological functions, i.e. authentication and data stream encryption. Microprocessor systems are also high end systems. The use of microprocessors facilitates the realisation of significantly more complex algorithms for the encryption and authentication than would be possible using the "hard-wired" logic of a state machine.

  High end systems are predominantly operated at the frequency 13.56 MHz. Because the pulse frequency of inductively coupled RFID transponders is derived from the transmission frequency of the reader, a pulse frequency that is higher by a factor of 100 than would be the case at 135 kHz is available. Therefore, even complex algorithms for authentication and encryption of the data stream can be executed within a reasonable period of time.

  The memory available in high end systems ranges from a few bytes to 16 kByte EEPROM (see Figure 2.14).

3

Fundamental Operating Principles

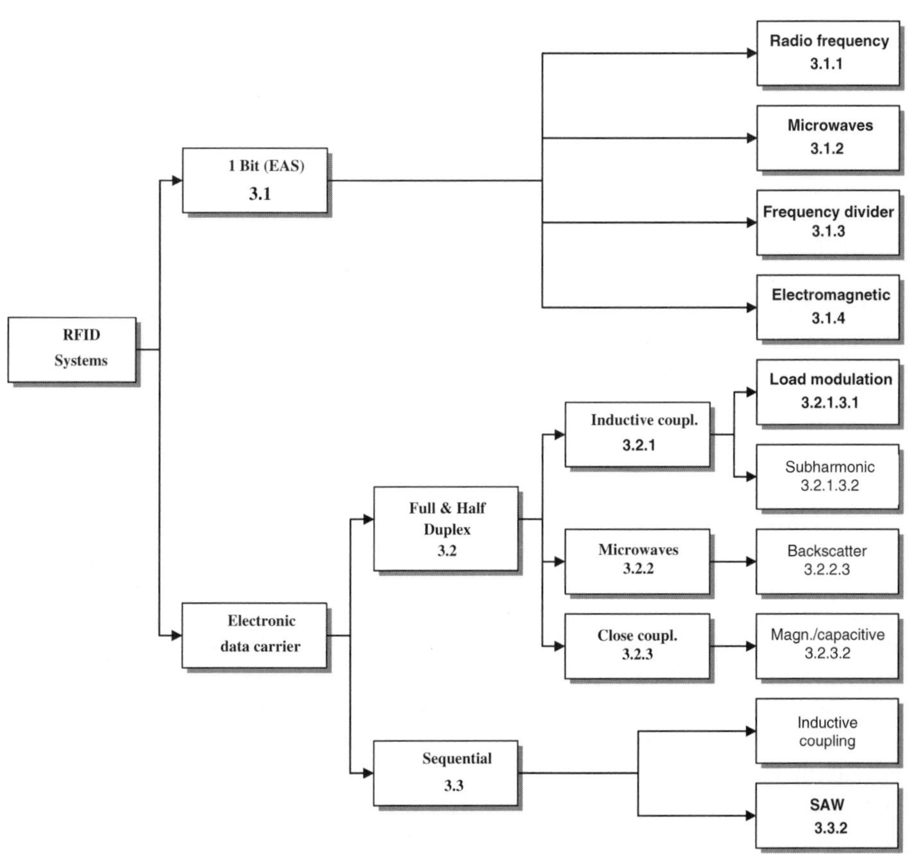

Figure 3.1: The allocation of the different operating principles of RFID systems into the sections of the chapter

This chapter describes the basic interaction between transponder and reader, in particular the power supply to the transponder and the data transfer between transponder and reader. For a more in-depth description of the physical interactions and mathematical models relating to inductive coupling or backscatter systems please refer to Chapter 4 "Physical Principles of RFID Systems".

3.1 1 Bit Transponder

A bit is the smallest unit of information that can be represented and has only two states: "1" or "0". This means that only two states can be represented by systems based upon a *1 bit transponder*: "transponder in interrogation zone" and "no transponder in interrogation zone". Despite this limitation, 1 bit transponders are very widespread – their main field of application is in electronic *anti-theft devices* in warehouses (*EAS* – electronic article surveillance).

An electronic article surveillance system is made up of the following components: the antenna of a "reader" or interrogator, the *security method* or *tag*, and an optional *deactivation device* for deactivating the tag after payment. Some systems also incorporate an *activator*, which is used to reactivate the tag after deactivation [gillert]. The main performance characteristic for all systems is the recognition or *detection rate*, in relation to the gate width (maximum distance between transponder and detector antenna).

The procedure for the inspection and testing of installed article surveillance systems is specified in the guideline *VDI 4470* entitled "Anti-theft systems for goods – detection gates. Inspection guidelines for customers". This guideline contains definitions and testing procedures for the calculation of the detection rate and false alarm ratio. It can be used by the retail trade as the basis for sales contracts or for monitoring the performance of installed systems on an ongoing basis. For the product manufacturer, the Inspection Guidelines for Customers represent an effective benchmark in the development and optimisation of integrated solutions for security projects [In accordance with VDI 4470].

3.1.1 Radio frequency

The *radio frequency (RF) procedure* is based upon LC resonant circuits adjusted to a defined resonant frequency f_R. Early versions employed inductive resistors made of wound enamelled copper wire with a soldered on capacitor in a plastic housing (*hard tag*). Modern systems employ coils etched between foils in the form of stick-on labels. To ensure that the damping resistance does not become too high and reduce the quality of the resonant circuit to an unacceptable level the thickness of the aluminium conduction tracks on the 25μm thick polyethylene foil must be at least 50μm [jörn].

Intermediate foils of 10μm thickness are used to manufacture capacitor plates. If a resonant circuit is moved into the vicinity of a magnetic alternating field, then energy from the alternating field can be induced in the resonant circuit via the coils of the resonant circuit (Faraday's law). If the frequency f_G of the alternating field corresponds with the resonant frequency f_R of the resonant circuit, then the resonant circuit produces a *sympathetic oscillation*. Due to the stimulation of oscillations in the transponder resonant circuit, the magnetic field provides the energy that is required for the stimulation of oscillations. The path of the oscillation can therefore be detected at the generator coil in the form of brief changes in voltage or current. This brief increase in coil current (or reduction in coil voltage) is known as a *dip*.

The relative magnitude of this dip is dependent upon the gap between the two coils, the quality of the induced resonant circuit (in the transponder) and above all the speed with which the resonant circuit is moved towards the magnetic field of the generator coil, which becomes weaker with increasing distance. In *anti-theft devices* (EAS) the maximum distance between the two coils is predetermined by constructional measures. The quality of the resonant circuit is also dependent upon the materials used.

It is not possible to influence the entry speed v into the generator field. A resonant circuit approaching at an infinitely lower speed would generate an infinitely small dip.

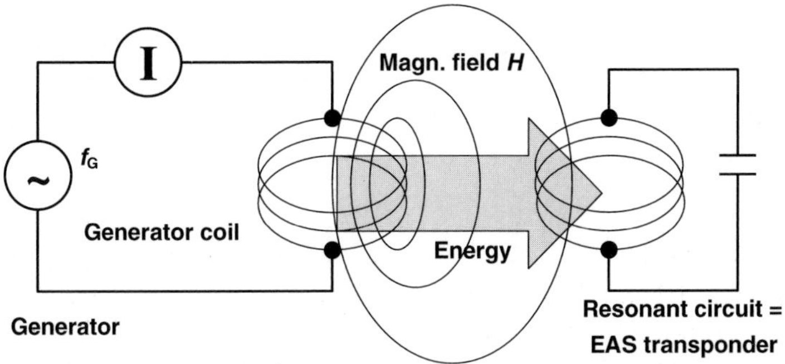

Figure 3.2: Operating principle of the EAS radio frequency procedure

However, the dip should be as clear as possible to guarantee reliable recognition of the attached resonant circuit (1 bit transponder),. This is achieved by using a small trick: the frequency of the magnetic field generated is not constant, it is "swept". This means that the generator frequency continuously crosses the range between a minimum and maximum frequency. The frequency range available to the swept systems is 8.2 MHz ± 10% [jörn].

Whenever the swept generator frequency exactly corresponds with the resonant frequency of the resonant circuit (in the transponder), the transponder begins to oscillate, producing a clear dip in the generator coil's supply current. In *sweep systems* the intensity of the dip is dependent upon the sweep frequency (speed of frequency change) rather than the transponder's speed of movement and can be adjusted to achieve an optimal recognition rate. Frequency tolerances of the transponder, which depend upon manufacturing tolerances or the presence of a metallic environment, therefore do not present a problem for reliable recognition.

Because the tag is not removed at the till, it must be altered so that it does not activate the anti-theft system. To achieve this, the cashier places the protected product into a device – the deactivator – that generates a sufficiently high magnetic field that the induced voltage destroys the foil capacitor of the transponder. The capacitors are designed with intentional

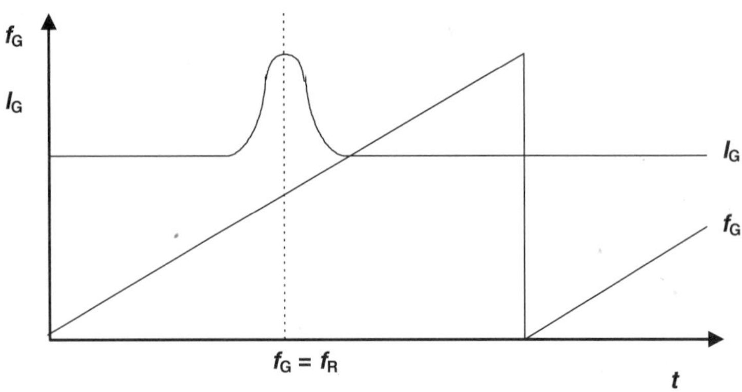

Figure 3.3: The occurrence of a "dip" in the resonant frequency. The generator frequency f_G is continuously swept between two cut-off frequencies. An RF tag in the generator field generates a clear dip at its resonant frequency f_R

short circuit points, so-called *dimples*. The breakdown of the capacitors is irreversible and disrupts the resonant circuit to such a degree that this can no longer be excited by the *sweep signal*.

Large area frame antennas are used to generate the required magnetic alternating field. The frame antennas are integrated into columns and combined to form gates. The classic design that can be seen in every large department store is illustrated in Figure 3.4. Gate widths of up to 2 m can be achieved using the RF procedure. The relatively low detection rate of 70% [gillert] is disproportionately influenced by certain product materials. Metals in particular (e.g. food tins) affect the resonant frequency of the tags and the coupling to the detector coil and thus have a negative effect on the detection rate. Tags of 50 x 50 mm must be used to achieve the gate width and detection rate mentioned above.

The range of products that have their own resonant frequencies (e.g. cable drums) present a great challenge for system manufacturers. If these resonant frequencies lie within the sweep frequency 8.2 MHz ± 10%, then they will always trigger false alarms.

Table 3.1: Typical system parameters [plotzke]

	System 1	System 2	System 3	System 4
Frequency / MHz:	1.86 – 2.18	7.44 – 8.73	7.30 – 8.70	7.40 – 8.60
Sweep frequency / Hz:	141	141	85	85

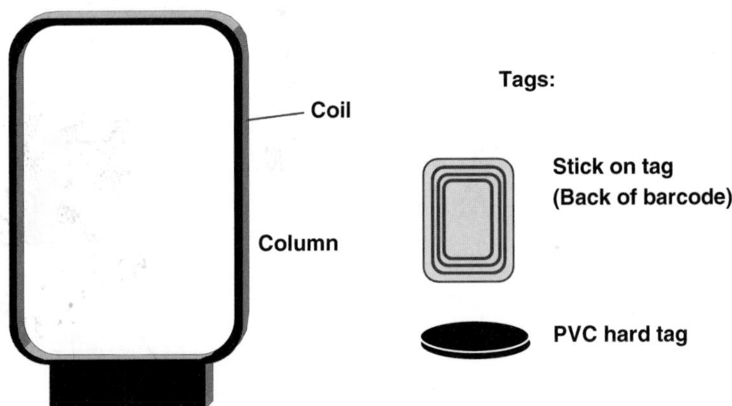

Figure 3.4: (Left) typical frame antenna of an RF system (height 1.20 – 1.60 m); (right) tag designs

3.1.2 Microwaves

EAS systems in the *microwave range* exploit the generation of harmonics at components with non-linear characteristic lines (e.g. diodes). The *harmonic* of a sinusoidal voltage *A* with a defined frequency f_A is a sinusoidal voltage *B*, whose frequency f_B is an integer multiple of the frequency f_A. The subharmonics of the frequency f_A are thus the frequencies $2f_A$, $3f_A$, $4f_A$ etc. The *N*th multiple of the output frequency is termed the *N*th harmonic (*N*th harmonic wave) in radio-engineering, the output frequency itself is termed the carrier wave or first harmonic.

In principle, every two-terminal network with a non-linear characteristic generates harmonics at the first harmonic. In the case of *non-linear resistances,* however, energy is consumed, so that only a small part of the first harmonic power is converted into the harmonic oscillation. Under favourable conditions, the multiplication of f to $n{\times}f$ occurs with an efficiency of $\eta = 1/n^2$. However, if non-linear energy storage is used for multiplication, then in the ideal case there are no losses [fleckner].

Capacitance diodes are particularly suitable non-linear energy stores for frequency multiplication. The number and intensity of the harmonics that are generated depend upon the capacitance diode's *dopant profile* and characteristic line gradient. The exponent n (also γ) is a measure for the gradient (= capacitance-voltage characteristic). For simple diffused diodes, this is 0.33 (e.g. BA110), for alloyed diodes it is 0.5 and for tuner diodes with a hyper-abrupt P-N junction it is around 0.75 (e.g. BB 141) [itt75].

The capacitance-voltage characteristic of alloyed capacitance diodes has a quadratic path and is therefore best suited for the doubling of frequencies. Simple diffused diodes can be used to produce higher harmonics [fleckner].

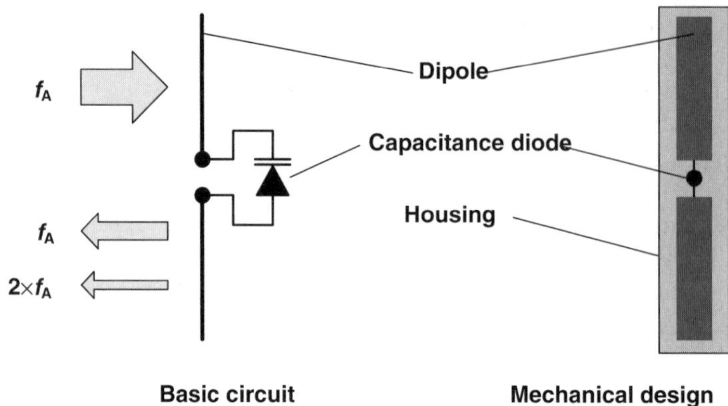

Figure 3.5: Basic circuit and typical construction format of a microwave tag

The layout of a 1 bit transponder for the generation of harmonics is extremely simple: a capacitance diode is connected to the base of a *dipole* adjusted to the carrier wave. Given a carrier wave frequency of 2.45 GHz the dipole has a total length of 6 cm. The carrier wave frequencies used are 915 MHz (outside Europe), 2.45 GHz or 5.6 GHz. If the transponder is located within the transmitter's range, then the flow of current within the diode generates and re-emits harmonics of the carrier wave. Particularly distinctive signals are obtained at two or three times the carrier wave, depending upon the type of diode used.

Transponders of this type cast in plastic (hard tags) are used mainly to protect textiles. The tags are removed at the till when the goods are paid for and they are subsequently reused.

Figure 3.6: Microwave tag in the interrogation zone of a detector

Figure 3.6 shows a transponder being placed within the range of a microwave transmitter operating at 2.45 GHz. The second harmonic of 4.90 GHz generated in the diode characteristic of the transponder is re-transmitted and detected by a receiver, which is adjusted to this precise frequency. The reception of a signal at the frequency of the 2nd harmonic can then trigger an alarm system.

If the amplitude or frequency of the carrier wave is modulated (ASK, FSK), then all harmonics incorporate the same modulation. This can be used to distinguish between "interference" and "useful" signals, preventing false alarms caused by external signals.

In the example above, the amplitude of the carrier wave is modulated with a signal of 1 kHz (100% ASK). The 2nd harmonic generated at the transponder is also modulated at 1 kHz ASK. The signal received at the receiver is demodulated and forwarded to a 1 kHz detector. Interference signals that happen to be at the reception frequency of 4.90 GHz cannot trigger false alarms because these are not normally modulated and, if they are, they will have a different modulation.

3.1.3 Frequency divider

This procedure operates in the long wave range at 100 – 135.5 kHz. The security tags contain a semiconductor circuit (microchip) and a resonant circuit coil made of wound enamelled copper. The resonant circuit is made to resonate at the operating frequency of the EAS system using a soldered capacitor. These transponders can be obtained in the form of hard tags (plastic) and are removed when goods are purchased.

The microchip in the transponder receives its power supply from the magnetic field of the security device (see Chapter 3, Section 3.2.1.1 "Power supply to passive transponders"). The frequency at the self inductive coil is divided by two by the microchip and sent back to the security device. The signal at half the original frequency is fed by a tap into the resonant circuit coil.

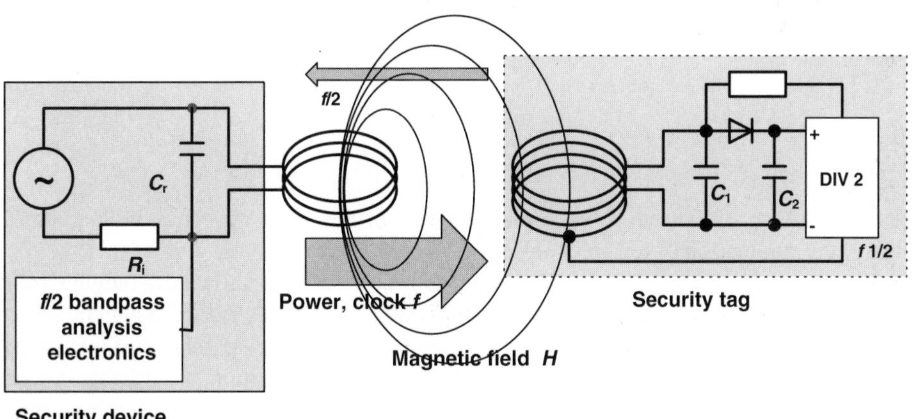

Figure 3.7: Basic circuit diagram of the EAS frequency division procedure: security tag (transponder) and detector (evaluation device)

The magnetic field of the security device is pulsed at a lower frequency (ASK modulated) to improve the detection rate. Similarly to the procedure for the generation of harmonics, the modulation of the carrier wave (ASK or FSK) is maintained at half the frequency (*subharmonic*). This is used to differentiate between "interference" and "useful" signals. This system almost entirely rules out false alarms.

Frame antennas, described in the section on RF systems, are used as sensor antennas.

Table 3.2: Typical system parameters [plotzke]

Frequency:	130 kHz
Modulation type:	100% ASK
Modulation frequency / modulation signal:	12.5 Hz or 25 Hz, rectangle 50%

3.1.4 *Electromagnetic types*

Electromagnetic types operate using strong magnetic fields in the *NF range* from 10 Hz to around 20 kHz. The security tag contains a soft magnetic *permalloy* strip with a steep flanked hysteresis curve, (see also Chapter 4 "Physical Principles of RFID Systems"). The polarity of the strips is periodically reversed by a strong magnetic alternating field. The sudden change of flux density B (non-linear characteristic: see 3.1.2 "Microwaves") in the strips in the vicinity of the zero crossover of the applied field strength H generates harmonics at the basic frequency of the security device, and these harmonics can be received and evaluated by the security device.

The electromagnetic type is optimised by superimposing additional signal sections with higher frequencies over the main signal. The marked nonlinearity of the strip's hysteresis curve generates not only harmonics but also signal sections with summation and differential frequencies of the supplied signals. Given a main signal of frequency f_S = 20 Hz and the additional signals f_1 = 3.5 and f_2 = 5.3 kHz the following signals are generated (1st order):

$$f_1 + f_2 = f_{1+2} = 8.80 \text{ kHz}$$
$$f_1 - f_2 = f_{1-2} = 1.80 \text{ kHz}$$
$$f_S + f_1 = f_{S+1} = 3.52 \text{ kHz} \quad \text{and so on ...}$$

The security device does not react to the harmonic of the basic frequency in this case, but rather to the summation or differential frequency of the extra signals.

The tags are available in the form of self-adhesive strips with lengths ranging from a few cm to 20 cm. Due to the extremely low operating frequency, electro-magnetic systems are the only system suitable for products containing metal. However, these systems have the disadvantage that the function of the tags is dependent upon position: for reliable detection the magnetic field lines of the security device must run vertically through the permalloy strips.

For deactivation, the tags are surrounded by a light magnetisable metal coating, which is magnetised at the till by a strong *permanent magnet* [plotzke]. Because of this permanent pre-magnetisation the polarity of the permalloy strips can no longer be reversed by the weaker alternating field of the security system and they are thus no longer detectable.

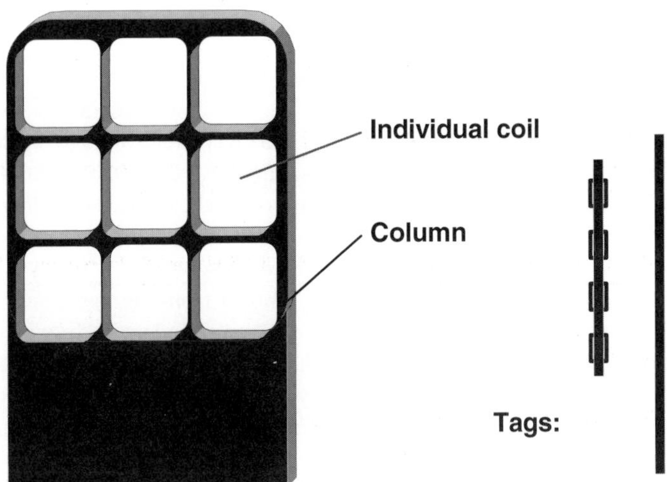

Figure 3.8: (Left) typical antenna design for a security system (height approximately 1.40 m); (right) possible tag designs

The tags can be reactivated at any time by demagnetisation. The process of deactivation and reactivation can be performed any number of times. For this reason, electro-magnetic goods protection systems were originally used mainly in lending libraries. Because the tags are small (min. 32 mm short strips) and cheap, these systems are now being used increasingly in the grocery industry.

In order to achieve the field strength necessary for demagnetisation of the permalloy strips, the field is generated by two coil systems in the columns at either side of a narrow passage. Several individual coils, typically 9 to 12, are located in the two pillars, and these generate weak magnetic fields in the centre and stronger magnetic fields on the outside [plotzke]. Gate widths of up to 1.50 m can now be realised using this method, whilst still achieving detection rates of 70 % [gillert].

Table 3.3: Typical system parameters [plotzke]

	System 1	System 2
Frequency:	215 Hz	21 Hz + 3.3 kHz + 5 kHz
Number of individual coils:	9	12
Max. Flux density B (individual coil):	1037 μT	118 μT

3.2 Full and Half Duplex Procedure

In contrast to 1 bit transponders, which normally exploit simple physical effects (oscillation stimulation procedures, stimulation of harmonics by diodes or the non-linear hysteresis curve of metals), the transponders described in this and subsequent sections use an electronic microchip as the data carrying device. This has a data storage capacity of up to a few Kbytes. To read from or write to the data carrying device it must be possible to transfer data between the transponder and a reader. This transfer takes place according to one of two main procedures: full and half duplex procedures, which are described in this section, and sequential systems, which are described in the following section.

In the *half duplex procedure* (HDX) the data transfer from the transponder to the reader alternates with data transfer from the reader to the transponder. At frequencies below 30 MHz this is most often used with the load modulation procedure, either with or without a subcarrier, which involves very simple circuitry. Closely related to this is the modulated reflected cross-section procedure that is familiar from radar technology and is used at frequencies above 100 MHz. Load modulation and modulated reflected cross-section procedures directly influence the magnetic or electromagnetic field generated by the reader and are therefore known as *"harmonic"* procedures.

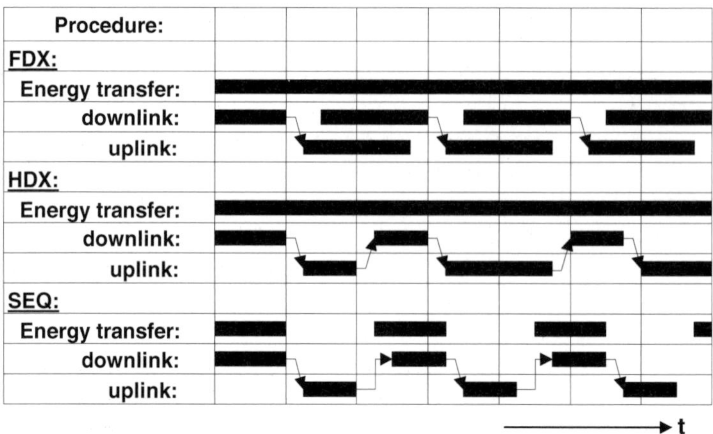

Figure 3.9: Representation of full duplex, half duplex and sequential systems over time. Data transfer from the reader to the transponder is termed downlink in the figure, whilst data transfer from the transponder to the reader is termed uplink

In the *full duplex procedure* (FDX) the data transfer from the transponder to the reader takes place at the same time as the data transfer from the reader to the transponder. This includes procedures in which data is transmitted from the transponder at a fraction of the frequency of the reader, i.e. a *"subharmonic"*, or at a completely independent, i.e. an *"anharmonic"* frequency.

However, both procedures have in common the fact that the transfer of energy from the reader to the transponder is continuous, i.e. it is independent of the direction of data flow. In sequential systems (SEQ), on the other hand, the transfer of energy from the transponder to the reader takes place for a limited period of time only (pulse operation → *pulsed system*). Data transfer from the transponder to the reader occurs in the pauses between the power supply to the transponder.

Unfortunately, the literature relating to RFID has not yet been able to agree a consistent nomenclature for these system variants. Rather, there has been a confusing and inconsistent classification of individual systems into full and half duplex procedures. Thus pulsed systems are often termed half duplex systems – this is correct from the point of view of data transfer – and all unpulsed systems are falsely classified as full duplex systems. For this reason, in this book pulsed systems – for differentiation from other procedures, and unlike most RFID literature(!) – are termed sequential systems (SEQ).

3.2.1 Inductive coupling

3.2.1.1 Power supply to passive transponders

An inductively coupled transponder comprises an electronic data carrying device, usually a single microchip and a large area coil that functions as an antenna.

Inductively coupled transponders are almost always operated passively. This means that all the energy needed for the operation of the microchip has to be provided by the reader. For this purpose, the reader's antenna coil generates a strong, high frequency electromagnetic field, which penetrates the cross-section of the coil area and the area around the coil. Because the wavelength of the frequency range used (< 135 kHz: 2400 m, 13.56 MHz: 22.1 m) is several times greater than the distance between the reader's antenna and the transponder, the electromagnetic field may be treated as a simple magnetic alternating field with regard to the distance between transponder and antenna (see Chapter 4, Section 4.2.1.1 "Transition from near to far field in magnetic antennas" for further details).

A small part of the emitted field penetrates the antenna coil of the transponder, which is some distance away from the coil of the reader. A voltage U_i is generated in the transponder's antenna coil by inductance. This voltage is rectified and serves as the power supply for the data carrying device (microchip). A capacitor C_r is connected in parallel with the reader's antenna coil, the capacitance of this capacitor being selected such that it works with the coil inductance of the antenna coil to form a parallel resonant circuit with a resonant frequency that corresponds with the transmission frequency of the reader. Very high currents are generated in the antenna coil of the reader by resonance step-up in the parallel resonant circuit, which can be used to generate the required field strengths for the operation of the remote transponder.

The antenna coil of the transponder and the capacitor C_1 form a resonant circuit tuned to the transmission frequency of the reader. The voltage U at the transponder coil reaches a maximum due to resonance step-up in the parallel resonant circuit.

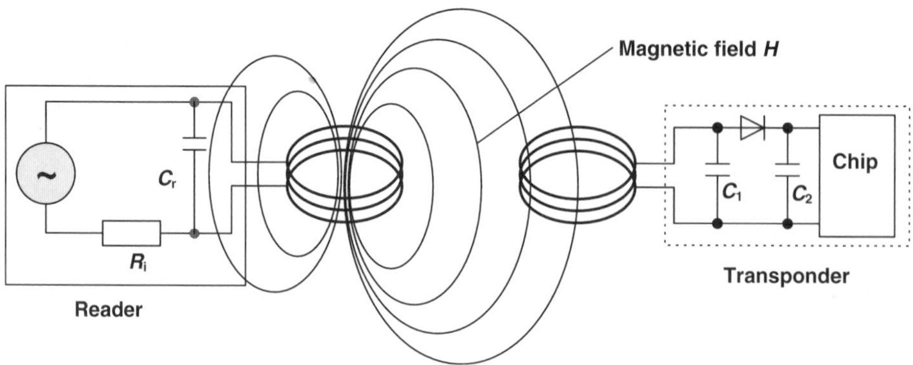

Figure 3.10: Power supply to an inductively coupled transponder from the energy of the magnetic alternating field generated by the reader

The layout of the two coils can also be interpreted as a transformer (*transformer coupling*), in which case there is only a very weak coupling between the two windings. The efficiency of power transfer between antenna coil of the reader and the transponder is proportional to the operating frequency f, the number of windings n, the area A enclosed by the transponder coil, the angle of the two coils relative to each other and the distance between the two coils.

Figure 3.11: Different designs of inductively coupled transponders. The photo shows half finished transponders, i.e. transponders before injection into a plastic housing (Reproduced by permission of AmaTech GmbH & Co. KG, D-Pfronten)

As frequency f increases, the required coil inductance of the transponder coil, and thus the number of windings n decreases (135 kHz: typical 100 – 1000 windings, 13.56 MHz: typical 3 – 10 windings). Because the voltage induced in the transponder is still proportional to frequency f (see Chapter 4 "Physical Principles of RFID Systems"), the reduced number of windings barely effects the efficiency of power transfer at higher frequencies.

Because they are nevertheless very inefficient, inductively coupled systems are only suited for the operation of low-current circuits. Only read only transponders (< 135 kHz) with extremely low power consumption are available for very great ranges > 1 m. Due to their high power consumption, transponders with a write function and complex security algorithms have a typical range of 15 cm, although there are some with ranges up to 80 cm.

Table 3.4: Overview of the power consumption of various RFID-ASIC building blocks [ATMEL]. The minimum supply voltage required for the operation of the microchip is 1.8 V, the maximum permissible voltage is 10 V

	Memory / Byte	Write / read distance	Power consumption	Frequency	Application
ASIC#1	6	15 cm	10 μA	120 kHz	Animal ID
ASIC#2	32	13 cm	600 μA	120 kHz	Goods flow, access check
ASIC#3	256	2 cm	6 μA	128 kHz	Public transport
ASIC#4	256	0.5 cm	< 1 mA	4 MHz*	Goods flow, public transport
ASIC#5	256	< 2 cm	~ 1 mA	4/13.56 MHz	Goods flow
ASIC#6	256	100 cm	500 μA	125 kHz	Access check
ASIC#7	2048	0.3 cm	< 10 mA	4.91 MHz*	Contactless chip cards
ASIC#8	1024	10 cm	~ 1 mA	13.56 MHz	Public transport
ASIC#9	8	100 cm	< 1 mA	125 kHz	Goods flow
ASIC#10	128	100 cm	< 1 mA	125 kHz	Access check

* Close coupling system

3.2.1.2 Data transfer transponder → reader

Load modulation As described above, inductively coupled systems are based upon a *transformer-type coupling* between the primary coil in the reader and the secondary coil in the transponder. This is true when the distance between the coils does not exceed 0.16 λ, so that the transponder is located in the *near field* of the transmitter antenna (for a more detailed definition of the near and far fields, please refer to Chapter 4 "Physical Principles of RFID Systems").

If a resonant transponder (i.e. the self-resonant frequency of the transponder corresponds with the transmission frequency of the reader) is placed within the magnetic alternating field of the reader's antenna, then this draws energy from the magnetic field. This additional power consumption can be measured as voltage drop at the internal resistance R_i in the reader through the supply current to the reader's antenna. The switching on and off of a load resistance at the transponder's antenna therefore effects voltage changes at the reader's antenna and thus has the effect of an amplitude modulation of the antenna voltage by the remote transponder. If the switching on and off of the load resistor is controlled by data, then this data can be transferred from the transponder to the reader. This type of data transfer is called *load modulation.*

To reclaim the data in the reader, the voltage measured at the reader's antenna is rectified. This represents the demodulation of an amplitude modulated signal. An example circuit is shown in Chapter 11, Section 11.3 "Low Cost Configuration – Reader IC U2270B".

Load modulation with subcarrier Due to the weak coupling between the reader antenna and the transponder antenna, the voltage fluctuations at the antenna of the reader that represent the useful signal are smaller by orders of magnitude than the output voltage of the reader.

In practice, for a 13.56 MHz system, given an antenna voltage of approximately 100V (voltage step-up by resonance) a useful signal of around 10 mV can be expected (= 80 dB signal/noise ratio). Because detecting this slight voltage change requires highly complicated circuitry, the modulation sidebands created by the amplitude modulation of the antenna voltage are utilised:

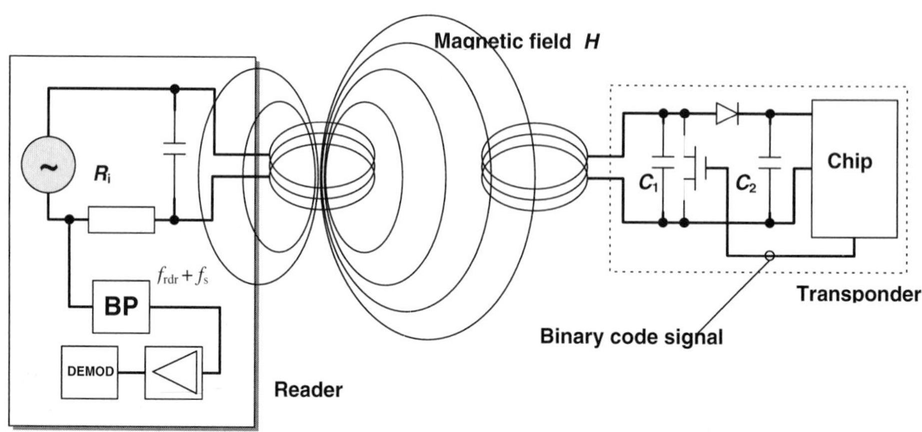

Figure 3.12: Generation of load modulation in the transponder by switching the drain-source-resistance of an FET on the chip. The reader illustrated is designed for the detection of a subcarrier

If the additional load resistor in the transponder is switched on and off at a very high elementary frequency f_S, then two spectral lines are created at a distance of $\pm f_S$ around the transmission frequency of the reader f_{READER}, and these can be easily detected (however f_S must be less than f_{READER}). In the terminology of radio technology the new elementary frequency is called a *subcarrier*). Data transfer is by ASK, FSK or PSK modulation of the subcarrier in time with the data flow. This represents an amplitude modulation of the subcarrier.

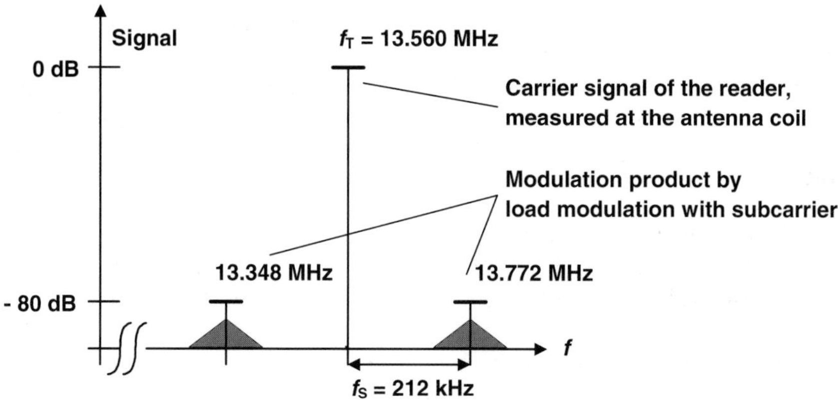

Figure 3.13: Load modulation creates two sidebands at a distance of the subcarrier frequency f_S around the transmission frequency of the reader. The actual information is carried in the sidebands of the two subcarrier sidebands, which are themselves created by the modulation of the subcarrier

Load modulation with a subcarrier creates two modulation sidebands at the reader's antenna at the distance of the subcarrier frequency around the operating frequency f_{READER}. These modulation sidebands can be separated from the significantly stronger signal of the reader by bandpass filtering (BP) on one of the two frequencies $f_{READER} \pm f_S$. Once it has been amplified, the subcarrier signal is now very simple to demodulate.

Because of the large bandwidth required for the transmission of a subcarrier, this procedure can only be used in the ISM frequency ranges for which this is permitted 6.78 MHz, 13.56 MHz and 27.125 MHz (see also Chapter 5 "Frequency Ranges and Radio Licensing Regulations").

Example circuit - load modulation with subcarrier Figure 3.14 shows an example circuit for a transponder using load modulation with a subcarrier. The circuit is designed for an operating frequency of 13.56 MHz and generates a subcarrier of 212 kHz.

The voltage induced at the antenna coil L1 by the magnetic alternating field of the reader is rectified using the bridge rectifier (D1 – D4) and after additional smoothing (C1) is available to the circuit as supply voltage. The parallel regulator (ZD 5V6) prevents the supply voltage from being subject to an uncontrolled increase when the transponder approaches the reader antenna.

Part of the high frequency antenna voltage (13.56 MHz) travels to the frequency divider's timing input (CLK) via the protective resistor (R1) and provides the transponder with the basis for the generation of an internal clocking signal. After division by 2^6 (= 64) a subcarrier clocking signal of 212 kHz is available at output Q7. The subcarrier clocking signal, controlled by a serial data flow at the data input (DATA), is passed to the switch (T1). If there is a logical HIGH signal at the data input (DATA), then the subcarrier clocking signal is passed to the switch (T1). The load resistor (R2) is then switched on and off in time with the subcarrier frequency.

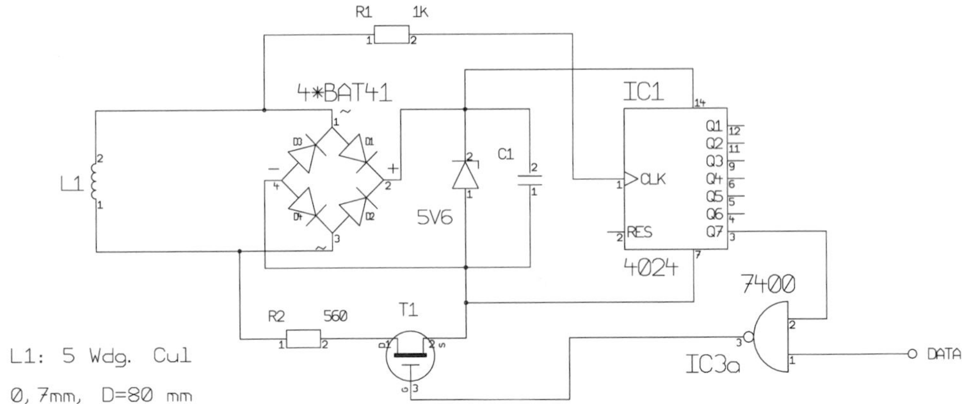

Figure 3.14: Example circuit for the generation of load modulation with subcarrier in an inductively coupled transponder

In the illustrated circuit an optional capacitor can be connected in parallel to coil L1 to make the transponder resonant circuit resonate at 13.56 MHz. The range of this "minimal transponder" can be significantly increased in this manner.

Subharmonic Procedure The subharmonic of a sinusoidal voltage *A* with a defined frequency f_A is a sinusoidal voltage *B*, whose frequency f_B is derived from an integer division of the frequency f_A. The subharmonics of the frequency f_A are therefore the frequencies $f_A/2, f_A/3, f_A/4$ ...

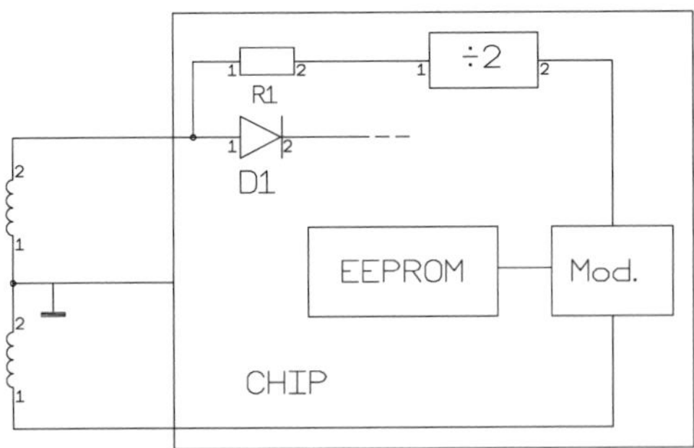

Figure 3.15: Basic circuit of a transponder with subharmonic back frequency. The received clocking signal is split into two, the data is modulated and fed into the transponder coil via a tap

In the subharmonic transfer procedure, a second frequency f_B, which is usually lower by a factor of two, is derived by digital division by two of the reader's transmission frequency f_A. The output signal f_B of a binary divider can now be modulated with the data stream from the transponder. The modulated signal is then fed back into the transponder's antenna via an output driver.

One popular operating frequency for subharmonic systems is 128 kHz. This gives rise to a transponder response frequency of 64 kHz.

The transponder's antenna consists of a coil with a central tap, whereby the power supply is taken from one end. The transponder's return signal is fed into the coil's second connection.

3.2.2 *Electromagnetic backscatter coupling*

3.2.2.1 Power supply to the transponder

RFID systems in which the gap between reader and transponder is greater than 1 m are called *long range systems*. In Europe these systems operate at frequencies between 433 MHz and 5.6 GHz. Outside Europe (Australia, North America) there are also individual systems in operation at 47 MHz or around 158 MHz.

The short wavelengths of these frequency ranges facilitate the construction of antennas with far smaller dimensions and greater efficiency than would be possible using the frequency ranges < 135 kHz and 13.56 MHz.

The efficiency of power transfer from the reader to the transponder can be estimated by calculating far field damping. Given a distance of 10 m between transmitter and receiver antenna, there is a path attenuation of approximately 60 dB at an operating frequency of 2.45 GHz. If we assume microchip power consumption of around 10 μW for the operation of a simple read only transponder (comparable to the power consumption of an inductively coupled transponder with the same functionality [sickert]), then power of 10 W (= 10μW + 60 dB) must be emitted from the reader (ERP) to supply the microprocessor. Unfortunately, statutory regulations in Europe only permit a maximum transmission power of 25 mW ERP. This is never enough to supply the microchip with energy for operation at the required range. *Microwave systems* therefore incorporate an additional battery to provide a power supply to the data carrying device. To prevent this battery from being subjected to unnecessary load, microchips generally have a power saving "power down" or "standby" mode. The power consumption of a chip in "power down" mode is not greater than a few μ*A*.

If the transponder moves out of range of a reader, then the chip automatically switches over to the power saving "power down" mode. The chip is not re-activated until a sufficiently strong signal from a reader is received, whereupon it switches back to normal operation.

3.2.2.2 Data transmission → reader

Modulated reflection cross-section We know from the field of *RADAR technology* that electromagnetic waves are reflected by objects with dimensions greater than around half the wavelength of the wave. The efficiency with which an object reflects electromagnetic waves is described by its *reflection cross-section*. Objects that are in resonance with the wave front that hits them, as is the case for antennas at the appropriate frequency for example, have a particularly large reflection cross-section.

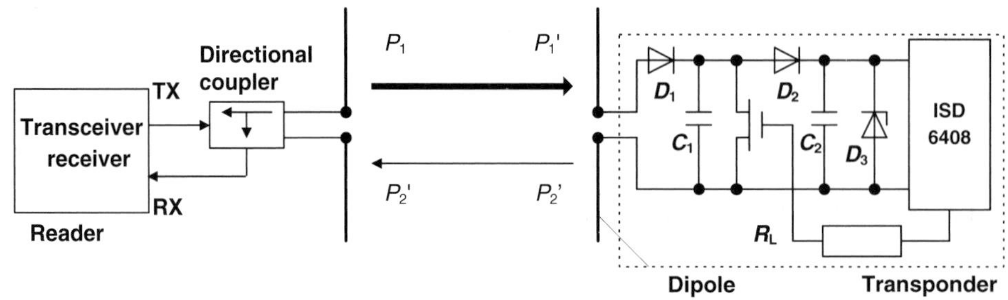

Figure 3.16: Operating principle of a backscatter transponder. The impedance of the chip is "modulated" by switching the chip's FET [isd]

Power P_1 is emitted from the reader's antenna, a small proportion of which (free space attenuation) reaches the transponder's antenna. The power P_1' is supplied to the antenna connections as HF voltage and after rectification by the diodes D_1 and D_2 this can be used as turn on voltage for the deactivation or activation of the power saving "power down" mode. The diodes used here are *low barrier Schottky diodes*, which have a particularly low threshold voltage. The voltage obtained may also be sufficient to serve as a power supply for short ranges.

A proportion of the incoming power P_1' is reflected by the antenna and returned as power P_2. The *reflection characteristics* (= reflection cross-section) of the antenna can be influenced by altering the load connected to the antenna. In order to transmit data from the transponder to the reader, a load resistor R_L connected in parallel with the antenna is switched on and off in time with the data stream to be transmitted. The amplitude of the power P_2 reflected from the transponder can thus be modulated (→ modulated backscatter).

The power P_2 reflected from the transponder is radiated into free space. A small proportion of this (free space attenuation) is picked up by the reader's antenna. The reflected signal therefore travels into the antenna connection of the reader in the "backwards direction" and can be decoupled using a *directional coupler* and transferred to the receiver input of a reader. The "forward" signal of the transmitter, which is stronger by powers of ten, is to a large degree suppressed by the directional coupler.

The ratio of power transmitted by the reader and power returning from the transponder (P_1 / P_2) can be estimated using the radar equation (for an explanation, refer to Chapter 4 "Physical Principles of RFID Systems").

3.2.3 Close coupling

3.2.3.1 Power supply to the transponder

Close coupling systems are designed for ranges between 0.1 cm and a maximum of 1 cm. The transponder is therefore inserted into the reader or placed onto a marked surface ("*touch & go*") for operation.

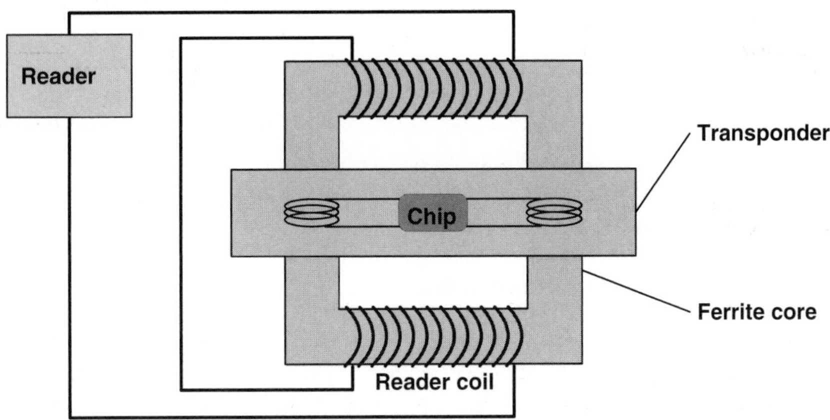

Figure 3.17: Close coupling transponder in an insertion reader with magnetic coupling coils

Inserting the transponder into the reader, or placing it on the reader, allows the transponder coil to be precisely positioned in the in the *air gap* of a ring-shaped or U-shaped core. The functional layout of the transponder coil and reader coil corresponds with that of a transformer. The reader represents the primary winding and the transponder coil represents the secondary winding of a transformer. A high frequency alternating current in the primary winding generates a high frequency magnetic field in the core and air gap of the arrangement, which also flows through the transponder coil. This power is rectified to provide a power supply to the chip.

Because the voltage U induced in the transponder coil is proportional to the frequency f of the exciting current, the frequency selected for power transfer should be as high as possible. In practice, frequencies in the range 1 – 10 MHz are used. In order to keep the losses in the transformer core low, a ferrite material that is suitable for this frequency must be selected as the core material.

Because, in contrast to inductively coupled or microwave systems, the efficiency of power transfer from reader to transponder is very good, close coupling systems are excellently suited for the operation of chips with a high power consumption. This includes microprocessors, which still require some 10 mW power for operation [sickert]. For this reason, the close coupling chip card systems on the market all contain microprocessors.

The mechanical and electrical parameters of contactless close coupling chip cards are defined in their own standard, ISO 10536. For other designs the operating parameters can be freely defined.

3.2.3.2 Data transfer transponder → reader

Magnetic coupling Load modulation with subcarrier is also used for magnetically coupled data transfer from the transponder to the reader in close coupling systems. Subcarrier frequency and modulation is specified in ISO 10536 for close coupling chip cards.

Capacitive coupling Due to the short distance between the reader and transponder, close coupling systems may also employ *capacitive coupling* for data transmission. Plate capacitors are constructed from coupling surfaces isolated from one another, and these are arranged in the transponder and reader such that when a transponder is inserted they are exactly parallel to one another.

This procedure is also used in close coupling smart cards. The mechanical and electrical characteristics of these cards are defined in ISO 10536.

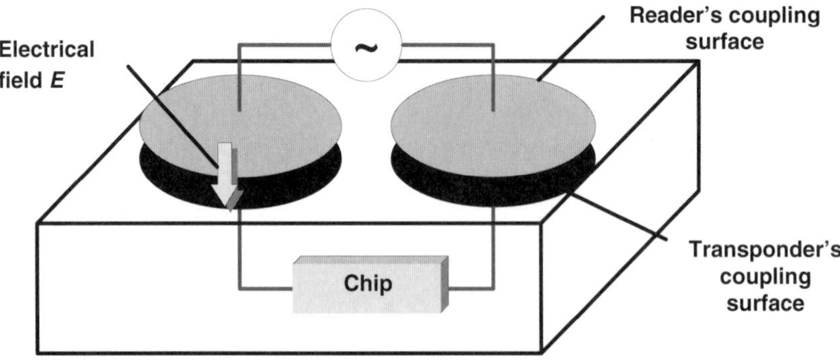

Figure 3.18: Capacitive coupling in close coupling systems occurs between two parallel metal surfaces positioned a short distance apart from each other

3.2.4 Data transfer reader → transponder

All known digital modulation procedures are used in data transfer from the reader to the transponder in full and half duplex systems, irrespective of the operating frequency or the coupling procedure. There are three basic procedures:

- ASK: Amplitude Shift Keying
- FSK: Frequency Shift Keying
- PSK: Phase Shift Keying

Because of the simplicity of demodulation, the majority of systems use ASK modulation.

3.3 Sequential Procedures

If the transmission of data and power from the reader to the data carrier alternates with data transfer from the transponder to the reader, then we speak of a *sequential procedure* (SEQ).

The characteristics used to differentiate between SEQ and other systems have already been described in the preceding section "Full and Half Duplex Procedures".

3.3.1 Inductive coupling

3.3.1.1 Power supply to the transponder

Sequential systems using inductive coupling are operated exclusively at frequencies below 135 kHz. A transformer type coupling is created between the reader's coil and the transponder's coil. The induced voltage generated in the transponder coil by the effect of an alternating field from the reader is rectified and can be used as a power supply.

In order to achieve higher efficiency of data transfer, the transponder frequency must be precisely matched to that of the reader, and the quality of the transponder coil must be carefully specified. For this reason the transponder contains an "*on-chip trimm capacitor*" to compensate for resonant frequency manufacturing tolerances.

However, unlike full and half duplex systems, in sequential systems the reader's transmitter does not operate on a continuous basis. The energy transferred to the transmitter during the transmission operation charges up a *charging capacitor* to provide an energy store. The transponder chip is switched over to standby or power saving mode during the charging operation, so that almost all of the energy received used to charge up the charging capacitor. After a fixed charging period the reader's transmitter is switched off again.

The energy stored in the transponder is used to send a reply to the reader. The minimum capacitance of the charging capacitor can be calculated from the necessary operating voltage and the chip's power consumption:

$$C = \frac{Q}{U} = \frac{I\,t}{\left[V_{\max} - V_{\min}\right]} \tag{3.1}$$

where V_{max}, V_{min} Limit values for operating voltage that may not be exceeded
I Power consumption of chip during operation
t Time required for the transmission of data from transponder to reader

For example, the parameters $I = 5\ \mu A$, $t = 20$ ms, $V_{max} = 4.5$ V and $V_{min} = 3.5$ V yield a charging capacitor of $C = 100$ nF [schürmann-93].

3.3.1.2 A comparison between FDX/HDX and SEQ systems

Figure 3.19 illustrates the different conditions arising from full / half duplex (FDX/HDX) and sequential (SEQ) systems.

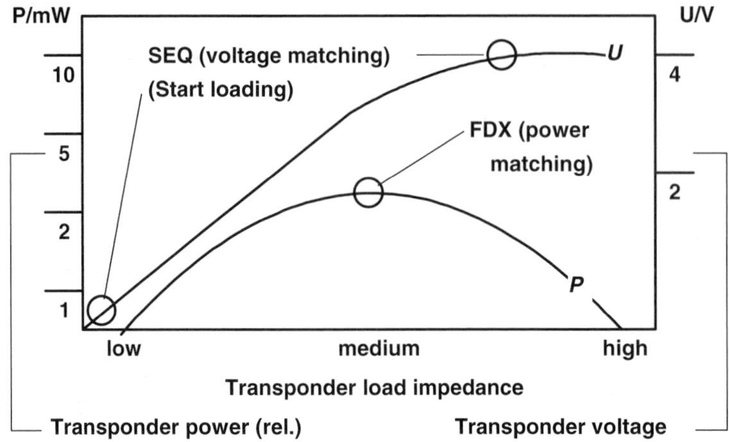

Figure 3.19: Comparison of induced transponder voltage in FDX/HDX and SEQ systems [schürmann-93]

Because the power supply from the reader to the transponder in full duplex systems occurs at the same time as data transfer in both directions, the chip is permanently in operating mode. *Power matching* between the transponder antenna (current source) and the chip (current consumer) is desirable to utilise the transmitted energy optimally. However, if precise power matching is used only half of the source voltage (= open circuit voltage of the coil) is available. The only option for increasing the available operating voltage is to increase the impedance (= load resistance) of the chip, however this is the same as decreasing the power consumption.

Therefore the design of full duplex systems is always a compromise between power matching (maximum power consumption P_{chip} at U_{chip} = ½ U_O) and voltage matching (minimum power consumption P_{chip} at maximum voltage $U_{chip} = U_O$).

The situation is completely different in sequential systems: during the charging process the chip is in standby or power saving mode, which means that almost no power is drawn through the chip.

The charging capacitor is fully discharged at the beginning of the charging process and therefore represents a very low ohmic load for the voltage source (Figure 3.19: Start loading). In this state, the maximum amount of current flows into the charging capacitor, whereas the voltage approaches zero (= *current matching*). As the charging capacitor is charged, the charging current starts to decrease according to an exponential function, and reaches zero when the capacitor is fully charged. The state of the charged capacitor corresponds with *voltage matching* at the transponder coil.

This achieves the following advantages for the chip power supply compared to a full/half duplex system:

- The full source voltage of the transponder coil is available for the operation of the chip. Thus the available operating voltage is up to twice that of a comparable full / half duplex system.
- The energy available to the chip is determined only by the capacitance of the charging capacitor and the charging period. Both values can in theory (!) be given any required magnitude. In full / half duplex systems the maximum power consumption of the chip is fixed by the power matching point (i.e. by the coil geometry and field strength H).

3.3.1.3 Data transmission transponder → reader

In sequential systems a full read cycle consists of two phases, the charging phase and the reading phase.

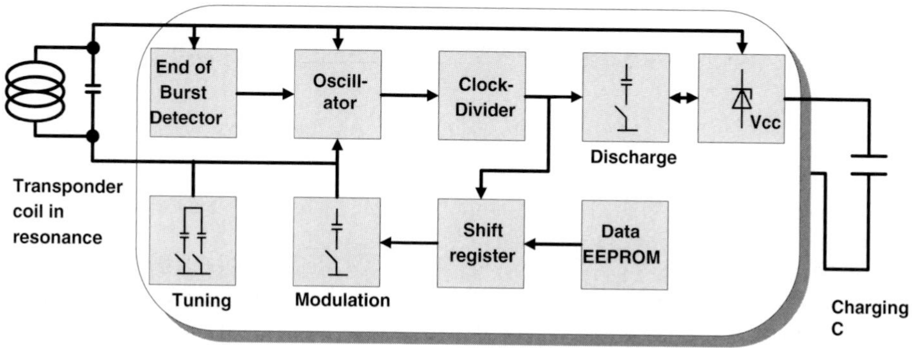

Figure 3.20: Block diagram of a sequential transponder by Texas Instruments TIRIS® Systems, using inductive coupling

The end of the charging phase is detected by an "*end of burst detector*", which monitors the path of voltage at the transponder coil and thus recognises the moment when the reader field is switched off. At the end of the charging phase an on-chip oscillator, which uses the resonant circuit formed by the transponder coil as a frequency determining component, is activated. A weak magnetic alternating field is generated by the transponder coil, and this can be received by the reader. This gives an improved signal-interference distance of typically 20 dB compared to full / half duplex systems, which has a positive effect upon the ranges that can be achieved using sequential systems.

The transmission frequency of the transponder corresponds with the resonant frequency of the transponder coil, which was adjusted to the transmission frequency of the reader when it was generated.

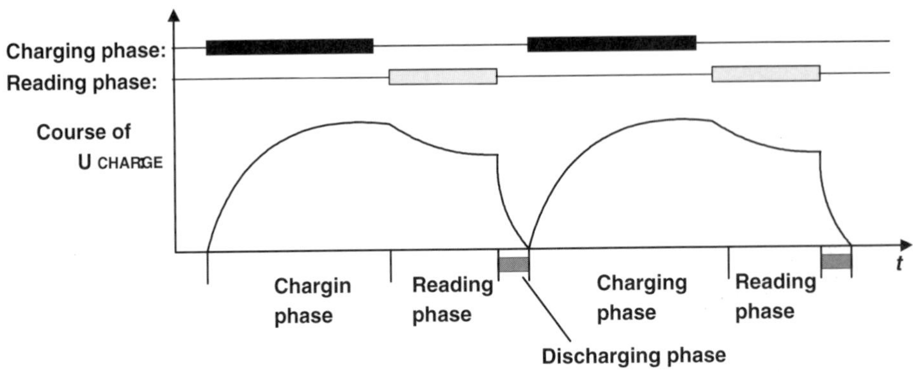

Figure 3.21: Voltage path of the charging capacitor of an inductively coupled SEQ transponder during operation

In order to be able to modulate the HF signal generated in the absence of a power supply, an additional modulation capacitor is connected in parallel with the resonant circuit in time with the data flow. The resulting frequency shift keying provides a *2 FSK modulation*.

After all the data has been transmitted, the discharge mode is activated to fully discharge the charging capacitor. This guarantees a safe Power-On-Reset at the start of the next charging cycle.

3.3.2 Surface acoustic wave transponder

Surface Acoustic Wave Devices (SAW) are based upon the piezoelectric effect and on the surface-related dispersion of elastic (= acoustic) waves at low speed. If an (ionic) crystal is elastically deformed in a certain direction, surface charges occur giving rise to electric voltages in the crystal (application: Piezo lighter). Conversely, the application of a surface charge to a crystal leads to an elastic deformation in the crystal grid (application: Piezo buzzer). Surface acoustic wave devices are operated at microwave frequencies, normally in the ISM range 2.45 GHz.

Electroacoustic converters (interdigital converters) and *reflectors* can be created using planar electrode structures on piezoelectric substrates. The normal substrate used for this application is *lithium niobate* or *lithium tantalate*. The electrode structure is created by a photolithographic procedure, similar to the procedure used in microelectronics for the manufacture of integrated circuits.

Figure 3.22 illustrates the basic layout of a surface wave transponder. A finger-shaped electrode structure – the *interdigital converter* – is positioned at the end of a long piezoelectrical substrate, a suitable *dipole antenna* for the operating frequency is attached to its busbar. The interdigital converter is used to convert between electrical signals and acoustic surface waves.

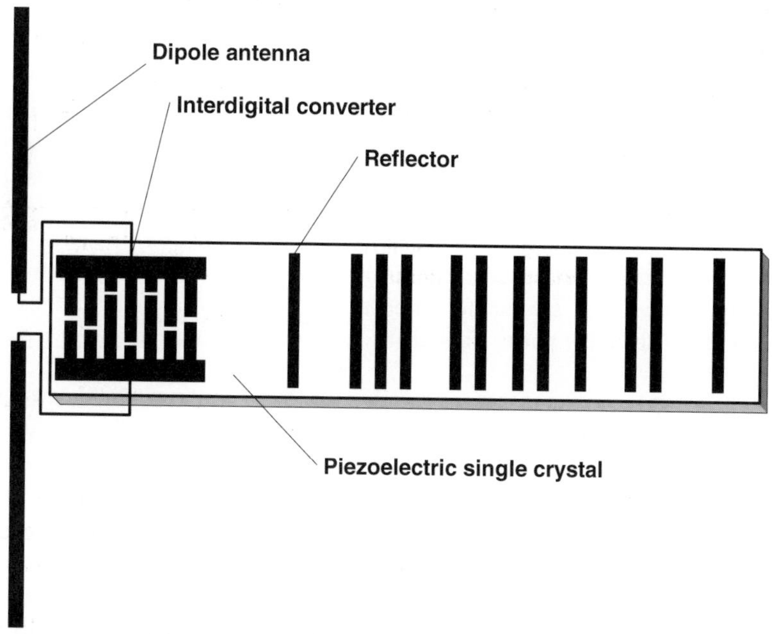

Figure 3.22: Basic layout of an SAW transponder. Interdigital converters and reflectors are positioned on the piezoelectric crystal

An electrical impulse applied to the busbar causes a mechanical deformation to the surface of the substrate due to the piezoelectrical effect between the electrodes (fingers), which disperses in both directions in the form of a surface wave (rayleigh wave). For a normal substrate the dispersion speed lies between 3000 and 4000 m/s. Similarly, a *surface wave* entering the converter creates an electrical impulse at the busbar of the interdigital converter due to the piezoelectric effect.

Individual electrodes are positioned along the remaining length of the surface wave transponder. The edges of the electrodes form a reflective strip and reflect a small proportion of the incoming surface waves. Reflector strips are normally made of aluminium, however some reflector strips are also in the form of etched grooves [meinke].

A high frequency *scanning pulse* generated by a reader is supplied from the dipole antenna of the transponder into the interdigital converter and is thus converted into an acoustic surface wave, which flows through the substrate in the longitudinal direction. The frequency of the surface wave corresponds with the carrier frequency of the sampling pulse (e.g. 2.45 GHz). The carrier frequency of the reflected and returned pulse sequence thus corresponds with the transmission frequency of the sampling pulse.

Part of the surface wave is reflected off each of the reflective strips that are distributed across the substrate, whilst the remaining part of the surface wave continues to travel to the end of the substrate and is absorbed there.

The reflected parts of the wave travel back to the interdigital converter, where they are converted into a high frequency pulse sequence and are emitted by the dipole antenna. This pulse sequence can be received by the reader. The number of pulses received corresponds with the number of reflective strips on the substrate. Likewise, the delay between the individual pulses is proportional to the spatial distance between the reflector strips on the substrate, and so the spatial layout of the reflector strips can represent a binary sequence of digits.

Due to the slow dispersion speed of the surface waves on the substrate the first response pulse is only received by the reader after a dead time of around 1.5 ms after the transmission of the scanning pulse. This gives decisive advantages for the reception of the pulse.

Reflections of the scanning pulse on the metal surfaces of the environment travel back to the antenna of the reader at the speed of light. A reflection over a distance of 100 m to the reader would arrive at the reader 0.6 ms after emission from the reader's antenna (travel time there and back, the signal is damped by > 160 dB). Therefore, when the transponder signal returns after 1.5 ms all reflections from the environment of the reader have long since died away, so they cannot lead to errors in the pulse sequence [dziggel].

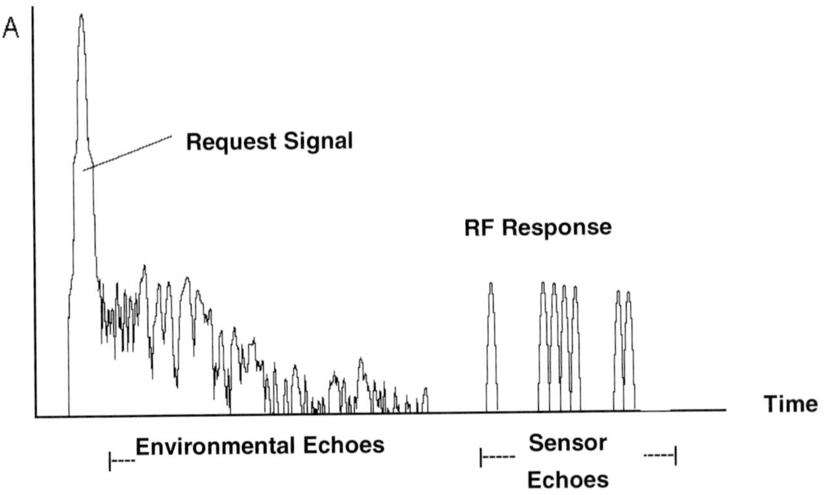

Figure 3.23: Only after the decay of the environmental echos do the sensor echoes of the surface acoustic wave transponder arrive (picture: Siemens AG, ZT KM, München)

Data storage capacity and data transfer speed of surface wave transponders depend upon the size of the substrate and the realisable minimum distance between the reflector strips on the substrate. In practice, around 16 – 32 bits are transferred at a data transfer rate of 500 kbit/s [sofis].

The range of a surface wave system depends mainly upon the transmission power of the scanning pulse and can be estimated using the radar equation (Chapter 4). At the permissible transmission power in the 2.45 GHz ISM frequency range a range of 1 – 2 m can be expected.

4

Physical Principles of RFID Systems

The vast majority of RFID systems operate according to the principle of *inductive coupling*. Therefore, understanding of the procedures of power and data transfer requires a thorough grounding in the physical principles of magnetic phenomena. This chapter therefore contains a particularly intensive study of the theory of magnetic fields from the point of view of RFID.

Electromagnetic fields – radio waves in the classic sense – are used in RFID systems that operate at above 30 MHz. To aid understanding of these systems we will investigate the propagation of waves in the far field and the principles of RADAR technology.

Electric fields play a secondary role and are only exploited for capacitive data transmission in close coupling systems. Therefore, this type of field will not be discussed further.

4.1 Magnetic Field

4.1.1 Magnetic field strength H

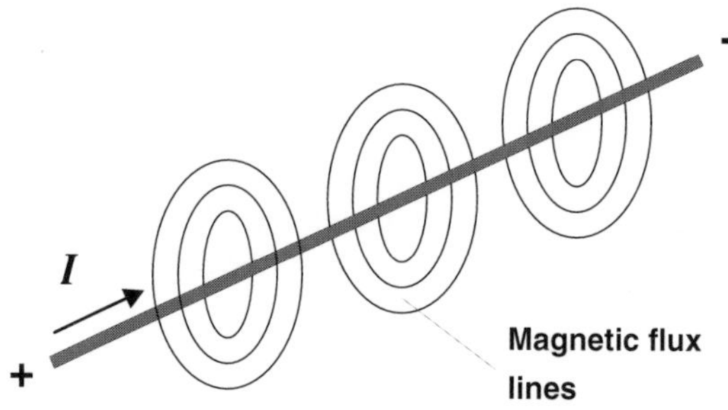

Figure 4.1: Lines of magnetic flux are generated around every current carrying conductor

Table 4.1: Constants used

Constant	Symbol	Value and unit
Electric field constant	ε_0	8.85×10^{-12} As/Vm
Magnetic field constant	μ_0	1.257×10^{-6} Vs/Am
Speed of light	c	299 792 km/s

Table 4.2: Units and abbreviations used

Variable	Symbol	Unit	Abbreviation
Magnetic field strength	H	Ampere per meter	A/m
Magnetic flux (n=number of windings)	Φ $\Psi = n\Phi$	Volt seconds	Vs
Magnetic inductance	B	Volt seconds per meter squared	Vs/m^2
Inductance	L	Henry	H
Mutual inductance	M	Henry	H
Electric field strength	E	Volts per metre	V/m
Electric current	I	Ampere	A
Electric voltage	U	Volt	V
Capacitance	C	Farad	F
Frequency	f	Hertz	Hz
Angular frequency	$\omega = 2\pi f$	1/seconds	1/s
Length	l	Metre	m
Area	A	Metre squared	m^2
Speed	v	Metres per second	m/s
Impedance	Z	Ohm	Ω
Wavelength	λ	Metre	m
Power	P	Watt	W
Power density	S	Watts per metre squared	W/m^2

Every moving charge (electrons in wires or in a vacuum), i.e. flow of current, is associated with a *magnetic field*. The intensity of the magnetic field can be demonstrated experimentally by the forces acting on a magnetic needle (compass) or a second electric current. The magnitude of the magnetic field is described by the *magnetic field strength H* regardless of the material properties of the space.

In the general form we can say that: "the contour integral of magnetic field strength along a closed curve is equal to the sum of the current strengths of the currents within it" [kuchling].

$$\sum I = \oint \vec{H} \cdot d\vec{s} \tag{4.1}$$

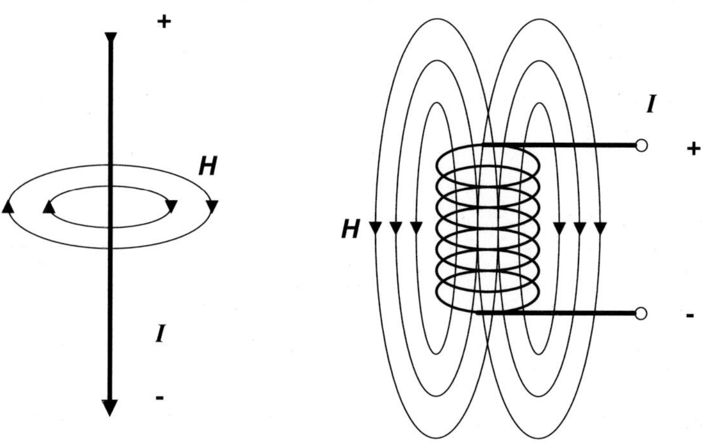

Figure 4.2: Lines of magnetic flux around a current carrying conductor and a current carrying cylindrical coil

We can use this formula to calculate the field strength H for different types of conductor:

In a straight conductor the field strength H along a circular *flux line* at a distance r is constant. The following is true [kuchling]:

$$H = \frac{1}{2\pi\, r} \tag{4.2}$$

4.1.1.1 Path of field strength H(x) in conductor loops

So-called "short cylindrical coils" or conductor loops are used as magnetic antennas to generate the *magnetic alternating field* in the write/read devices of inductively coupled RFID systems.

If the measuring point is moved away from the centre of the coil along the coil axis (x axis), then the strength of the field H will decrease as the distance x is increased. A more in-depth investigation shows that the field strength in relation to the radius (or area) of the coil remains constant up to a certain distance and then falls rapidly (see Figure 4.4). In free space, the decay of field strength is approximately 60 dB per decade in the near field of the coil, and flattens out to 20 dB per decade in the far field of the electromagnetic wave that is generated (a more precise explanation of these effects can be found in the section entitled "The generation of electromagnetic waves").

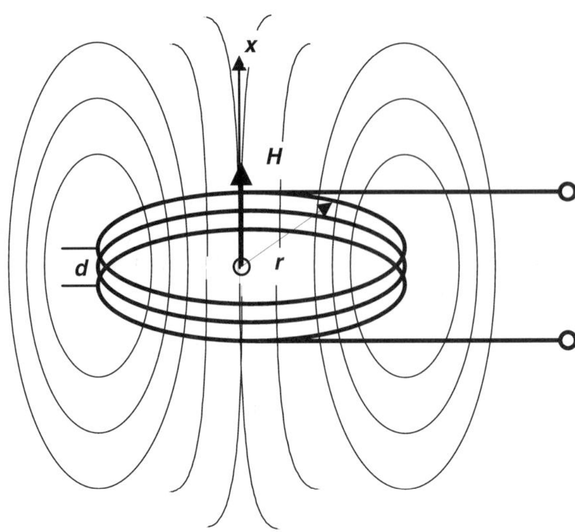

Figure 4.3: The path of the lines of magnetic flux around a "short" cylindrical coil, or conductor loop, similar to those employed in the transmitter antennas of inductively coupled RFID systems

The following equation can be used to calculate the path of field strength along the x axis of a round coil (= conductor loop) similar to those employed in the transmitter antennas of inductively coupled RFID systems [paul]:

$$H = \frac{I \cdot N \cdot R^2}{2\sqrt{\left(R^2 + x^2\right)^3}} \qquad (4.3)$$

where N: number of windings, R: circle radius r, x: distance from the centre of the coil in the x direction. The following boundary condition applies to this equation: $d \ll R$ and $x < \lambda/2\pi$ (the transition into the electromagnetic far field begins at a distance $> 2\pi$, see the section entitled "Generation of electromagnetic waves").

At distance 0 or, in other words, at the centre of the antenna, the formula can be simplified to [kuchling]:

$$H = \frac{I \cdot N}{2R} \qquad (4.4)$$

We can calculate the *field strength path* of a rectangular conductor loop with edge length $a{\cdot}b$ at a distance of x using the following equation. This format is often used as a transmitter antenna.

$$H = \frac{N \cdot I \cdot ab}{4\pi\sqrt{\left(\frac{a}{2}\right)^2 + \left(\frac{b}{2}\right)^2 + x^2}} \cdot \left(\frac{1}{\left(\frac{a}{2}\right)^2 + x^2} + \frac{1}{\left(\frac{b}{2}\right)^2 + x^2} \right) \quad (4.5)$$

The following figure shows the calculated field strength path $H(x)$ for three different antennas at a distance 0 – 20 m. The number of windings and the antenna current are constant in each case, the antennas differ only in radius R. The calculation is based upon the following values: $H1{:}R = 55$ cm, $H2{:}R = 7.5$ cm, $H3{:}R = 1$ cm.

The calculation results confirm that the increase in field strength flattens out at short distances ($x < R$) from the antenna coil. Interestingly, the smallest antenna exhibits a significantly higher field strength at the centre of the antenna (distance = 0), but at greater distances ($x > R$) the largest antenna generates a significantly higher field strength. It is vital that this effect is taken into account in the design of antennas for inductively coupled RFID systems.

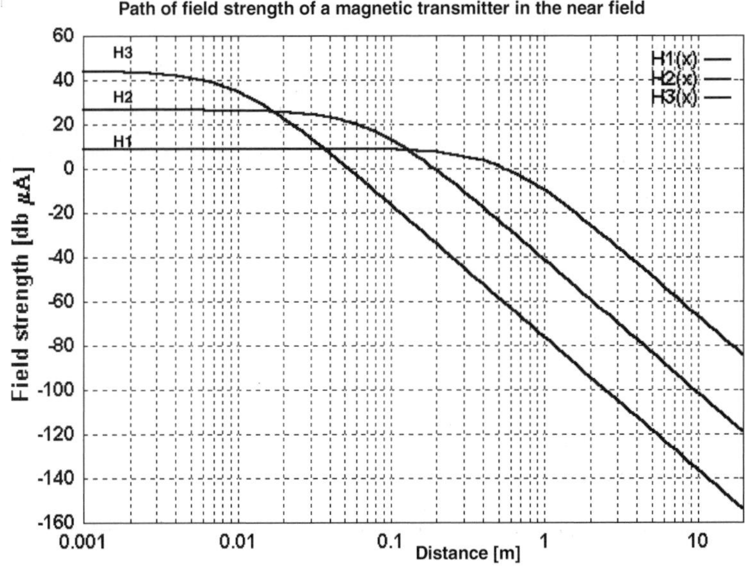

Figure 4.4: Path of magnetic field strength H in the near field of "short" cylinder coils, or conductor coils, as the distance in the x direction is increased

4.1.1.2 Optimal antenna diameter

If the radius R of the transmitter antenna is varied at a constant distance x from the transmitter antenna under the simplifying assumption of constant coil current I in the transmitter antenna, then field strength H is found to be at its highest where $R \approx x$. This is not surprising because even where distance $x > 0$, an increase in radius R brings about a proportionate decrease in the field strength. A reduction in the radius R of the transmitter antenna where distance $x > 0$ brings about a supraproportional reduction in the field strength $H(x) \sim x^{-3}$ in the case where the distance x is greater than the radius R. A change in the coil radius R at a constant distance x gives rise to the path of field strength H shown in Figure 4.5: field strength H increases rapidly where $x > R$; field strength H reaches a maximum where $x = R$, thereafter decreasing in proportion to the distance x.

As a rule of thumb, we can assume that the *read range* of a reader, i.e. the interrogation zone within which a transponder can be read, roughly corresponds with the radius of the transmitter antenna. From this we find that the optimal antenna radius R for a given system range x_{MAX} can be expressed as $R \approx x_{MAX}$.

However, an accurate estimate of a system's maximum read range requires knowledge of the interrogation field strength of the transponder in question (the minimum field strength H required for operation). If the antenna radius R is too great, then the field strength H, calculated according to equation 4.3, may be too low to supply the transponder with sufficient operating energy, even where distance $x = 0$.

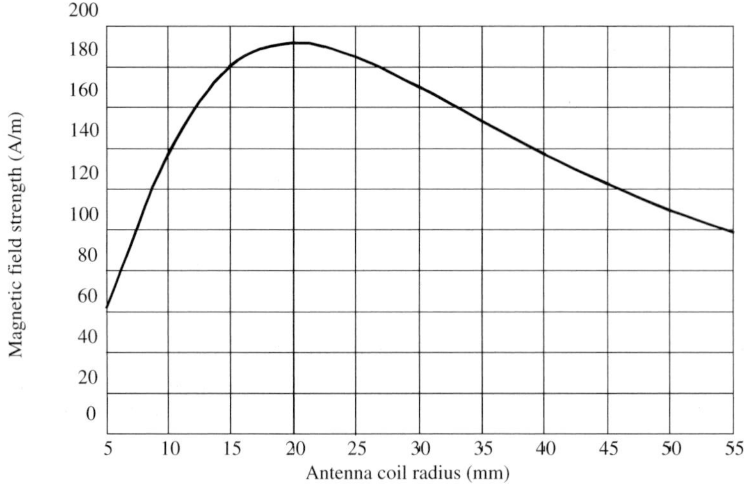

Figure 4.5: Path of magnetic field strength H at a distance $x = 20$ mm, where the coil radius $R = 5 - 55$ (Reproduced by permission of TEMIC Semiconductor GmbH, Heilbronn)

4.1.2 Magnetic flux and magnetic flux density

The magnetic field of a (cylindrical) coil will exert a force on a magnetic needle. If a soft iron core is inserted into a (cylindrical) coil – all other things remaining equal – then the force acting on the magnetic needle will increase. The quotient $I \times N$ (Section 4.1.1) remains constant and therefore so does field strength. However, the flux density – the total number of flux lines – which is decisive for the force generated (cf. [Pauls]), has increased.

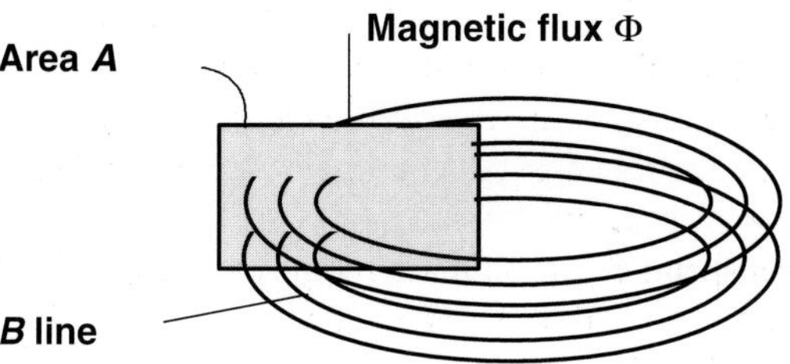

Figure 4.6: Relationship between magnetic flux Φ and flux density *B*

The total number of lines of magnetic flux that pass through the inside of a cylindrical coil, for example, is denoted by *magnetic flux* Φ. Magnetic flux density *B* is a further variable related to area *A* (this variable is often referred to as "magnetic inductance *B* in the literature") [reichel]. Magnetic flux is expressed as

$$\Phi = B \cdot A \tag{4.6}$$

The material relationship between flux density *B* and field strength *H* is expressed by the material equation:

$$B = \mu_0 \mu_r H = \mu H \tag{4.7}$$

The constant μ_0 is the magnetic field constant ($\mu_0 = 4\pi \times 10^{-6}$ Vs/Am) and describes the permeability (= magnetic conductivity) of a vacuum. The variable μ_r is called relative permeability and indicates how much greater than or less than μ_0 the permeability of a material is.

4.1.3 Inductance L

A magnetic field, and thus a magnetic flux Φ, will be generated around a conductor of any shape. This will be particularly intense if the conductor is in the form of a loop (coil). Normally, there is not one conduction loop, but *N* loops of the same area *A*, through which the same current *I* flows. Each of the conduction loops contributes the same proportion Φ to the total flux ψ [paul].

$$\Psi = \sum_N \Phi_N = N \cdot \Phi = N \cdot \mu \cdot H \cdot A \tag{4.8}$$

The ratio of the interlinked flux ψ that arises in an area enclosed by current *I*, to the current in the conductor that encloses it (conductor loop) is denoted by *inductance L*.

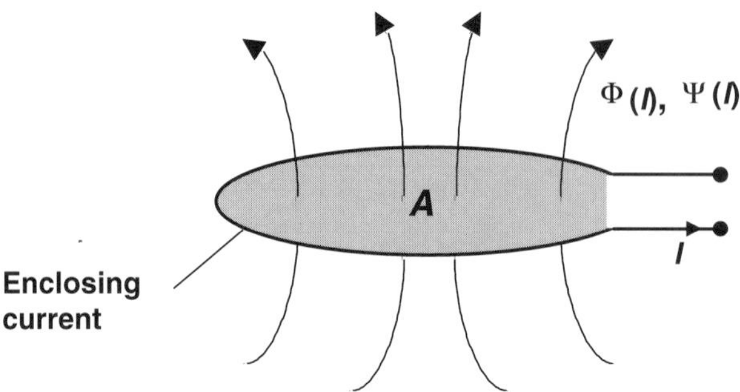

Figure 4.7: Definition of inductance L

$$L = \frac{\Psi}{I} = \frac{N \cdot \Phi}{I} = \frac{N \cdot \mu \cdot H \cdot A}{I} \tag{4.9}$$

Inductance is one of the characteristic variables of conductor loops (coils). The inductance of a conductor loop (coil) depends totally upon the material properties (permeability) of the space that the flux flows through and the geometry of the layout.

Inductance of a conductor loop

If we assume that the diameter d of the wire used is very small compared to the diameter D of the conductor coil ($d/D < 0.0001$) a very simple approximation can be used:

$$L = N^2 \mu_0 R \cdot \ln\left(\frac{2R}{d}\right) \tag{4.10}$$

where R is the radius of the conductor loop and d is the diameter of the wire used.

4.1.4 *Mutual inductance M*

If a second conductor loop 2 (area A_2) is located in the vicinity of conductor loop 1 (area A_1), through which a current is flowing, then this will be subject to a proportion of the total magnetic flux Φ flowing through A_1. The two circuits are connected together by this partial flux or coupling flux. The magnitude of the coupling flux ψ_{21} depends upon the geometric dimensions of both conductor loops, the position of the conductor loops in relation to one another, and the magnetic properties of the medium (e.g. permeability) in the layout.

Similarly to the definition of the (self) inductance L of a conductor loop, the *mutual inductance* M_{21} of conductor loop 2 in relation to conductor loop 1 is defined as the ratio of the partial flux ψ_{21} enclosed by conductor loop 2, to the current I_1 in conductor loop 1 [paul]:

$$M_{21} = \frac{\Psi_{21}(I_1)}{I_1} = \oint_{A2} \frac{B_2(I_1)}{I_1} \cdot dA_2 \tag{4.11}$$

Similarly, there is also a mutual inductance M_{12}. Here, current I_2 flows through the conductor loop 2, thereby determining the coupling flux ψ_{12} in loop 1. The following relationship applies:

$$M = M_{12} = M_{21} \tag{4.12}$$

Mutual inductance describes the coupling of two circuits via the medium of a magnetic field. Mutual inductance is always present between two electric circuits. Its dimension and unit are the same as for inductance.

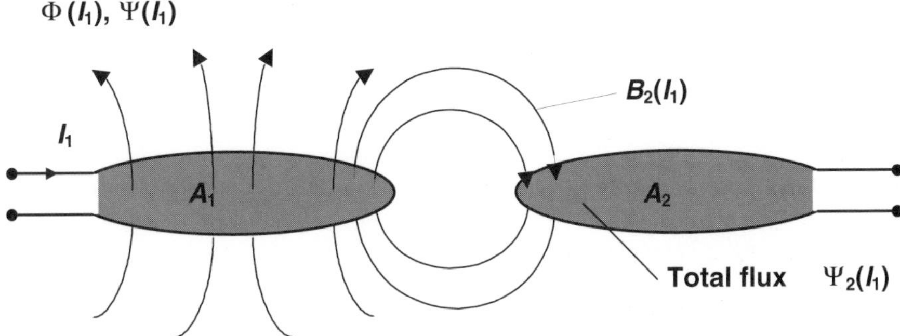

Figure 4.8: The definition of mutual inductance M_{21} by the coupling of two coils via a partial magnetic flow

The coupling of two electric circuits via the magnetic field is the physical principle upon which inductively coupled RFID systems are based. Figure 4.9 shows a calculation of the mutual inductance between a transponder antenna and three different reader antennas, which differ only in diameter. The calculation is based upon the following values: M_1: R = 55 cm, M_2: R = 7.5 cm, M_3: R = 1 cm, transponder: R = 3.5 cm. N = 1 for all reader antennas.

The path of mutual inductance shows a strong similarity to the path of the magnetic field strength in the near field of a conductor loop (see Section 4.1.1). However, this is not surprising because $\Psi_{A2}(I_1) = N \cdot B_2(I_1) \cdot A_2 = \mu_0 \cdot N \cdot H(x, I_1) \cdot A_2$. Mutual inductance M_3, however, is a little lower overall because the extremely small reader antenna (1 cm) means that the reader antenna field can only flow through a small proportion of area A_2 of the transponder antenna.

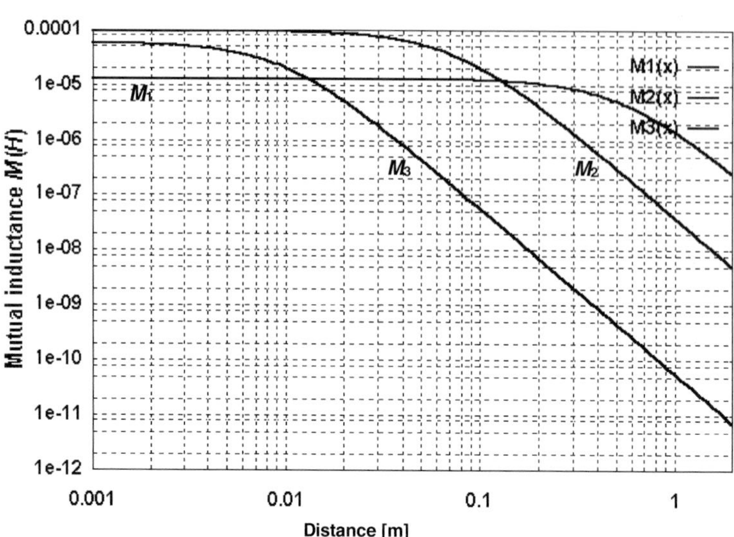

Figure 4.9: Path of mutual inductance between reader and transponder antenna as the distance in the *x* direction increases

4.1.5 Coupling coefficient k

Mutual inductance is a quantitative description of the flux coupling of two conductor loops. The *coupling coefficient k* is introduced so that we can make a qualitative prediction about the coupling of the conductor loops independent of their geometric dimensions. The following applies

$$k = \frac{M}{\sqrt{L_1 \cdot L_2}} \tag{4.13}$$

The coupling coefficient always varies between the two extreme cases $0 \leq k \leq 1$.

- k=0: Full decoupling due to great distance or magnetic shielding.
- k=1: Total coupling. Both coils are subject to the same magnetic flux Φ. The transformer is a technical application of total coupling, whereby two or more coils are wound onto a highly permeable iron core.

An analytic calculation is only possible for very simple antenna configurations. For two parallel conductor loops centred on a single x-axis the coupling coefficient according to [roz] can be approximated from the following equation. However, this only applies where

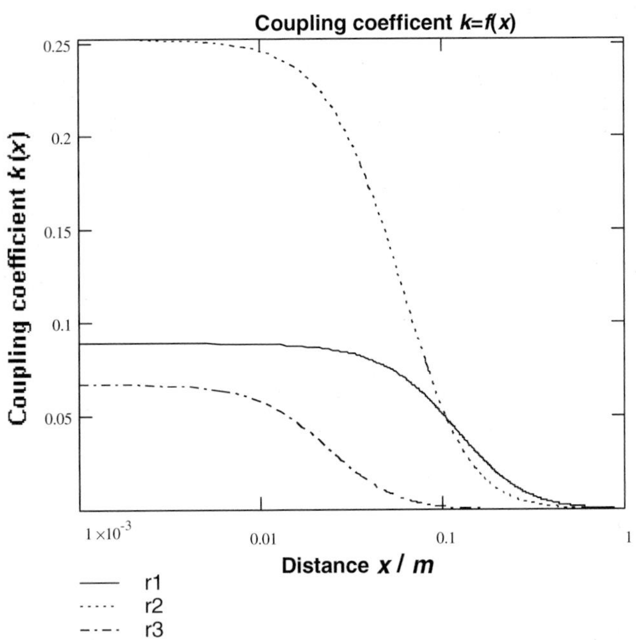

Figure 4.10: Path of the coupling coefficient for different sized conductor loops. Transponder antenna: $r_{Transp} = 2$ cm, reader antenna: $r_1 = 10$ cm, $r_2 = 7.5$ cm, $r_3 = 1$ cm

the radii of the conductor loops fulfil the condition $r_{Transp} \leq r_{Reader}$. The distance between the conductor loops on the x-axis is denoted by x.

$$k(x) \approx \frac{r_{Transp}^2 \cdot r_{Reader}^2}{\sqrt{r_{Transp} \cdot r_{Reader}} \cdot \left(\sqrt{x^2 + r_{Reader}^2}\right)^3} \tag{4.14}$$

Due to the permanent connection between the coupling coefficient and mutual inductance M, and because of the relationship $M = M_{12} = M_{21}$, the formula is also applicable to transmitter antennas smaller than the transponder antenna. Where $r_{Transp} \geq r_{Reader}$, we write

$$k(x) \approx \frac{r_{Transp}^2 \cdot r_{Reader}^2}{\sqrt{r_{Transp} \cdot r_{Reader}} \cdot \left(\sqrt{x^2 + r_{Transp}^2}\right)^3} \tag{4.15}$$

The coupling coefficient $k(x)$ is 1 (= 100%) is achieved where the distance between the conductor loops is zero (x=0) and the antenna radii are the same ($r_{Transp} = r_{Reader}$), because in this case the conductor loops are in the same place and are exposed to exactly the same magnetic flux ψ.

In practice, inductively coupled transponder systems operate with coupling coefficients that may be as low as 0.01 (< 1%).

4.1.6 Faraday's law

Any change to the magnetic flux Φ generates an electric field strength E_i. This characteristic of the magnetic field is described by *Faraday's law.*

The effect of the electric field generated in this manner depends upon the material properties of the surrounding area. Figure 4.11 shows some of the possible effects [paul]:

- Vacuum: In this case, the field strength E gives rise to an *electric rotational field.* Periodic changes in magnetic flux (high frequency current in an antenna coil) generate an electromagnetic field that propagates itself into the distance.
- Open conductor loop: An open circuit voltage builds up across the ends of an almost closed conductor loop, which is normally called induced voltage or *induced voltage.* This voltage corresponds with the line integral (path integral) of the field strength E that is generated along the path of the conductor loop in space.

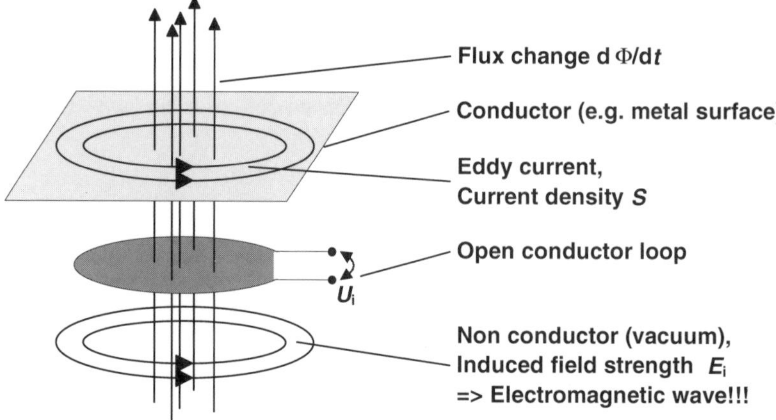

Figure 4.11: Induced electric field strength E in different materials. From above: Metal surface, conductor loop and vacuum

- Metal surface: An electric field strength E is also induced in the metal surface. This causes free charge carriers to flow in the direction of the electric field strength. Currents flowing in circles are created, so-called *eddy currents.* This works against the exciting magnetic flux (Lenz's law), which may significantly damp the magnetic flux in the vicinity of *metal surfaces.* However, this effect is undesirable in inductively coupled RFID systems (installation of a transponder or reader antenna on a metal surface) and must therefore be prevented by suitable countermeasures (see Section 4.1.12.3 "Ferrite shielding in a metallic environment").

In its general form Faraday's law is written follows:

$$u_{\mathrm{i}} = \oint E_{\mathrm{i}} \cdot \mathrm{d}s = -\frac{\mathrm{d}\Psi(t)}{\mathrm{d}t} \qquad (4.16)$$

For a conductor loop configuration with N windings, we can also say that $u_{\mathrm{i}} = N \cdot \mathrm{d}\Psi/\mathrm{d}t$. (The value of the contour integral $\int E_{\mathrm{i}} \cdot \mathrm{d}s$ can be increased N-times if the closed integration path is carried out N-times [paul]).

To improve our understanding of inductively coupled RFID systems we will now consider the effect of inductance on magnetically coupled conduction loops.

A time variant current $i_1(t)$ in conduction loop L_1 generates a time variant magnetic flux $\mathrm{d}\Phi(i_1)/\mathrm{d}t$. In accordance with the inductance law, a voltage is induced in the conductor loops L_1 and L_2 that through which some degree of magnetic flux is flowing. We can differentiate between two cases:

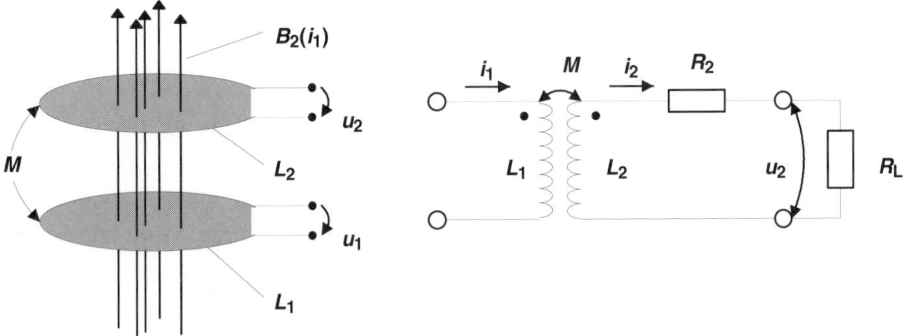

Figure 4.12: Left: Magnetically coupled conductor loops. Right: Equivalent circuit diagram for magnetically coupled conductor loops

- *Self inductance* The flux change generated by the current change $\mathrm{d}i_n/\mathrm{d}t$ induces a voltage u_n in the same conductor circuit
- *Mutual inductance*: The flux change generated by the current change $\mathrm{d}i_n/\mathrm{d}t$ induces a voltage in the adjacent conductor circuit L_m. Both circuits are coupled by mutual inductance.

Figure 4.12 shows the equivalent circuit diagram for coupled conductor loops. In an inductively coupled RFID system L_1 would be the transmitter antenna of the reader. L_2 represents the antenna of the transponder, where R_2 is the *coil resistance* of the transponder antenna. The current consumption of the data memory is symbolised by the load resistor R_L.

A time varying flux in the conductor loop L_1 induces voltage u_{2i} in the conductor loop L_2 due to mutual inductance M. The flow of current creates an additional voltage drop across the coil resistance R_2, meaning that the voltage u_2 can be measured at the terminals. The current through the load resistor R_L is calculated from the expression u_2/R_L. The current through L_2 also generates an additional magnetic flux, which opposes the magnetic flux $\Psi_1(i_1)$. The above is summed up in the following equation:

$$u_2 = +\frac{d\Psi_2}{dt} = M\frac{di_1}{dt} - L_2\frac{di_2}{dt} - i_2 R_2 \tag{4.17}$$

Because, in practice, i_1 and i_2 are sinusoidal (HF) alternating currents we write equation 4.17 in the more appropriate complex notation (where $\omega = 2\pi f$):

$$u_2 = j\omega M \cdot i_1 - j\omega L_2 \cdot i_2 - i_2 R_2 \tag{4.18}$$

If i_2 is replaced by u_2/R_L in equation 4.18, then we can solve the equation for u_2

$$u_2 = \frac{j\omega M \cdot i_1}{1 + \frac{j\omega L_2 + R_2}{R_L}} \quad \begin{cases} R_L \to \infty: & u_2 = j\omega M \cdot i_1 \\ R_L \to 0: & u_2 \to 0 \end{cases} \tag{4.19}$$

4.1.7 *Resonance*

The voltage u_2 induced in the transponder coil is used to provide the power supply to the data memory (*microchip*) of a passive transponder (see Section 4.1.8.1 "Power supply to the transponder"). In order to significantly improve the efficiency of the equivalent circuit diagram illustrated in Figure 4.12, a capacitor C_2 is connected in parallel with the transponder coil L_2 to form a parallel resonant circuit with a resonant frequency that corresponds with the operating frequency of the RFID system in question. However, in 13.56 MHz systems with anticollision procedures, the resonant frequency selected for the transponder is often 1–5 MHz higher to minimise the effect of the interaction of transponders on overall performance. This is because the overall resonant frequency of two transponders directly adjacent to one another is always lower than the resonant frequency of a single transponder. The resonant frequency of the parallel resonant circuit can be calculated using the Thomson equation:

$$f = \frac{1}{2\pi\sqrt{L_2 \cdot C_2}} \tag{4.20}$$

In practice, C_2 is made up of a parallel capacitor C_2' and a parasitic capacitance C_p from the real circuit. $C_2 = (C_2' + C_p)$. The required capacitance for the parallel capacitor C_2' is found using the Thomson equation, taking into account the parasitic capacitance C_p:

$$C_2' = \frac{1}{(2\pi f)^2 L_2} - C_p \tag{4.21}$$

Figure 4.13 shows the equivalent circuit diagram of a real transponder. R_2 is the natural resistance of the transponder coil L_2, while the current consumption of the data carrier (chip) is represented by the load resistor R_L.

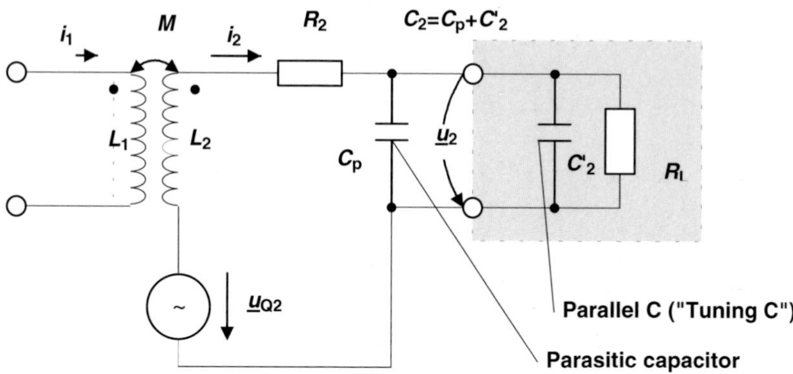

Figure 4.13: Equivalent circuit diagram for magnetically coupled conductor loops. Transponder coil L_2 and parallel capacitor C_2 form a parallel resonant circuit to improve the efficiency of voltage transfer. The transponder's data carrier is represented by the grey box

If a voltage $u_{Q2} = u_i$ is induced in the coil L_2, the following voltage u_2 can be measured at the data carrier load resistor R_L in the equivalent circuit diagram shown in Figure 4.13:

$$u_2 = \frac{ju_{Q2}}{1+\left(j\omega\ L_2 + R_2\right)\cdot\left(\frac{1}{R_L} + j\omega\ C_2\right)} \tag{4.22}$$

We now replace the induced voltage $u_{Q2} = ui$ with the factor responsible for its generation

$$u_{Q2} = u_i = j\omega M\, i_1 = \omega k\sqrt{L_1 L_2} \tag{4.23}$$

thus obtaining the relationship between voltage u_2 and the magnetic coupling of transmitter and transponder coil:

$$u_2 = \frac{j\omega M \cdot i_1}{1+\left(j\omega\ L_2 + R_2\right)\cdot\left(\frac{1}{R_L} + j\omega\ C_2\right)} \tag{4.24}$$

and

$$u_2 = \frac{j\omega \cdot k \cdot \sqrt{L_1 \cdot L_2} \cdot i_1}{1 + \left(j\omega\ L_2 + R_2\right) \cdot \left(\frac{1}{R_L} + j\omega\ C_2 \right)} \tag{4.25}$$

or in the non complex form [jurisch]:

$$u_2 = \frac{\omega \cdot k \cdot \sqrt{L_1 L_2} \cdot i_1}{\sqrt{\left(\frac{\omega L_2}{R_L} + \omega R_2 C_2 \right)^2 + \left(1 - \omega^2 L_2 C_2 + \frac{R_2}{R_L} \right)^2}} \tag{4.26}$$

where $C_2 = C_2' + C_p$.

Figure 4.14 shows the simulated path of u_2 over a large frequency range for a possible transponder system, with and without resonance. The current i_1 in the transmitter antenna (and thus also $\Phi(i_1)$), inductance L_2, mutual inductance M and R_2 and R_2 are held constant over the entire frequency range.

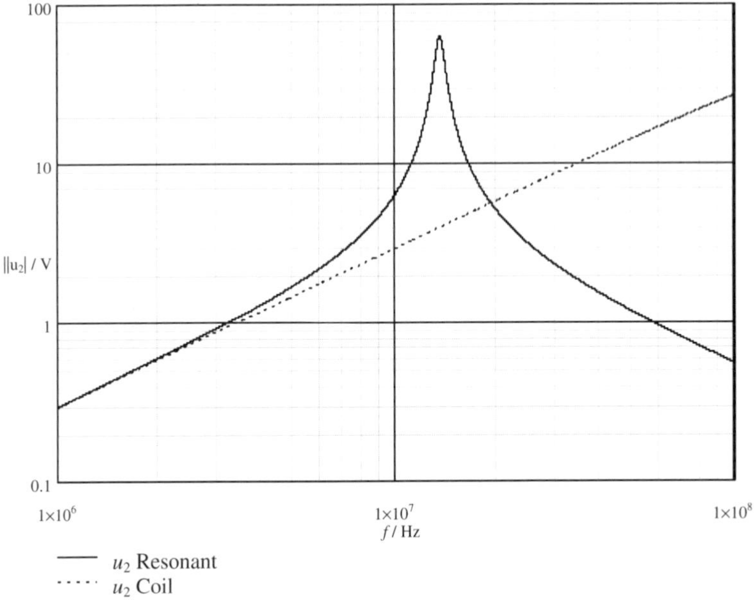

Figure 4.14: Voltage path at a transponder coil in the frequency range 1 to 100 MHz, given a constant magnetic field strength H or constant current i_1. A transponder coil with a parallel capacitor shows a clear voltage step-up when excited at its resonant frequency (f_{RES} = 13.56 MHz)

We see that the path of voltage u_2 for the circuit with the coil alone (circuit from Figure 4.12) is almost identical to that of the parallel resonant circuit (circuit from Figure 4.13) below its resonant frequency, but that when the resonant frequency is reached, voltage u_2 increases by more than a power of ten in the parallel resonant circuit compared to the voltage u_2 for the coil alone. Above the resonant frequency, however, the voltage u_2 falls rapidly in the parallel resonant circuit, even falling below the value for the coil alone.

For transponders in the frequency range below 135 kHz, the transponder coil L_2 is generally connected in parallel with a chip capacitor ($C_2' = 20 - 220$ pF), to achieve the desired resonant frequency. At the higher frequencies 13.56 MHz and 27.125 MHz, the required capacitance C_2 is usually so low that it is provided by the input capacitance of the data carrier together with the parasitic capacitance of the transponder coil.

Let us now investigate the influence of the circuit elements R_2, R_L and L_2 on voltage u_2. To gain a better understanding of interactions between the individual parameters we will now introduce the Q factor (the Q factor crops up again when we investigate the connection of transmitter antennas in section 11.4.1.3 – "The influence of the Q factor"). We will refrain from deriving formulae because the electric resonant circuit is dealt with in detail in the background reading.

The Q factor is a measure for the voltage and current step-up in a resonant circuit at resonant frequency. Its reciprocal 1/Q denotes the circuit damping d. The Q factor is very simple to calculate for the equivalent circuit in Figure 4.13. In this case ω is the circuit frequency ($\omega = 2\pi f$) of the transponder resonant circuit:

$$Q = \frac{1}{R_2 \cdot \sqrt{\frac{C_2}{L_2}} + \frac{1}{R_L} \cdot \sqrt{\frac{L_2}{C_2}}} = \frac{1}{\frac{R_2}{\omega\ L_2} + \frac{\omega\ L_2}{R_L}} \tag{4.27}$$

A glance at equation 4.27 shows that when $R_2 \rightarrow \infty$ and $R_L \rightarrow 0$, the Q factor also tends towards zero. On the other hand, when the transponder coil has a very low coil resistance $R_2 \rightarrow 0$ and there is a very high load resistor $R_L >> 0$ (in accordance with very low transponder chip power consumption), very high Q factors can be achieved. The voltage u_2 is now proportional to the quality of the resonant circuit, which means that the relationship between voltage u_2 and R_2 and R_L is clearly defined.

Voltage u_2 thus tends towards zero where $R_2 \rightarrow \infty$ and $R_L \rightarrow 0$. At a very low transponder coil resistance $R_2 \rightarrow 0$ and a high value load resistor $R_L >> 0$, a very high voltage u_2 can be achieved (compare equation 4.26).

It is interesting to note the path of voltage u_2 when the inductance of the transponder coil L_2 is unchanged, in which case the resonance condition is maintained (i.e. $C_2 = 1/\omega_2 L_2$ for all values of L_2). We see that for certain values of L_2, voltage u_2 reaches a clear peak (Figure 4.15).

If we now consider the path of the Q factor as a function of L_2 (Figure 4.16), then we observe a maximum at the same value for transponder inductance L_2. The maximum voltage $u_2 = f(L_2)$ is therefore derived from the maximum Q factor, $Q = f(L_2)$, at this point.

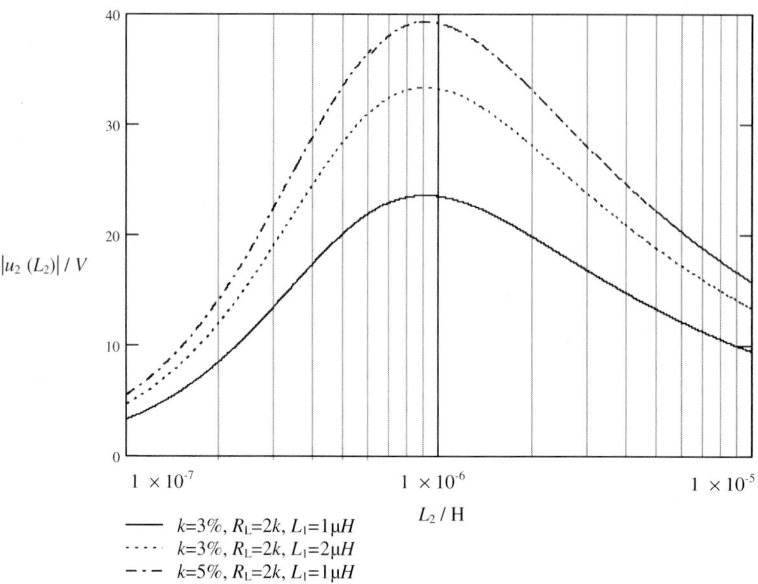

Figure 4.15: Path of voltage u_2 for different values of transponder inductance L_2. The resonant frequency of the transponder is equal to the transmission frequency of the reader for all values of L_2 (i_1 = 0.5 A, f = 13.56 MHz, R_2 = 1Ω)

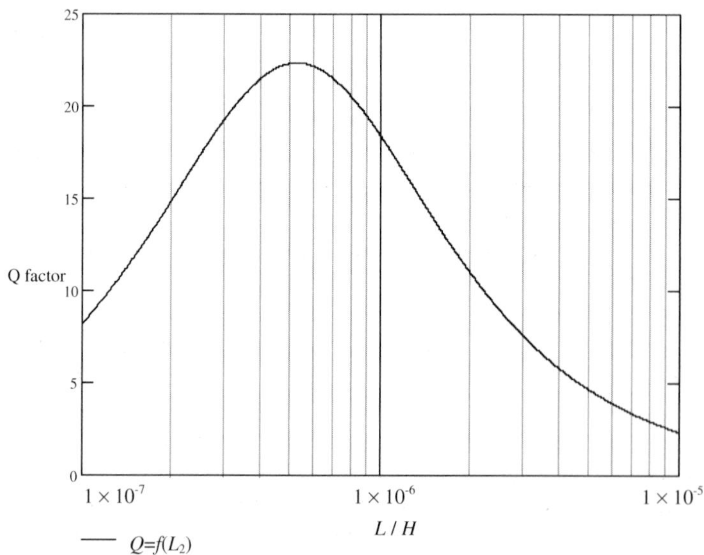

Figure 4.16: Path of the Q factor as a function of transponder inductance L_2, where the resonant frequency of the transponder is constant (f = 13.56 MHz, R_2 = 1Ω)

This indicates that for every pair of parameters (R_2, R_L), there is an inductance value L_2 at which the Q factor, and thus also the supply voltage u_2 to the data carrier, is at a maximum. This should always be taken into consideration when designing a transponder, because this effect can be exploited to optimise the energy range of an inductively coupled RFID system. However, we must also bear in mind that the influence of component tolerances in the system also reaches a maximum in the Q_{max} range. This is particularly important in systems designed for mass production, where the whole system should be designed so that reliable operation is still possible in the range $Q << Q_{max}$ at the maximum distance between transponder and reader.

R_L should be set at the same value as the input resistance of the data carrier after setting the "power on" reset, i.e. after the activation of the voltage regulator, as is the case for the maximum energy range of the system.

4.1.8 Practical operation of the transponder

4.1.8.1 Power supply to the transponder

Transponders are classified as active or passive depending upon the type of power supply they use.

Active transponders incorporate their own battery to provide the power supply to the data carrier. In these transponders, the voltage u_2 is generally only required to generate a "wake up" signal. As soon as the voltage u_2 exceeds a certain limit this signal is activated and puts the data carrier into operating mode. The transponder returns to the power saving "sleep" or "standby mode" after the completion of a transaction with the reader, or when the voltage u_2 falls below a minimum value.

In passive transponders the data carrier has to obtain its power supply from the voltage u_2. To achieve this, the voltage u_2 is converted into direct current using a low loss bridge rectifier and then smoothed. A simple basic circuit for this application is shown in Figure 3.14 (Chapter 3, Section 3.2.1.2 "Data transfer transponder reader").

4.1.8.2 Voltage regulation

The induced voltage u_2 in the transponder coil very rapidly reaches high values due to resonance step-up in the resonant circuit. Considering the example in Figure 4.14, if we increase the coupling coefficient k – possibly by reducing the gap between reader and transponder – or the value of the load resistor R_L, then voltage u_2 will reach a level much greater than a hundred volts. However, the operation of a data carrier requires a constant *operating voltage* of 3–5 V (after rectification).

In order to regulate voltage u_2 independently of the coupling coefficient k or other parameters, and to hold it constant in practice, a voltage-dependent shunt resistor R_S is connected in parallel with the load resistor R_L.

As induced voltage $u_{Q2} = u_i$ increases, the value of the shunt resistor R_S falls, thus reducing the quality of the transponder resonant circuit to such a degree that the voltage u_2 remains constant. To calculate the value of the shunt resistor for different variables, we refer back to equation 4.25 and introduce the parallel connection of R_L and R_S in place of the constant load resistor R_L. The variable voltage u_2 is replaced by the constant voltage u_{Transp} – the desired input voltage of the data carrier – giving the following equation for R_S.

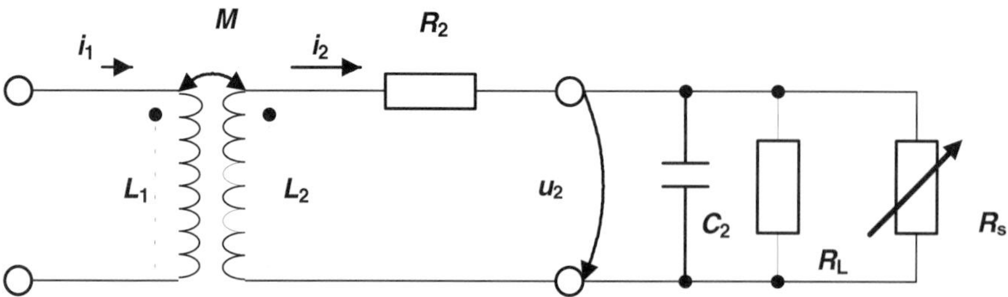

Figure 4.17: Operating principle of the voltage regulation in the transponder using a shunt regulator

$$R_S = \left| \frac{1}{\dfrac{\left(\dfrac{j\omega \cdot k \cdot \sqrt{L_1 L_2} \cdot i_1}{u_{\text{Transp}}}\right) - 1}{j\omega L_2 + R_2} - j\omega C_2 - \dfrac{1}{R_L}} \right|_{u_{2-\text{unreg}} > u_{\text{Transp}}} \qquad (4.28)$$

for lower values of $u_{2\text{ - unreg}}$, $R_S = \infty$.

Figure 4.18 shows the path of voltage u_2 with an "ideal" shunt regulator. Voltage u_2 initially increases in proportion with the coupling coefficient k. When u_2 reaches its desired value, the value of the shunt resistor begins to fall inversely to k, thus maintaining a constant value for voltage u_2.

Figure 4.19 shows the variable value of the shunt resistor R_S as a function of the coupling coefficient. In this example the value range for the shunt resistor covers several powers of ten. This can only be achieved using a semiconductor circuit, therefore so-called *shunt* or *parallel regulators* are used in inductively coupled transponders. This is an electronic regulator circuit, the internal resistance of which falls disproportionately sharply when a threshold voltage is exceeded. A simple shunt regulator based upon a zener diode [Nürmann] is shown in Figure 4.20.

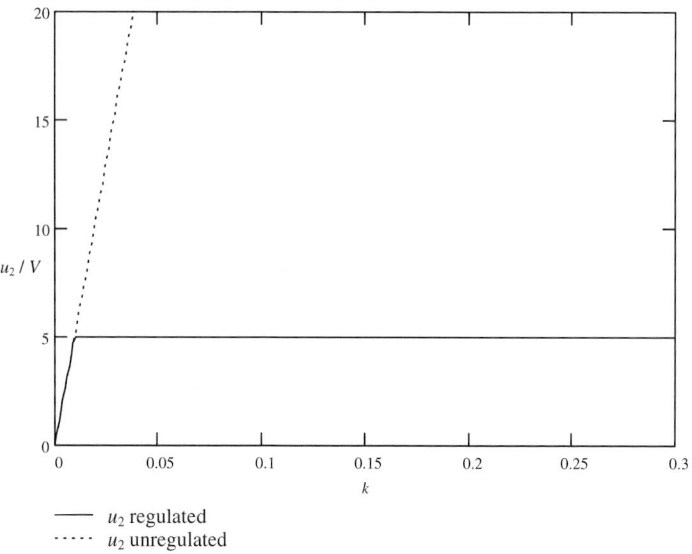

Figure 4.18: Example of the path of voltage u_2 with and without shunt regulation in the transponder, where the coupling coefficient k is varied by altering the distance between transponder and reader antenna. (The calculation is based upon the following parameters: $i_1 = 0.5$ A, $L_1 = 1$ μH, $L_2 = 3.5$ μH, $R_L = 2$ kΩ, $C_2 = 1/\omega_2 L_2$)

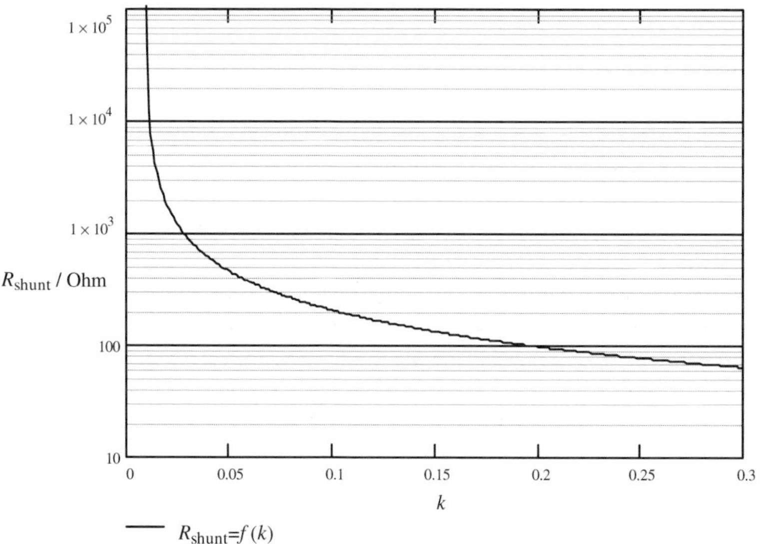

Figure 4.19: The value of the shunt resistor R_S must be adjustable over a wide range to keep voltage u_2 constant regardless of the coupling coefficient k. All parameters are the same as for the previous Figure (parameters as Figure 4.18)

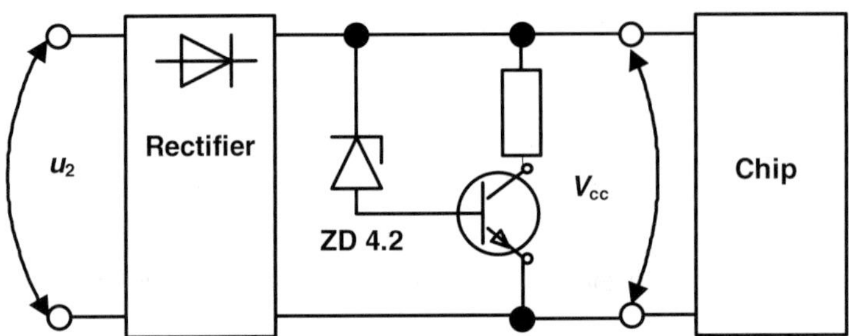

Figure 4.20: Example circuit for a simple shunt regulator

4.1.9 Interrogation field strength H_{min}

We can now use the results obtained in Section 4.1.7 to calculate the interrogation field strength of a transponder. This is the minimum field strength H_{min} (at a maximum distance x between transponder and reader) at which the supply voltage u_2 is just high enough for the operation of the data carrier.

However, u_2 is not the internal operating voltage of the data carrier here (3V or 5V), it is the input of the chip. The voltage regulator (shunt regulator) should not yet be active at this supply voltage. R_L corresponds with the input resistance of the data carrier after the "power on reset", C_2 is made up of the input capacitance C_p of the data carrier (chip) and the parasitic capacitance of the transponder layout C_2': $C_2 = (C_2' + C_p)$.

The inductive voltage (source voltage $u_{Q2} = u_i$) of a transponder coil can be calculated using equation 4.29 for the general case. If we assume a homogeneous, sinusoidal magnetic field in air (permeability constant = μ_0) we can derive the following, more appropriate, formula:

$$u_i = \mu_0 \cdot A \cdot N \cdot \omega \cdot H_{eff} \tag{4.29}$$

where H_{eff} is the effective field strength of a sinusoidal magnetic field, ω is the angular frequency of the magnetic field, N is the number of windings of the transponder coil L_2, and A is the cross sectional area of the transponder coil.

We now replace $u_{Q2} = u_i = j\omega M \cdot i_1$ from equation 4.24 with formula 4.29 and thus obtain the following equation for the circuit in Figure 4.13:

$$u_2 = \frac{j\omega \cdot \mu_0 \cdot H_{eff} \cdot A \cdot N}{1 + (j\omega L_2 + R_2)\left(\frac{1}{R_L} + j\omega C_2\right)} \tag{4.30}$$

Multiplying out the denominator:

$$u_2 = \frac{j\omega \cdot \mu_0 \cdot H_{\text{eff}} \cdot A \cdot N}{j\omega\left(\frac{L_2}{R_L} + R_2 C_2\right) + \left(1 - \omega^2 L_2 C_2 + \frac{R_2}{R_L}\right)} \tag{4.31}$$

We now solve this equation for H_{eff} and obtain the value of the complex form. This yields the following relationship for the interrogation field $H_{\min}$ in the general case:

$$H_{\min} = \frac{u_2 \cdot \sqrt{\left(\frac{\omega L_2}{R_L} + \omega R_2 C_2\right)^2 + \left(1 - \omega^2 L_2 C_2 + \frac{R_2}{R_L}\right)^2}}{\omega \cdot \mu_0 \cdot A \cdot N} \tag{4.32}$$

A more detailed analysis of equation 4.32 shows that the interrogation field strength is dependent upon the frequency $\omega = 2\pi f$ in addition to the antenna area A, the number of windings N (the transponder coil), the minimum voltage u_2 and the input resistance R_2. This is not surprising, because we have determined a resonance step-up of u_2 at the resonant frequency of the transponder resonant circuit. Therefore, when the transmission frequency of the reader corresponds with the resonant frequency of the transponder, the interrogation field strength $H_{\min}$ is at its minimum value.

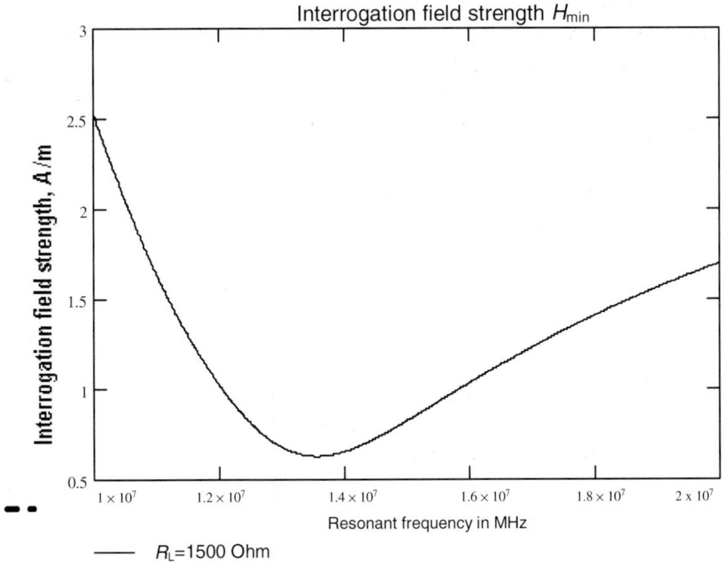

Figure 4.21: Interrogation sensitivity of a contactless smart card where the transponder resonant frequency is detuned in the range 10–20 MHz. ($N = 4$, $A = 0.05 \times 0.08\ \text{m}^2$, $u_2 = 5\text{V}$, $L_2 = 3.5\ \mu\text{H}$, $R_2 = 5\Omega$, $R_L = 1.5\ \text{k}\Omega$). If the transponder resonant frequency deviates from the transmission frequency (13.56 MHz) of the reader an increasingly high field strength is required to address the transponder. In practical operation this results in a reduction of the read range

To optimise the interrogation sensitivity of an inductively coupled RFID system, the resonant frequency of the transponder should be matched precisely to the transmission frequency of the reader. Unfortunately, this is not always possible in practice. Firstly, tolerances occur during the manufacture of a transponder, which lead to a deviation in the transponder resonant frequency. Secondly, there is also a technical reason for setting the resonant frequency of the transponder a few percent higher than the transmission frequency of the reader (for example in systems using anticollision procedures to keep the interaction of nearby transponders low).

Some semiconductor manufacturers incorporate additional smoothing capacitors into the transponder chip to smooth out frequency deviations in the transponder (see Figure 3.20, "tuning-C"). During manufacture the transponder is adjusted to the desired frequency by switching individual smoothing capacitors on and off [schürmann-93].

In equation 4.32 the resonant frequency of the transponder is expressed as the product L_2C_2. This is not recognisable at first glance. In order to make a direct statement about the frequency dependency of the interrogation sensitivity, we rearrange equation 4.20 to obtain:

$$L_2C_2 = \frac{1}{(2\pi\ f_0)^2} = \frac{1}{\omega_0{}^2} \tag{4.33}$$

By substituting the right hand term into the root of equation 4.32 we obtain a function, in which the dependence of the interrogation field strength H_{min} on the relationship between the transmission frequency of the reader (ω) and the resonant frequency of the transponder (ω_0) is now clearly expressed. This is based upon the assumption that the change in the resonant frequency of the transponder is caused by a change in the capacitance of C_2 (e.g. due to temperature dependence or manufacturing tolerances of this capacitance), whereas the inductance L_2 of the coil remains constant. To express this, the capacitor C_2 in the left hand term below the root of equation 4.32 is replaced by $C_2 = (\omega_{02} \cdot L_2)$-1:

$$H_{min} = \frac{u_2 \cdot \sqrt{\omega^2 \left(\frac{L_2}{R_L} + \frac{R_2}{\omega_0^2 L_2}\right)^2 + \left(\frac{\omega_0^2 - \omega^2}{\omega_0^2} + \frac{R_2}{R_L}\right)^2}}{\omega\mu_0 \cdot A \cdot N} \tag{4.34}$$

Therefore a deviation of the transponder resonant frequency from the transmission frequency of the reader will lead to a higher transponder interrogation field strength and thus to a lower read range.

4.1.9.1 "Energy range" of transponder systems

If the interrogation field of a transponder is known, then we can also estimate the energy range associated with a certain reader. The energy range of a transponder is the distance from the reader antenna at which there is just enough energy to operate the transponder (defined by u_{2min} and R_L). However, whether the energy range obtained also corresponds with the maximum functional range of the system also depends upon whether the data transmitted from the transponder can be detected by the reader at the distance in question.

Given a known antenna current I, radius R and number of windings N_1 of the transmitter antenna, the path of the field strength in the x-direction can be calculated using equation 4.5 (see Section 4.1.1.1 "Path of field strength $H(x)$ in conductor loops"). (If the antenna current of the transmitter antenna is not known it can be calculated from the measured field strength $H(x)$ at a distance x where the antenna radius R and the number of windings N_1 are known (see Section 4.1.1.1. "Path of field strength H(x) in conductor loops")). If we solve the equation for x we obtain the following relationship between the energy range and interrogation field H_{min} of a transponder for a given reader:

$$x = \sqrt{\sqrt[3]{\left(\frac{I \cdot N_1 \cdot R^2}{2 \cdot H_{min}}\right)^2} - R^2} \tag{4.35}$$

As an example (see Figure 4.22), let us now consider the energy range of a transponder as a function of the power consumption of the data carrier ($R_L = u_2/i_2$). The reader in this example generates a field strength of 0.115 A/m at a distance of 80 cm from the transmitter antenna (radius R of transmitter antenna: 40 cm), this is a typical value for RFID systems in accordance with ISO 15693.

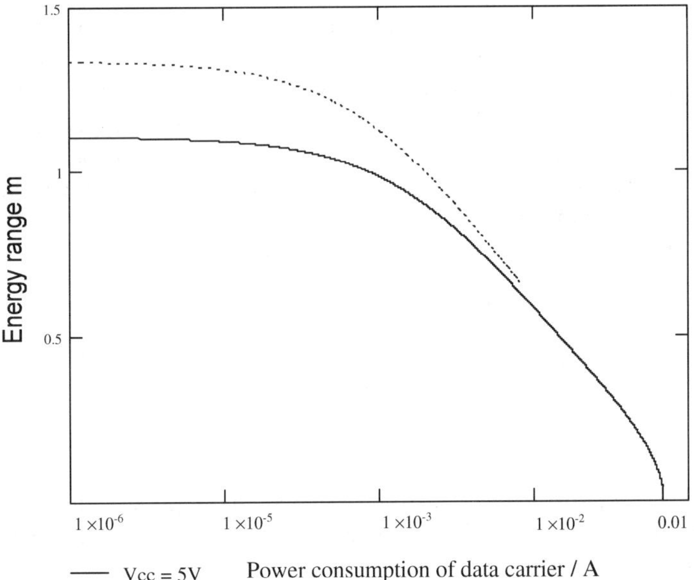

Figure 4.22: The energy range of a transponder also depends upon the power consumption of the data carrier (R_L). The transmitter antenna of the simulated system generates a field strength of 0.115 A/m at a distance of 80 cm, a value typical for RFID systems in accordance with ISO 15693 (transmitter: I = 1A, N_1 = 1, R = 0.4 m. Transponder: $A = 0.048 \times 0.076$ m^2 (smart card), N = 4, L_2 = 3.6 μH, u_{2min} = 5V / 3V)

As the current consumption of the transponder (lowering R_L) increases, the interrogation sensitivity of the transponder also increases and the energy range falls.

The maximum energy range of the transponder is determined by the distance between transponder and reader at which the minimum power supply u_{2min} required for the operation of the data carrier exists (that is $i_2 \to 0$, $R_L \to \infty$), even with an unloaded transponder resonant circuit. Where distance $x = 0$ the maximum current i_2 represents a limit, above which the supply voltage for the data carrier falls below u_{2min}, which means that the reliable operation of the data carrier can no longer be guaranteed in this operating state.

4.1.10 Total transponder – reader system

Up to this point we have considered the characteristics of inductively coupled systems primarily from the point of view of the transponder. In order to analyse in more detail the interaction between transponder and reader in the system, we need to take a slightly different view and first examine the electrical properties of the reader so that we can then go on to study the system as a whole.

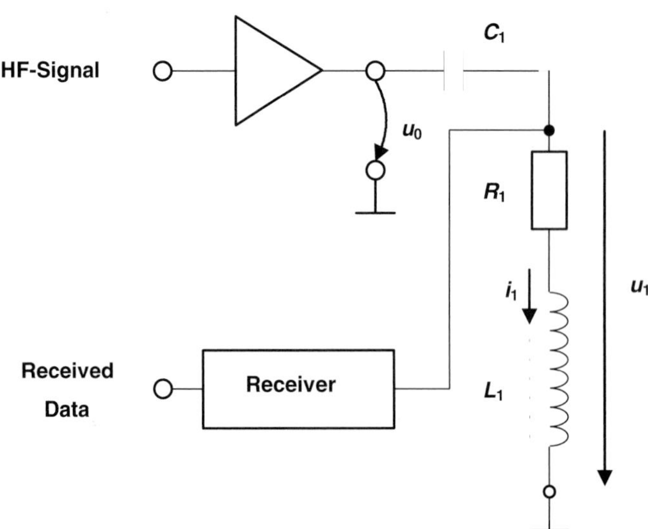

Figure 4.23: Equivalent circuit diagram of a reader with antenna L_1. The transmitter output branch of the reader generates the HF voltage u_0. The receiver of the reader is directly connected to the antenna coil L_1

Figure 4.23 shows the equivalent circuit diagram for a reader (the practical realisation of this circuit configuration can be found in Section 11.4 “Connection of Antennas”, Figures 11.13 and 11.15). The conductor loop necessary to generate the magnetic alternating field is represented by the coil L_1. The series resistor R_1 corresponds with the ohmic losses of the wire resistance in the conductor loop L_1. In order to obtain maximum current in the conductor coil L_1 at the reader operating frequency f_{TX}, a series resonant circuit with the resonant frequency $f_{RES} = f_{TX}$ is created by the serial connection of the capacitor C_1. The resonant frequency of the series resonant circuit can be very easily calculated using the Thomson equation (4.20). The operating state of the reader can be described by:

$$f_{TX} = f_{RES} = \frac{1}{2\pi\sqrt{L_1 \cdot C_1}} \tag{4.36}$$

Because of the series configuration, the total impedance Z_1 of the series resonant circuit is the sum of individual impedances, i.e.:

$$Z_1 = R_1 + j\omega \; L_1 + \frac{1}{j\omega \; C_1} \tag{4.37}$$

At the resonant frequency f_{RES}, however, the impedances of L_1 and C_1 cancel each other out. The total impedance Z_1 is determined by R_1 only and thus reaches a minimum.

$$\left. j\omega L_1 + \frac{1}{j\omega C_1} = 0 \right|_{f_{res}} \Rightarrow Z_1(f_{res}) = R_1 \tag{4.38}$$

The antenna current i_1 reaches a maximum at the resonant frequency and is calculated (under the assumption of an ideal voltage source where $R_i = 0$) from the source voltage u_0 of the transmitter high level stage, and the ohmic coil resistance R_1.

$$i_1(f_{res}) = \frac{u_0}{Z_1(f_{res})} = \frac{u_0}{R_1} \tag{4.39}$$

The two voltages u_1, across the conductor loop L_1, and u_{C1}, across the capacitor C_1, are in antiphase and cancel each other out at the resonant frequency because current i_1 is the same. However, the individual values may be very high. Despite the low source voltage u_0, which is usually just a few volts, figures of a few hundred volts can easily be reached at L_1 and C_1. Designs for conductor loop antennas for high currents must therefore incorporate sufficient voltage resistance in the components used, in particular the capacitors, because otherwise these would easily be destroyed by arcing. Figure 4.24 shows an example of voltage step-up at resonance.

Despite the fact that the voltage may reach very high levels, it is completely safe to touch the voltage-carrying components of the reader antenna. Because of the additional capacitance of the hand, the series resonant circuit is rapidly detuned, thus reducing the resonance step-up of voltage.

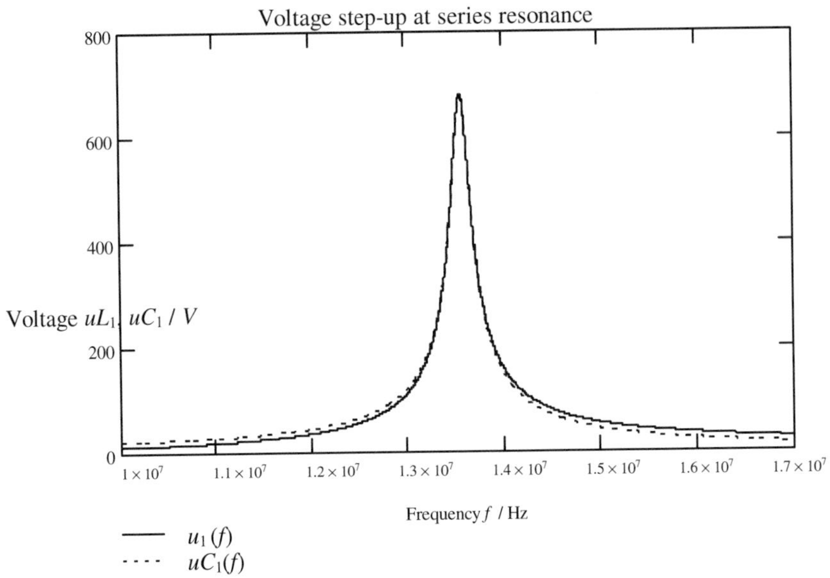

Figure 4.24: Voltage step-up at the coil and capacitor in a series resonant circuit in the frequency range 10 – 17 MHz (f_{RES} = 13.56 MHz, u_0 = 10V(!), R_1 = 2.5 Ω, L_1 = 2 μH, C_1 = 68.8 pF). The voltage at the conductor coil and series capacitor reaches a maximum of above 700 V at the resonant frequency. Because the resonant frequency of the reader antenna of an inductively coupled system always corresponds with the transmission frequency of the reader, components should be sufficiently voltage resistant

4.1.10.1 Transformed transponder impedance Z_T'

If a transponder enters the magnetic alternating field of the conductor coil L_1 a change can be detected in the current i_1. The current i_2 induced in the transponder coil acts upon current i_1 responsible for its generation via the magnetic mutual inductance M. This is in accordance with Lenz's law which states that "the direction of an induced e.m.f. is always such that it tends to set up a current opposing the motion or the change of flux responsible for inducing that e.m.f." [paul].

In order to simplify the mathematical description of the mutual inductance on the current i_1, let us now introduce an imaginary impedance, the "complex transformed transponder impedance" Z_T'. The electrical behaviour of the series resonant circuit of the reader in the presence of mutual inductance is as if the imaginary impedance Z_T' were actually present as a discrete component: Z_T' takes on a finite value $|Z_T'| > 0$. If the mutual inductance is removed, e.g. by withdrawing the transponder from the field of the conductor loop, then $|Z_T'| = 0$. We will derive the calculation of this transformed impedance step by step:

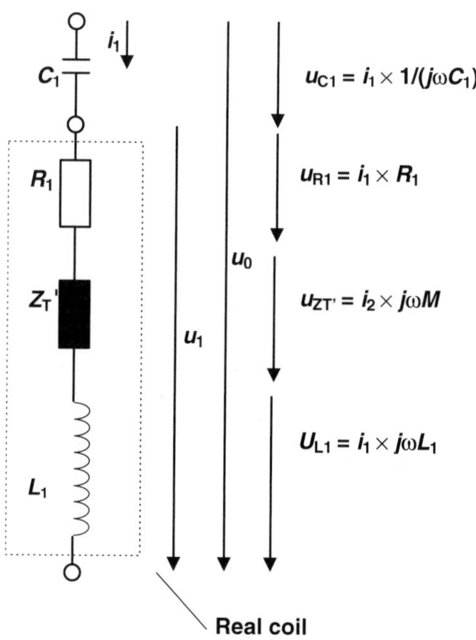

Figure 4.25: Equivalent circuit diagram of the series resonant circuit – the change in current i_1 in the conductor loop of the transmitter due to the influence of a magnetically coupled transponder is represented by the impedance Z_T'

The source voltage u_0 of the reader can be divided into the individual voltages u_{C1}, u_{R1}, u_{L1} and u_{ZT} in the series resonant circuit, as is illustrated in Figure 4.25. Figure 4.26 shows the vector diagram for the individual voltages in this circuit at resonance.

Because of the constant current i_1 in the series resonant circuit, the source voltage u_0 can be represented as the sum of the products of individual impedances with the current i_1. The transformed impedance Z_T' is expressed by the product $j\omega M \cdot i_2$:

$$u_0 = \frac{1}{j\omega C_1} \cdot i_1 + j\omega L_1 \cdot i_1 + R_1 \cdot i_1 - j\omega M \cdot i_2 \tag{4.40}$$

Because the series resonant circuit is operated at its resonant frequency, the individual impedances $(j\omega C_1)$-1 and $j\omega L_1$ cancel each other out. The voltage u_0 is therefore only divided between the resistance R_1 and the transformed transponder impedance Z_T', as we can see from the vector diagram (Figure 4.26). Equation 4.40 can be further simplified to:

$$u_0 = R_1 \cdot i_1 - j\omega M \cdot i_2 \tag{4.41}$$

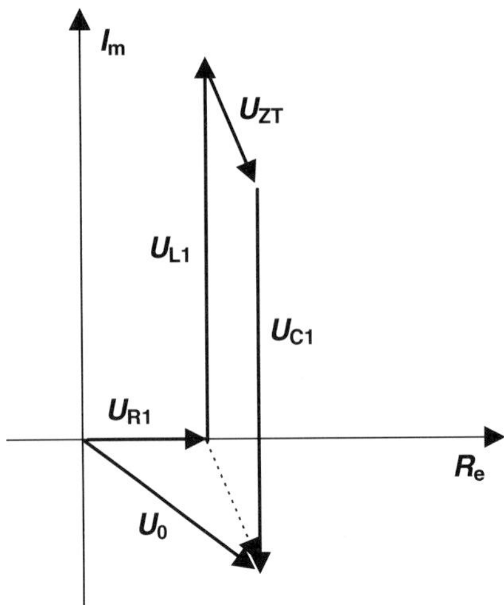

Figure 4.26: The vector diagram for voltages in the series resonance circuit of the reader antenna at resonant frequency. The figures for individual voltages u_{L1} and u_{C1} reach much higher levels than the total voltage u_0

We now require an expression for the current i_2 in the coil of the transponder, so that we can calculate the value of the transformed transponder impedance. Figure 4.27 gives an overview of the currents and voltages in the transponder.

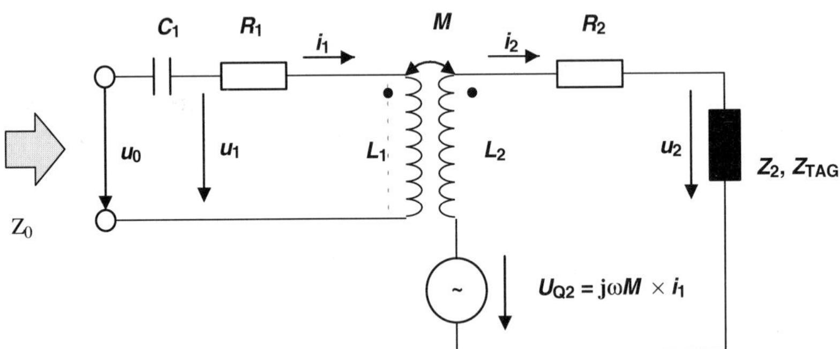

Figure 4.27: Simple equivalent circuit diagram of a transponder in the vicinity of a reader. The impedance Z_2 of the transponder is made up of the load resistor R_L (data carrier) and the capacitor C_2

The source voltage u_{Q2} is induced in the transponder coil L_2 by mutual inductance M. Current i_2 in the transponder is calculated from the quotient of the voltage u_2 divided by the sum of the individual impedances $j\omega L_2$, R_2 and Z_2 (here Z_2 represents the total input impedance of the data carrier and the parallel capacitor C_2). In the next step, we replace the voltage u_{Q2} by the voltage responsible for its generation $u_{Q2} = j\omega M \cdot i_1$, yielding the following expression for u_0:

$$u_0 = R_1 \cdot i_1 - j\omega M \cdot \frac{u_{Q2}}{R_2 + j\omega L_2 + Z_2} = R_1 \cdot i_1 - j\omega M \cdot \frac{j\omega M \cdot i_1}{R_2 + j\omega L_2 + Z_2} \tag{4.42}$$

Because it is generally impractical to work with the mutual inductance M, in a final step we replace M with

$$M = k\sqrt{L_1 \cdot L_2}$$

because the values k, L_1 and L_2 of a transponder are generally known. We write:

$$u_0 = R_1 \cdot i_1 + \frac{\omega^2 k^2 \cdot L_1 \cdot L_2}{R_2 + j\omega\ L_2 + Z_2} \cdot i_1 \tag{4.43}$$

Dividing both sides of equation 4.43 by i_1 yields the total impedance $Z_0 = u_0/i_1$ of the series resonant circuit in the reader as the sum of R_1 and the transformed transponder impedance Z_T'. Thus Z_T' is found to be:

$$Z_T' = \frac{\omega^2 k^2 \cdot L_1 \cdot L_2}{R_2 + j\omega\ L_2 + Z_2} \tag{4.44}$$

Impedance Z_2 represents the parallel connection of C_2 and R_L in the transponder. We replace Z_2 with the complete expression for C_2 and R_L and thus finally obtain an expression for Z_T' that incorporates all components of the transponder and is thus applicable in practice:

$$Z_T' = \frac{\omega^2 k^2 \cdot L_1 \cdot L_2}{R_2 + j\omega\ L_2 + \frac{R_L}{1 + j\omega\ R_L C_2}} \tag{4.45}$$

4.1.10.2 Influencing variables of Z_T'

Let us now investigate the influence of individual parameters on the transformed transponder impedance Z_T'. In addition to line diagrams, locus curves are also suitable for this: there is precisely one vector in the complex Z level for every parameter value in the function $Z_T' = f$("parameter") and thus exactly one point on the curve.

All line diagrams and locus curves from Section 4.1.10 are – unless stated otherwise – calculated using the following constant parameter values:

$L_1 = 1\mu H$	$L_2 = 3.5\ \mu H$
$C_1 = 1/(\omega_{TX})2 \cdot L_1$ (resonance)	$R_2 = 5\ \Omega$
$C_2 = 1/(\omega_{RX})2 \cdot L_2$ (resonance)	$R_L = 5\ k\Omega$
$f_{RES} = f_{TX} = 13.56$ MHz	$k = 15\%$

Transmission frequency f_{TX} We first change – at constant transponder resonant frequency f_{RES} – the transmission frequency f_{TX} of the reader. Although this case does not occur in practice, it is very useful as a theoretical experiment to help us to understand the principles behind the transformed transponder impedance Z_T'.

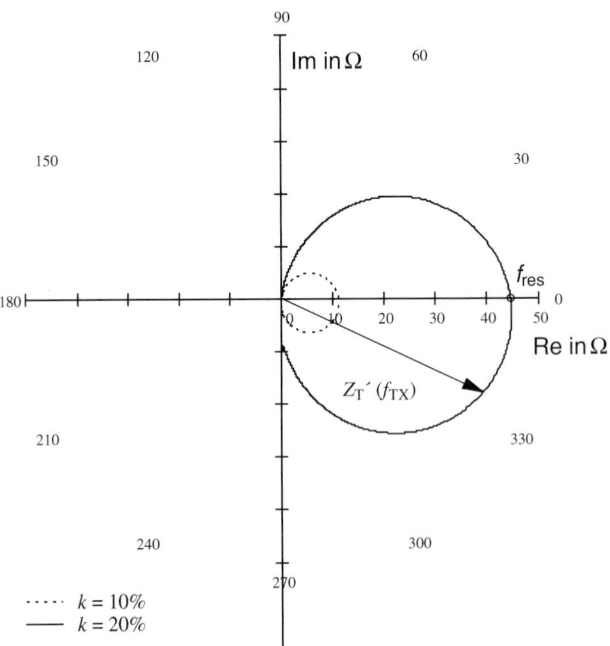

Figure 4.28: The impedance locus curve of the complex transformed transponder impedance Z_T' as a function of transmission frequency ($f_{TX} = 1 - 30$ MHz) of the reader corresponds with the impedance locus curve of a parallel resonant circuit

Figure 4.28 shows the locus curve $Z_T' = f(f_{TX})$ for this case. The impedance vector Z_T' traces a circle in the clockwise direction in the complex Z plane as transmission frequency f_{TX} increases.

In the frequency range below the transponder resonant frequency ($f_{TX} < f_{RES}$) the impedance vector Z_T' is initially found in quadrant I of the complex Z level. The transformed transponder impedance Z_T' has an inductive effect in this frequency range.

If the transmission frequency precisely corresponds with the transponder resonant frequency ($f_{TX} = f_{RES}$) then the reactive impedances for L_2 and C_2 in the transponder cancel each other out. Z_T' acts as an ohmic (real) resistor – the locus curve thus intersects the real x-axis of the complex Z plane at this point.

In the frequency range above the transponder resonant frequency ($f_{TX} > f_{RES}$), the locus curve passes through quadrant IV of the complex Z plane – Z_T' has a capacitive effect in this range.

The impedance locus curve of the complex transformed transponder impedance Z_T' corresponds with the impedance locus curve of a damped parallel resonant circuit with a parallel resonant frequency equal to the resonant frequency of the transponder. Figure 4.29 shows an equivalent circuit diagram for this. The complex current i_2 in the coil L_2 of the transponder resonant circuit is transformed by the magnetic mutual inductance M in the antenna coil L_1 of the reader and acts as a parallel resonant circuit with the (frequency dependent) impedance Z_T'. The value of the real resistance R' in the equivalent circuit diagram corresponds with the point of intersection of the locus curve Z_T' with the real axis in the Z plane.

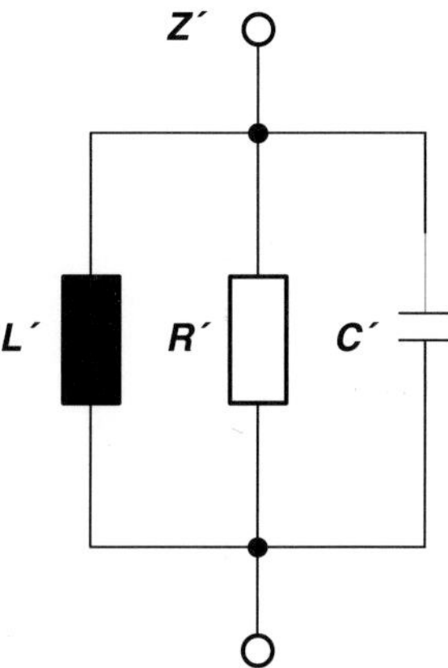

Figure 4.29: The equivalent circuit diagram of complex transformed transponder impedance Z_T' is a damped parallel resonant circuit

Coupling coefficient *k* Given constant geometry of the transponder and reader antenna, the coupling coefficient is defined by the distance and angle of the two coils in relation to each other (see Section 4.1.5 "Coupling coefficient *k*"). The influence of metals in the vicinity of the transmitter or transponder coil on the coupling coefficient should not be disregarded (e.g. shielding effect caused by eddy current losses). In practice, therefore, the coupling coefficient is the parameter that varies the most. Figure 4.30 shows the

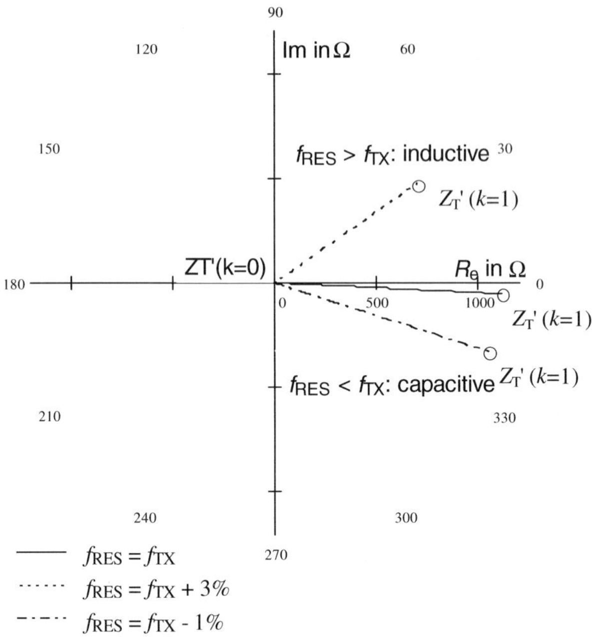

Figure 4.30: The locus curve of $Z_T'(k = 0 - 1)$ in the complex impedance level as a function of the coupling coefficient k is a straight line

locus curve of the complex transformed transponder impedance for the range $0 \leq k \leq 1$. We differentiate between three ranges:

- $k = 0$: If the transponder coil L_2 is removed from the field of the reader antenna L_1 entirely, then no mutual inductance occurs. For this limit case, the transformed transponder impedance is no longer effective, that is $Z_T'(k = 0) = 0$.
- $0 < k < 1$: If the transponder coil L_2 is slowly moved towards the reader antenna L_1, then the coupling coefficient, and thus also the mutual inductance M between the two coils, increases continuously. The value of complex transformed transponder impedance decreases proportionately, whereby $Z_T' \sim k_2$. When f_{TX} exactly corresponds with f_{RES}, $Z_T'(k)$ remains real for all values of k. (The low angular deviation in the locus curve in Figure 4.30 for $f_{RES} = f_{TX}$ is therefore due to the fact that the resonant frequency calculated according to equation 4.20 is only valid without limitations for the undamped parallel resonant circuit. Given damping by R_L and R_2, on the other hand, there is a slight detuning of the resonant frequency. However, this effect can be largely disregarded in practice and thus will not be considered further here.) Given a detuning of the transponder resonant frequency ($f_{RES} \neq f_{TX}$) on the other hand, Z_T' also has an inductive or capacitive component.

- $k = 1$: This case only occurs if the construction formats of both coils are exactly identical, so that the windings of the two coils L_1 and L_2 lie directly on top of each other at distance $d = 0$. $Z_T'(k)$ reaches a maximum in this case. In general the following applies: $|Z_T'(k)_{max}| = |Z_T'(K_{max})|$

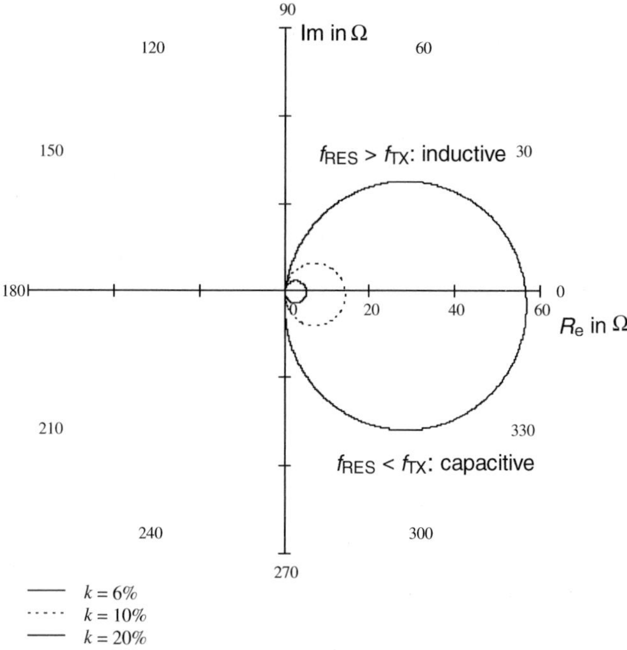

Figure 4.31: The locus curve of Z_T' ($C_2 = 10 - 110$ pF) in the complex impedance level as a function of the capacitance C_2 in the transponder is a circle in the complex Z plane. The diameter of the circle is proportional to k_2

Transponder capacitance C_2 We will now change the value of transponder capacitance C_2, whilst keeping all other parameters constant. This naturally detunes the resonant frequency f_{RES} of the transponder in relation to the transmission frequency f_{TX} of the reader. In practice, different factors may be responsible for a change in C_2:

- Manufacturing tolerances lead to a statistical deviation from the target value.
- A dependence of the data carrier's input capacitance on the input voltage u_2 due to effects in the semiconductor: $C_2 = f(u_2)$.
- Intentional variation of the capacitance of C_2 for the purpose of data transmission (we will deal with so-called "capacitive load modulation" in more detail in Section 4.1.10.3.)
- Detuning due to environmental influences such as metal, temperature, moisture, "hand capacitance" when the smart card is touched.

Figure 4.31 shows the locus curve for $Z_T'(C_2)$ in the complex impedance level. As expected, the locus curve obtained is the circle in the complex Z plane that is typical of a parallel resonant circuit. Let us now consider the extreme values for C_2:

- $C_2 = 1/\omega_{TX2}L_2$: The resonant frequency of the transponder in this case precisely corresponds with the transmission frequency of the reader (see equation 4.20). The current i_2 in the transponder coil reaches a maximum at this value due to resonance step-up and is real. Because $Z_T' \sim j\omega M \cdot i_2$ the value for impedance Z_T' also reaches a maximum – the locus curve intersects the real axis in the complex Z plane. The following applies: $|Z_T'(C_2)|_{max} = |Z_T'(C_2 = 1/(\omega_{TX})^2 \cdot L_2)|$.
- $C_2 \neq 1/\omega_{TX2}L_2$: If the capacitance C_2 is less than or greater than $C_2 = 1/\omega_{TX2}L_2$ then the resonant frequency of the transponder will be detuned and will be vary significantly from the transmission frequency of the reader. The polarity of the current i_2 in the resonant circuit of the transponder varies when the resonant frequency is exceeded, as we can see from Figure 4.32. Similarly, the locus curve of Z_T' describes the familiar circular path in the complex Z plane. The following is true for both extreme values:

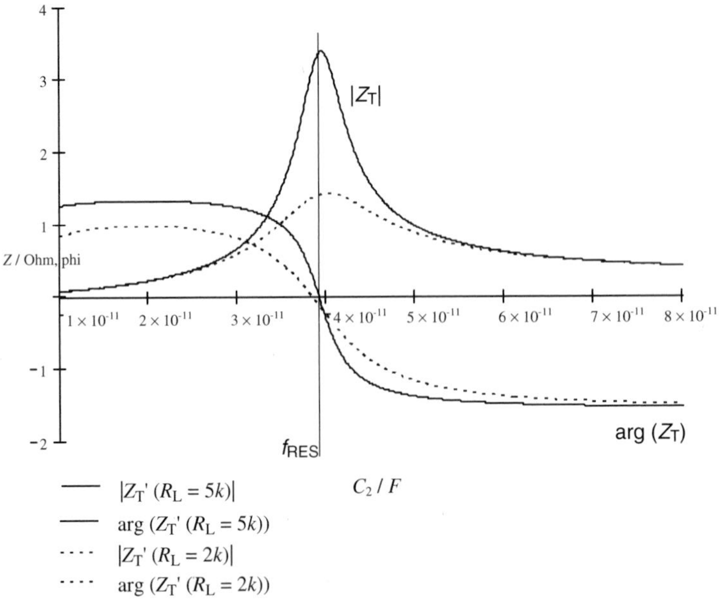

Figure 4.32: Magnitude and phase of the transformed transponder impedance Z_T' as a function of C_2. The maximum value of the magnitude of Z_T' is reached when the transponder resonant frequency matches the transmission frequency of the reader. The polarity of the phase angle of Z_T' varies

$$Z_T'(C_2 \to 0) = \frac{\omega^2 k^2 \cdot L_1 \cdot L_2}{j\omega\ L_2 + R_2 + R_L} \tag{4.46}$$

(no resonance step-up)

$$Z_t'(C_2 \to \infty) = \frac{\omega^2 k^2 \cdot L_1 \cdot L_2}{R_2 + j\omega\ L_2} \tag{4.47}$$

("short circuited" transponder coil)

Load resistance R_L The load resistance R_L is an expression for the power consumption of the data carrier (microchip) in the transponder. In practice, the load resistance is not constant, but falls as the coupling coefficient increases due to the influence of the shunt regulator (voltage regulator). The power consumption of the data carrier during the read or write operation also varies. Furthermore, the magnitude of the load resistance is often intentionally altered in order to transmit data to the reader (see Section 4.1.10.3 "Load modulation").

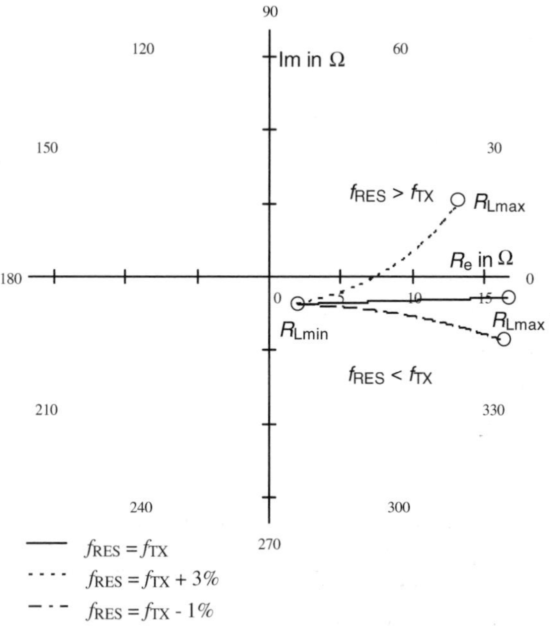

Figure 4.33: Locus curve of Z_T' ($R_L = 0.3$ - 3kΩ) in the impedance level as a function of the load resistance R_L in the transponder for different transponder resonant frequencies

Figure 4.33 shows the corresponding locus curve for $Z_T' = f(R_L)$. This shows that the transformed transponder impedance is proportional to R_L. Increasing load resistance R_L, corresponding with a lower(!) current in the data carrier, thus also leads to a greater transformed transponder impedance Z_T'. This can be explained by the influence of the load resistance R_L on the Q factor: A high-ohmic load resistance R_L leads to a high Q factor in the resonant circuit and thus to a greater current step-up in the transponder resonant circuit. Due to the proportionality $Z_T' \sim j\omega M \cdot i_2$ – and not to i_{RL} – we obtain a correspondingly high magnitude for the transformed transponder impedance.

If the transponder resonant frequency is detuned we obtain a curved locus curve for the transformed transponder impedance Z_T'. This can also be traced back to the influence of the Q factor, because the phase angle of a detuned parallel resonant circuit also increases as the Q factor increases ($R_L \uparrow$), as we can see from a glance at Figure 4.32.

Let us reconsider the two extreme values of R_L. The following is true:

$$Z_t'(R_L \to 0) = \frac{\omega^2 k^2 \cdot L_1 \cdot L_2}{R_2 + j\omega\ L_2} \tag{4.48}$$

("short-circuited" transponder coil)

$$Z_T'(R_L \to \infty) = \frac{\omega^2 k^2 \cdot L_1 \cdot L_2}{j\omega\ L_2 + R_2 + \frac{1}{j\omega\ C_2}} \tag{4.49}$$

(unloaded transponder resonant circuit)

Transponder inductance L_2 Let us now investigate the influence of inductance L_2 on the transformed transponder impedance, whereby the resonant frequency of the transponder is again held constant, so that $C_2 = 1/\omega_{TX2} L_2$.

Transformed transponder impedance reaches a clear peak at a given inductance value, as a glance at the line diagram (Figure 4.34) shows. This behaviour is reminiscent of the path of voltage $u_2 = f(L_2)$, which we have already investigated in Section 4.1.7. Here too the peak transformed transponder impedance occurs where the Q factor, and thus the current i_2 in the transponder, is at a maximum ($Z_T' \sim j\omega M \cdot i_2$). Please refer to Section 4.1.7 "Resonance" for an explanation of the mathematical relationship between load resistance and the Q factor.

4.1.10.3 Load modulation

In addition to some other principles (see Chapter 3 "Fundamental Operating Principles"), so-called load modulation is the most frequently used procedure for data transmission from transponder to reader by some margin. By varying the circuit parameters of the transponder resonant circuit in time with the data stream, the magnitude and phase of the transformed transponder impedance can be influenced (modulation) such that an appropriate evaluation procedure in the reader can be used to reconstruct the data transmitted from the transponder (demodulation).

Of all the circuit parameters in the transponder resonant circuit, only two can be altered by the data carrier: the load resistance R_L and the parallel capacitance C_2. Therefore RFID literature distinguishes between ohmic (or real) and capacitive load modulation.

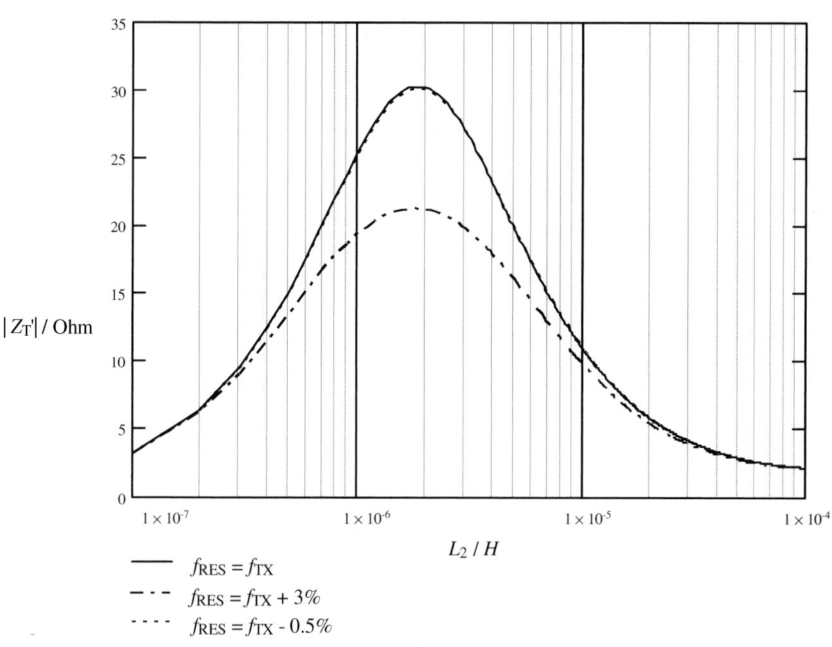

Figure 4.34: The magnitude of Z_T' as a function of the transponder inductance L_2 at a constant resonant frequency f_{RES} of the transponder. The maximum value of Z_T' coincides with the maximum value of the Q factor in the transponder

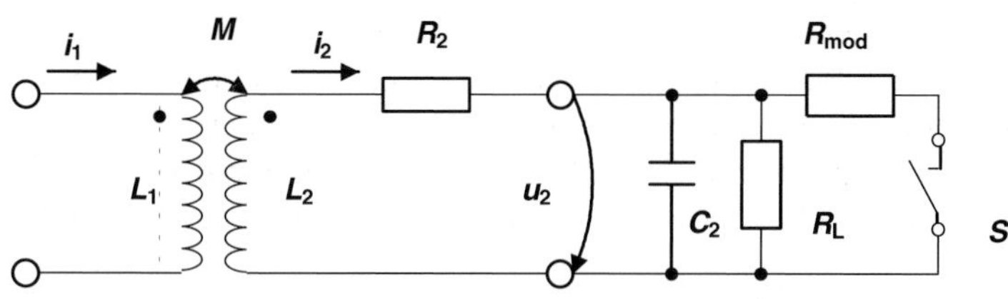

Figure 4.35: Equivalent circuit diagram for a transponder with load modulator. Switch S is closed in time with the data stream – or a modulated subcarrier signal – for the transmission of data

Ohmic load modulation In this type of load modulation a parallel resistor R_{mod} is switched on and off within the data carrier of the transponder in time with the data stream (or in time with a modulated subcarrier). We already know from the previous section that the parallel connection of R_{mod} (→ reduced total resistance) will reduce the magnitude of the Q factor and thus also the transformed transponder impedance Z_T'. This is also visible from the locus curve for the ohmic load modulator: the magnitude Z_T' is switched between the values Z_T' (R_L) and Z_T' (R_L ||R_{mod}) by the load modulator in the transponder. The phase of Z_T' remains almost constant during this process (assuming $f_{TX} = f_{RES}$).

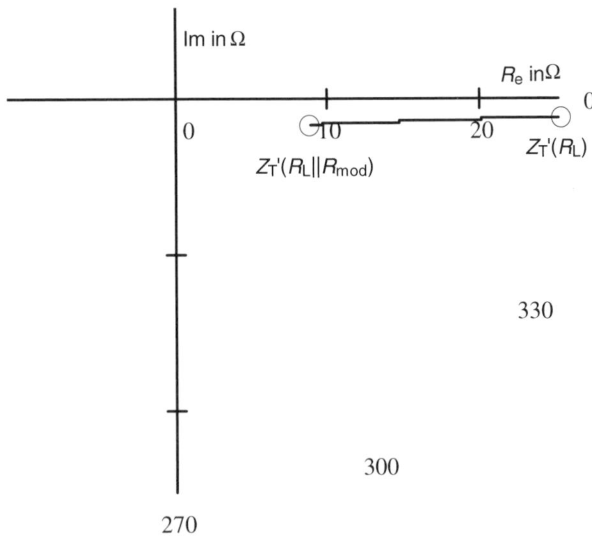

Figure 4.36: Locus curve of the transformed transponder impedance with ohmic load modulation (R_L ||R_{mod} = 1.5 – 5 kΩ) of an inductively coupled transponder. The parallel connection of the modulation resistor R_{mod} results in a lower magnitude for Z_T'

In order to be able to reconstruct (i.e. demodulate) the transmitted data, the falling voltage u_{ZT} at Z_T' must be sent to the receiver (RX) of the reader. Unfortunately, Z_T' is not accessible in the reader as a discrete component, because the voltage u_{ZT} is induced in the real antenna coil L_1. However, the voltages u_{L1} and u_{R1} also occur at the antenna coil L_1, and they can only be measured at the terminals of the antenna coil as the total voltage u_{RX}. The summed voltage is available to the receiver branch of the reader (see also Figure 4.23).

The vector diagram in Figure 4.37 shows the magnitude and phase of the voltage components u_{ZT}, u_{L1} and u_{R1} which make up the compound voltage u_{RX}. The magnitude and phase of u_{RX} is varied by the modulation of the voltage component u_{ZT} by the load modulator in the transponder. *Load modulation* in the transponder thus generates *amplitude modulation* of the reader antenna voltage u_{RX}. The transmitted data is therefore

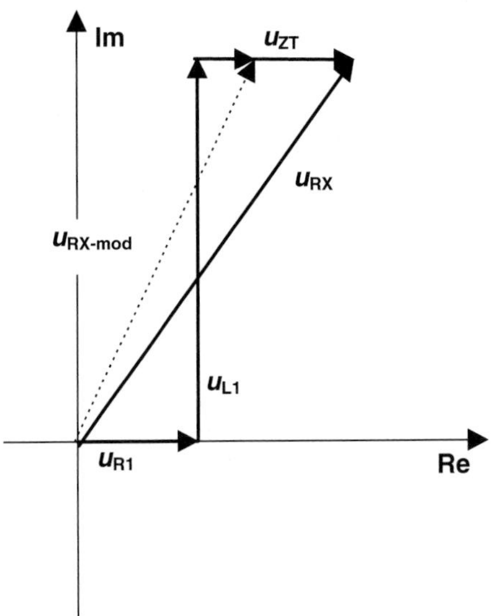

Figure 4.37: Vector diagram for the compound voltage u_{RX} that is available to receiver of a reader. The magnitude and phase of u_{RX} is modulated at the antenna coil of the reader (L_1) by ohmic load modulation

not available in the baseband at L_1, but is located in the modulation products (= modulation sidebands) of the (load) modulated voltage u_{RX} (see Chapter 6 "Coding and Modulation").

Capacitive load modulation In capacitive load modulation it is an additional capacitor C_{mod}, rather than a modulation resistance, that is switched on and off in time with the data stream (or in time with a modulated subcarrier). This causes the resonant frequency of the transponder to be switched between two frequencies. We know from the previous section that the detuning of the transponder resonant frequency markedly influences the magnitude and phase of transformed transponder impedance Z_T'.

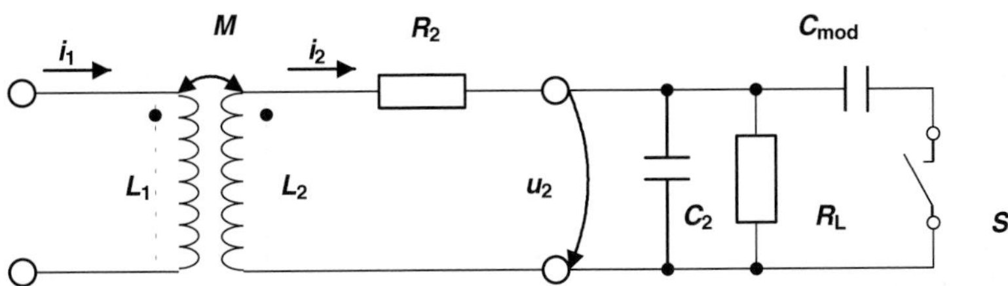

Figure 4.38: Equivalent circuit diagram for a transponder with capacitive load modulator. To transmit data the switch S is closed in time with the data stream – or a modulated subcarrier signal

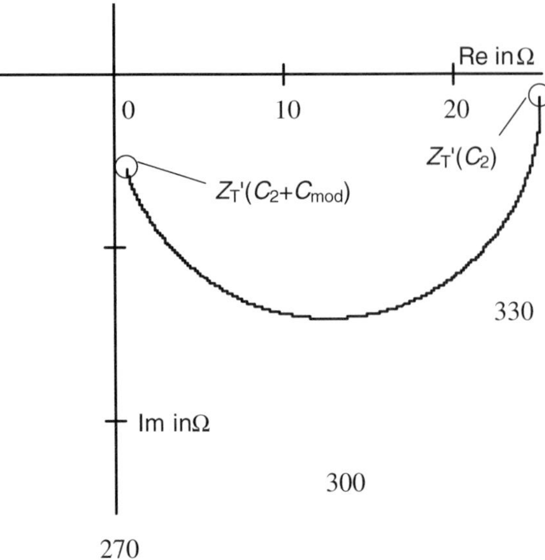

Figure 4.39: Locus curve of transformed transponder impedance for capacitive load modulation ($C_2 \| C_{mod}$ = 40 – 60 pF) of an inductively coupled transponder. The parallel connection of a modulation capacitor C_{mod} results in a modulation of the magnitude and phase of the transformed transponder impedance Z_T`

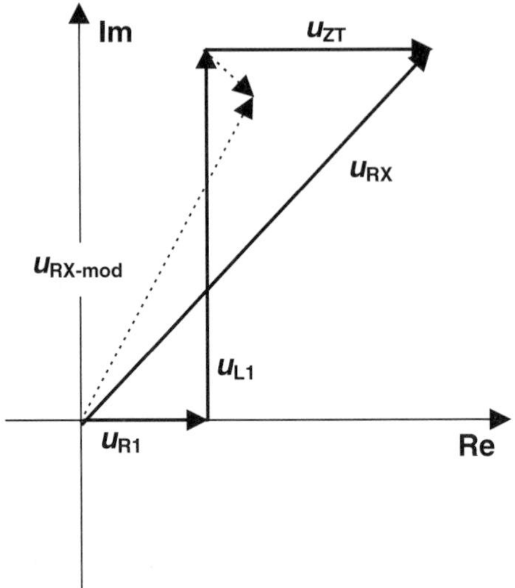

Figure 4.40: Vector diagram of the compound voltage u_{RX} available to the receiver of the reader. The magnitude and phase of this voltage is modulated at the antenna coil of the reader (L_1) by a capacitive load modulator

This is also clearly visible from the locus curve for the capacitive load modulator (Figure 4.39): the magnitude of Z_T' is switched between the values $Z_T'(\omega_{RES1})$ and $Z_T'(\omega_{RES2})$. The locus curve for Z_T' thereby passes through a segment of the circle in the complex Z plane that is typical of the parallel resonant circuit.

Demodulation of the data signal is similar to the procedure used with ohmic load modulation. The capacitive *load modulation* generates a combination of *amplitude and phase modulation* of voltage u_{RX} at the reader antenna and should therefore be processed in an appropriate manner in the receiver branch of the reader. The relevant vector diagram is shown in Figure 4.40.

Demodulation in the reader For transponders in the frequency range < 135 kHz the load modulator is generally controlled directly by a serial data stream encoded in the baseband, e.g. a Manchester encoded bit sequence. The modulation signal from the transponder can be recreated by rectification of the amplitude modulated voltage at the antenna coil of the reader (see Section 11.3 "Low Cost Configuration").

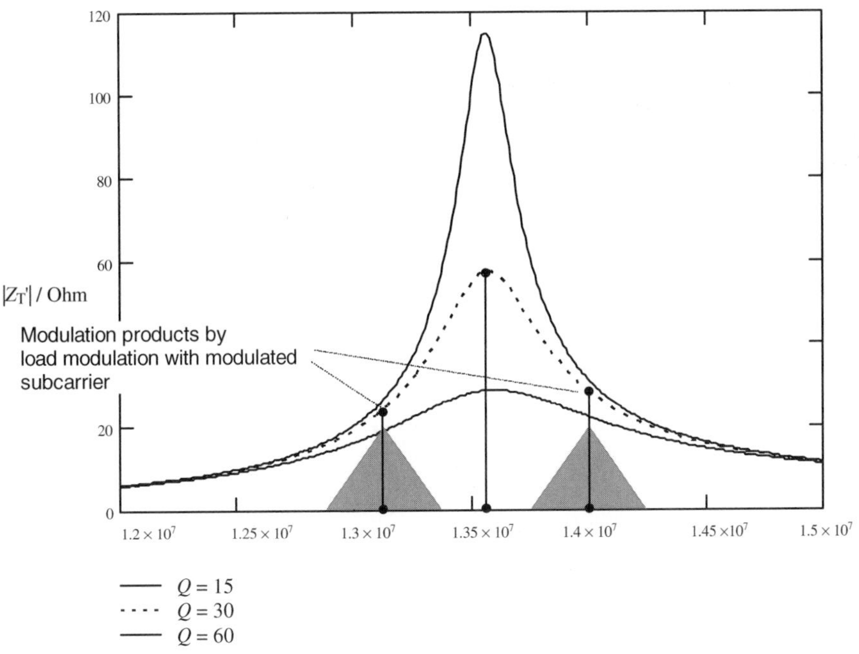

Figure 4.41: The transformed transponder impedance reaches a maximum at the resonant frequency of the transponder. The modulation sidebands of the current i_2 are damped in amplitude due to the influence of the bandwidth B. (Example for f_H = 440 kHz)

In higher frequency systems operating at 6.78 MHz or 13.56 MHz, on the other hand, the transponder's load modulator is controlled by a modulated subcarrier signal (see Section 6.2.4 "Modulation procedures with subcarrier"). The subcarrier frequency f_H is normally 212 kHz or 847 kHz (in accordance with ISO 14443-2).

Load modulation generates two sidebands at a distance of $\pm f_H$ to either side of the transmission frequency (see Figure 6.13). The information to be transmitted is contained uniformly in each of the two sidebands. One of the two sidebands is filtered in the reader and finally demodulated to reclaim the baseband signal of the modulated data stream.

The influence of the Q factor As we know from the preceding section, we attempt to maximise the Q factor in order to also maximise the energy range and the retroactive transformed transponder impedance. From the point of view of the energy range, a high Q factor in the transponder resonant circuit is definitely desirable. If we want to transmit data from or to the transponder, then there will be a certain minimum bandwidth that is required for the transmission from the data carrier in the transponder to the receiver in the reader. However, the bandwidth B of the transponder resonant circuit is inversely proportional to the Q factor.

$$B = \frac{f_{\text{RES}}}{Q} \tag{4.50}$$

Each load modulation operation in the transponder causes a corresponding amplitude modulation of the current i_2 in the transponder coil. The modulation sidebands of the current i_2 that this generates are limited to some degree by the bandwidth of the transponder resonant circuit, which is limited by practical considerations. The bandwidth B determines a frequency range around the resonant frequency f_{RES}, at the limits of which the modulation sidebands of the current i_2 in the transponder reach a damping of 3 dB relative to the resonant frequency. If the Q factor of the transponder is too high, then the modulation sidebands of the current i_2 are damped to such a degree due to the low bandwidth that the range is reduced (transponder signal range). This can rapidly become critical, particularly in transponder systems with a modulated subcarrier.

As mentioned above, transponders used in 13.56 MHz systems that support an anticollision algorithm are adjusted to a resonant frequency of 15 – 18 MHz to minimise the mutual influence of transponders. Due to the marked detuning of the transponder resonant frequency relative to the transmission frequency of the reader, the two modulation sidebands of a load modulation system with subcarrier are transmitted at a different level (see Figure 4.42).

The term 'bandwidth' is problematic here (the frequencies of the reader and the modulation sidebands may even lie outside the bandwidth of the transponder resonant circuit). However, the selection of the correct Q factor for the transponder resonant circuit is still important, because it can influence the transient effects during load modulation.

Ideally, the "mean Q factor" of the transponder will be selected such that the energy range and transponder signal range of the system are identical. However, the calculation of

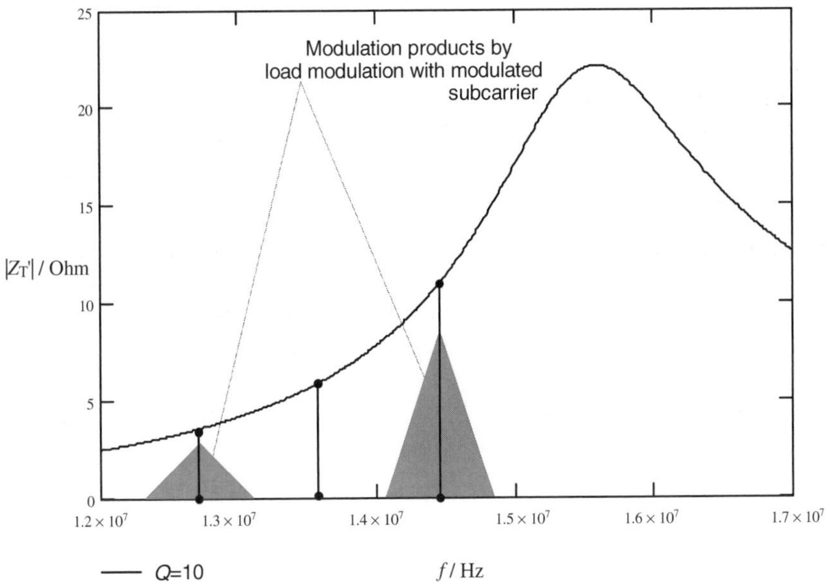

Figure 4.42: If the transponder resonant frequency is markedly detuned compared to the transmission frequency of the reader, then the two modulation sidebands will be transmitted at different levels. (f_H = 847 kHz)

an ideal Q factor is non-trivial and should not be underestimated because the Q factor is also strongly influenced by the shunt regulator (in connection with the distance *d* between transponder and reader antenna) and the load modulator itself. Furthermore, the influence of the bandwidth of the transmitter antenna (series resonant circuit) on the level of the load modulation sidebands should not be underestimated.

Therefore, the development of an inductively coupled RFID system is always a compromise between the system's range and its data transmission speed (baud rate / subcarrier frequency). Systems that require a short transaction time (that is rapid data transmission and large bandwidth) often only have a range of a few centimetres, whereas systems with relatively long transaction times (that is slow data transmission and low bandwidth) can be designed to achieve a greater range. A good example of the former case is provided by contactless smart cards for local public transport applications, which carry out authentication with the reader within a few 100 ms and must also transmit booking data. Contactless smart cards for "hands free" access systems that transmit just a few bytes – usually the serial number of the data carrier – within 1 – 2 seconds are an example of the latter case. A further consideration is that in systems with "large" transmission antenna, the data rate of the reader is restricted by the fact that only small sidebands may be generated because of the need to comply with the radio licensing regulations (ETS, FCC). Table 4.3 gives a brief overview of the relationship between range and bandwidth in inductively coupled RFID systems.

Table 4.3: Relationship between range and bandwidth in 13.56 MHz systems. An increasing Q factor in the transponder permits a greater range in the transponder system, however this is at the expense of the bandwidth and thus also the data transmission speed (baud rate) between transponder and reader

System	Baud rate	Subcarrier	f_{TX}	Range
ISO1444	3106 kBd	847 kHz	13.56 MHz	0 – 10 cm
ISO 15693 short	26.48 kBd	484 kHz	13.56 MHz	0 – 30 cm
ISO 15693 long	6.62 kBd	484 kHz	13.56 MHz	0 – 70 cm
Long range system	9.0 kBd	212 kHz	13.56 MHz	0 – 1 m
LF system	2 – 4 kBd	no subcarrier	< 125 kHz	0 – 1.5 m

4.1.11 Measuring the coupling coefficient k

The coupling coefficient k and the associated mutual inductance M are the most important parameters for the design of an inductively coupled RFID system. These parameters are difficult to determine from the – often complicated – field path. Mathematics may be fun, but has its limits. The necessary software to calculate a numeric simulation is often unavailable – or it may simply be that the time or patience is lacking.

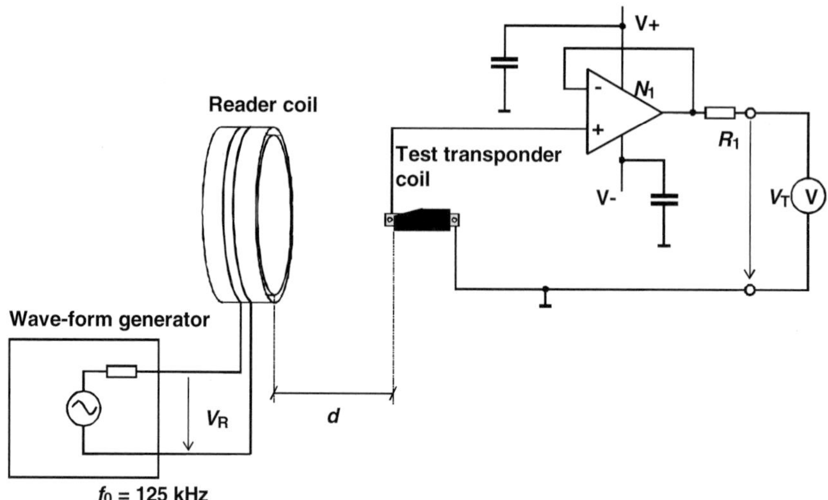

Figure 4.43: Measurement circuit for the measurement of the magnetic coupling coefficient k. N1: TL081 or LF 356N, R1: 100 – 500 Ω. (Reproduced by permission of TEMIC Semiconductor GmbH, Heilbronn)

However, the coupling coefficient k can be quickly calculated for an existing system by means of a simple measurement. This requires a test transponder coil, whose electrical and mechanical parameters correspond with those of the real transponder. The coupling coefficient can be simply calculated from the measured voltages U_R at the reader coil and U_T at the transponder coil:

$$k = A_k \cdot \frac{U_T}{U_R} \cdot \sqrt{\frac{L_R}{L_T}} \tag{4.51}$$

where U_T = voltage at the transponder coil, U_R = voltage at the reader coil, L_T and L_R = inductance of the coils, A_K = coupling coefficient (<1).

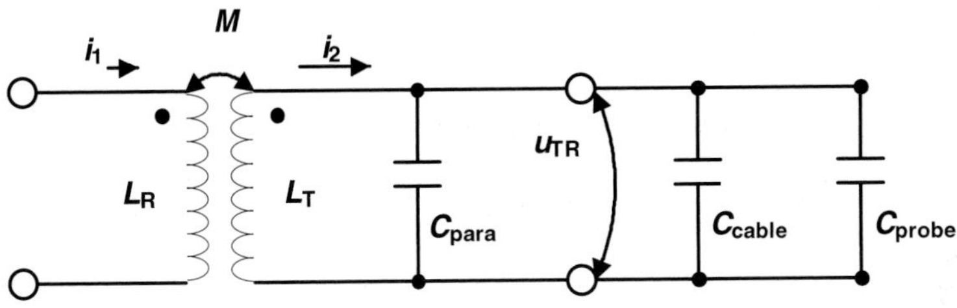

Figure 4.44: Equivalent circuit diagram of the test transponder coil with the parasitic capacitances of the measuring circuit

The parallel, parasitic capacitances of the measuring circuit and the test transponder coil itself influence the result of the measurement because of the undesired current i_2. To compensate for this effect Equation 4.52 contains a correction factor A_K. Where $C_{TOT} = C_{PARA} + C_{CABLE} + C_{PROBE}$ (see Figure 4.44) the correction factor is defined as:

$$A_k = 2 - \frac{1}{1 - \left(\omega^2 \cdot C_{TOT} \cdot L_T\right)} \tag{4.52}$$

In practice, the correction factor in the low capacitance layout of the measuring circuit is $A_K \sim 0.99 - 0.8$ [temic].

4.1.12 *Magnetic materials*

Materials with a relative *permeability* >1 are termed ferromagnetic materials. These materials are iron, cobalt, nickel, various alloys and ferrite.

4.1.12.1 Properties of magnetic materials and ferrite

One important characteristic of a magnetic material is the *magnetisation characteristic* or *hysteresis curve*. This describes $B = f(H)$, which displays a typical path for all ferromagnetic materials.

Starting from the unmagnetised state of the ferromagnetic material, the virgin curve *A* → *B* is obtained as the magnetic field strength *H* increases. During this process, the molecular magnets in the material align themselves in the *B* direction. (Ferromagnetism is based upon the presence of molecular magnetic dipoles. In these, the electron circling the atomic core represents a current and generates a magnetic field. In addition to the movement of the electron along its path, the rotation of the electron around itself, the spin, also generates a magnetic moment, which is of even greater importance for the material's magnetic behaviour.) Because there is a finite number of these molecular magnets, the number that remain to be aligned falls as the magnetic field increases, thus the gradient of the hysteresis curve falls. When all molecular magnets have been aligned, *B* rises in proportion to *H* only to the same degree as in a vacuum.

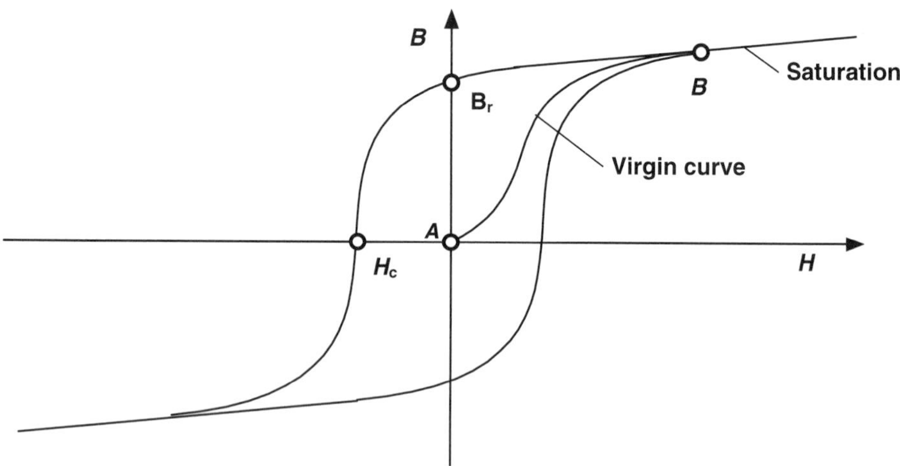

Figure 4.45: Typical magnetisation or hysteresis curve for a ferromagnetic material

When the field strength *H* falls to $H = 0$, the flux density *B* falls to the positive residual value B_R, the remanence. Only after the application of a opposing field (-*H*) does the flux density *B* fall further and finally return to zero. The field strength necessary for this is termed the coercive field strength H_C.

Ferrite is the main material used in high frequency technology. This is used in the form of soft magnetic ceramic materials (low B_r), composed mainly of mixed crystals or compounds of iron oxide (Fe_2O_3) with one or more oxides of bivalent metals (NiO, ZnO, MnO etc.) [Vogt]. The manufacturing process is similar to that for ceramic technologies (sintering).

The main characteristic of ferrite is its high specific electrical resistance, which varies between 1 and 10^6 Ω depending upon the material type, compared to the range for metals, which vary between 10^{-5} and 10^{-4} Ωm. Because of this, eddy current losses are low and can be disregarded over a wide frequency range.

The relative permeability of ferrites can reach the order of magnitude of $\mu_r = 2000$.

An important characteristic of ferrite materials is their material-dependent limit frequency, which is listed in the datasheets provided by the ferrite manufacturer. Above the limit frequency increased losses occur in the ferrite material, and therefore ferrite should not be used outside the specified frequency range.

4.1.12.2 Ferrite antennas in LF transponders

Some applications require extremely small transponder coils. In transponders for animal identification, typical dimensions for cylinder coils are $d \times l$ = 5 mm × 0.75 mm. The mutual inductance that is decisive for the power supply of the transponder falls sharply due to its proportionality with the cross sectional area of the coil ($M \sim A$, equation 4.11). By inserting a ferrite material with a high permeability μ into the coil ($M \sim \Psi \quad M \sim \mu \cdot H \cdot A$, equation 4.11), the mutual inductance can be significantly increased, thus compensating for the small cross sectional area of the coil.

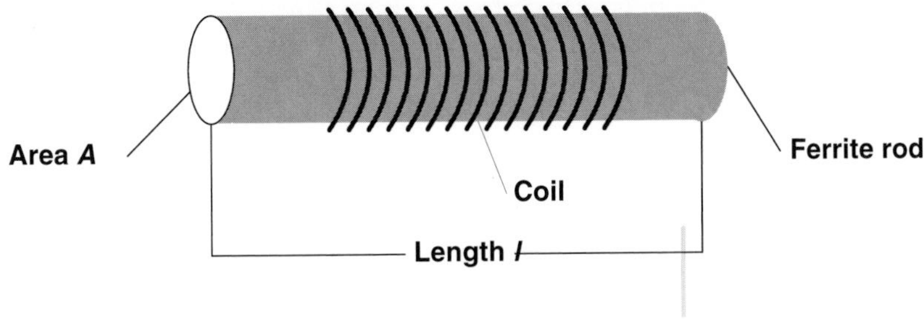

Figure 4.46: Configuration of a ferrite antenna in a 135 kHz glass transponder

The inductance of a *ferrite antenna* can be calculated according to the following equation [philmag]:

$$L = \frac{\mu_0 \mu_{\text{Ferrite}} \cdot n^2 \cdot A}{l} \tag{4.53}$$

4.1.12.3 Ferrite shielding in a metallic environment

The use of (inductively coupled) RFID systems often requires that reader or transponder antenna be mounted directly upon a metallic surface. This might be the reader antenna of an automatic ticket dispenser or a transponder for mounting on gas bottles.

However, it is not possible to fit a magnetic antenna directly onto a metallic surface. The magnetic flux through the metal surface induces eddy currents within the metal, which oppose the field responsible for their creation, i.e. the reader's field (Lenz's law), thus damping the magnetic field in the surface of the metal to such a degree that communication between reader and transponder is no longer possible. It makes no difference here whether the magnetic field is generated by the coil mounted upon the metal surface (reader antenna) or the field approaches the metal surface from "outside" (transponder on metal surface).

By inserting highly permeable ferrite between the coil and metal surface it is possible to largely prevent the occurence of eddy currents. This makes it possible to mount the antenna on metal surfaces.

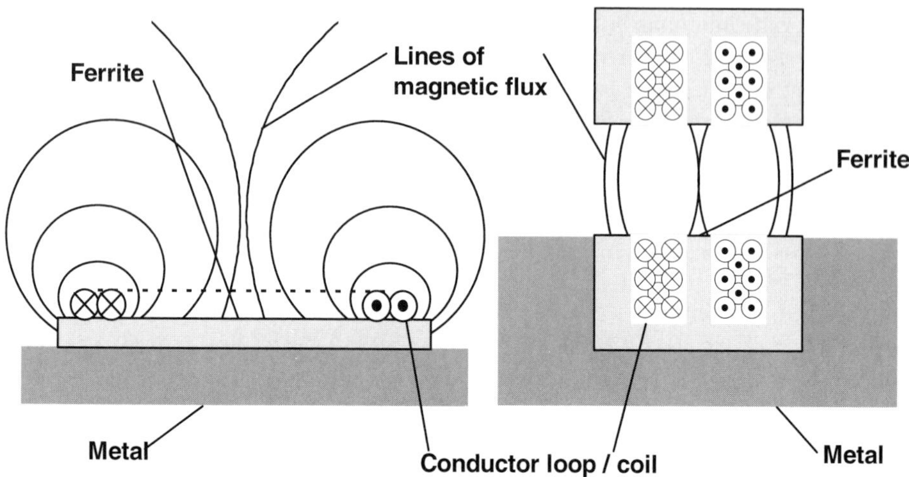

Figure 4.47: Reader antenna (left) and gas bottle transponder in a u-shaped core with read head (right) can be mounted directly upon or within metal surfaces using a ferrite shielding

When fitting antennas onto ferrite surfaces it is necessary to take into account the fact that the inductance of the conductor loop or coils may be significantly increased by the permeability of the ferrite material, and it may therefore be necessary to readjust the resonant frequency or even redimension the matching network (in readers) altogether (see Chapter 11, Section 11.4 entitled "Connection of Antennas".)

4.2 Electromagnetic Waves

4.2.1 The creation of electromagnetic waves

Earlier in the book we described how a varying magnetic field in space induces an electric field with closed field lines (rotational field) (see also Figure 4.11). The electric field surrounds the magnetic field and itself varies over time. Due to the variation of the electric rotational field, a magnetic field with closed field lines occurs in space (rotational field). It surrounds the electric field and itself varies over time, thus generating another electric field. Due to the mutual dependence of the time varying fields there is a chain effect of electric and magnetic fields in space [fricke].

Radiation can only occur given a finite propagation speed ($c \approx$ 300 000 km/s; *speed of light*) for the electromagnetic field, which prevents a change in the voltage at the antenna from being followed immediately by the field in the vicinity of the change. Because of this, even at the alternating voltage's zero crossover, the field lines remaining in space from the previous half wave cannot end at the antenna, but close into themselves forming eddies. The eddies in the opposite direction that occur in the next half wave propel the existing eddies, and thus the energy stored in this field, away from the emitter at the speed of light c. The magnetic field is interlinked with the varying electrical field that propagates at the same time. When a certain distance is reached, the fields are released from the emitter, and this point represents the beginning of electromagnetic radiation (→ far field). At high frequencies, that is small wavelengths, the radiation generated is particularly effective, because in this case the separation takes place in the direct vicinity of the emitter, where high field strengths still exist. The wavelength λ of an electromagnetic wave is calculated from the quotient of the speed of light and the frequency of the radiation:

$$\lambda = \frac{c}{f} \tag{4.54}$$

In the far field there is a fixed relationship between the electric and magnetic field strengths of the electromagnetic field. Both field strengths are linked with each other by the *field characteristic impedance* Z_F.

$$Z_F = \frac{E}{F} = \sqrt{\mu_0 \varepsilon_0^{-1}} = 376.73\Omega \tag{4.55}$$

In radiotechnology *field strengths* are usually quoted as a relative figure. The conversion between the electric and magnetic field strength is thus simplified:

$$H[\text{dB } \mu\text{A/m}] = E\,[\text{dB } \mu\text{V/m}] - 51.5 \text{ dB} \tag{4.56}$$

Electromagnetic waves propagate quasi-optically in the far field (free space). The field strength of the electromagnetic wave falls as the distance from the radiation source increases. Given unimpeded propagation, the following equation describes the so-called *free space attenuation* a_F from a source to a point at a distance d:

$$a_F\,/\text{dB} = -147.6 + 20\log(d) + 20\log(f) \tag{4.57}$$

4.2.1.1 Transition from near to far field in magnetic antennas

The primary magnetic field begins at the antenna and induces electric field lines in space. This area is termed the *near field* of an antenna. At a distance of $\lambda/2\pi$ the electromagnetic field separates from the antenna and wanders into space as an electromagnetic field. The separated electromagnetic field is termed the *far field.*

A separated electromagnetic wave can no longer have a direct effect, i.e. by inductive or capacitive coupling, on the antenna from which it was generated. For inductively coupled RFID systems this means that when the far field has begun, a transformer-type (inductive) coupling is no longer possible. The radius $r_F = \lambda/2\pi$ around the antenna therefore represents an insurmountable range limit for inductively coupled systems.

Table 4.4: r_F and λ for different frequency ranges

Frequency	Wavelength λ	$\lambda/2\pi$
< 135 kHz	> 2222 m	> 353 m
6.78 MHz	44.7 m	7.1 m
13.56 MHz	22.1 m	3.5 m
27.125 MHz	11.0 m	1.7 m

The field strength path of a magnetic antenna along the coil x axis follows the relationship $1/d^3$ in the near field, as demonstrated above. This corresponds with a damping of 60 dB per decade (the distance). Upon the transition to the far field, on the other hand, the damping path flattens out, because after the separation of the field from the antenna only free space attenuation is relevant to the field strength path. The field strength then decreases according to the relationship $1/d$ as distance increases. This corresponds with a damping of just 20 dB per decade (the distance).

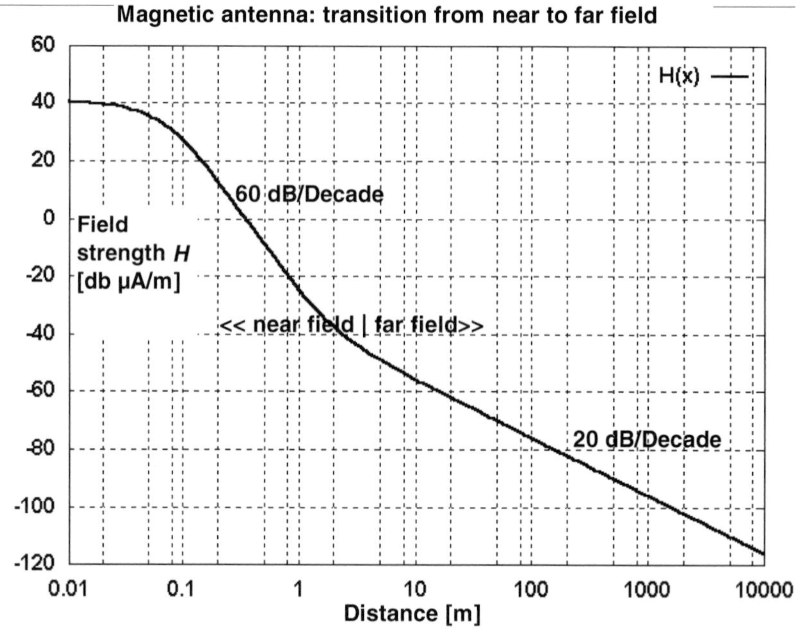

Figure 4.48: Path of the magnetic field strength H in the transition from near to far field at a frequency of 13.56 MHz

4.2.2 *Reflection of electromagnetic waves*

An *electromagnetic wave* emitted into the surrounding space by the antenna encounters various objects. Part of the high frequency energy that reaches the object is absorbed by the object and converted into heat, the other part is scattered in many directions with varying intensity. A small part of the reflected energy finds its way back to the transmitter antenna. *RADAR technology* uses this reflection to measure the distance and position of distant objects.

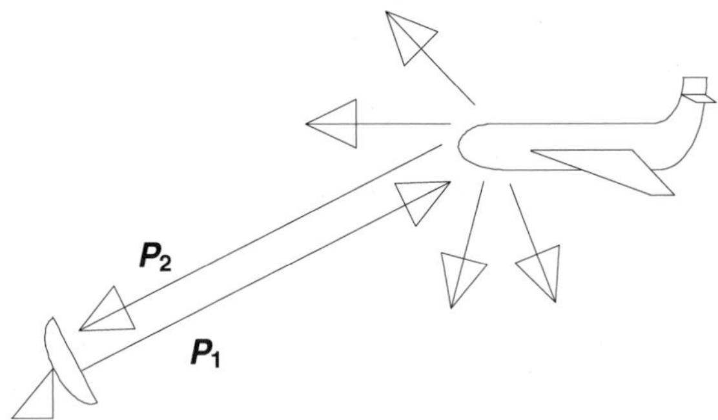

Figure 4.49: Reflection off a distant object is also used in RADAR technology

In RFID systems the reflection of electromagnetic waves (*backscatter system, modulated radar cross-section*) is used for the transmission of data from a transponder to a reader. Because the *reflective properties* of objects generally increase with increasing frequency, these systems are used mainly at frequency ranges 915 MHz (USA), 2.45 GHz and higher.

Let us now consider the relationships in an RFID system in more detail. The high frequency generated by the transmitter is propagated uniformly in all directions by the antenna of the reader. The *power density* S at the location of the transponder is thus:

$$S = \frac{P \cdot G}{4\,\pi\, R^2} \tag{4.58}$$

where P is the transmission power of the reader, G is the gain of the transmitter antenna and R is the distance between reader and antenna.

The transponder reflects a power P_2 that is proportional to the power density S:

$$P_2 = \sigma \cdot S \tag{4.59}$$

The *radar cross section* σ (RCS) is a measure of an object's ability to reflect electromagnetic waves. The radar cross section depends upon a range of parameters, such as object size, shape, material, surface structure, but also wavelength and polarisation.

The following equation describes the power density that finally returns to the antenna of the reader.

$$S_{\text{Back}} = \frac{P \cdot G \cdot \sigma}{(4\pi)^2 \cdot R^4} \tag{4.60}$$

We can find the reception power of the receiver antenna from its effective absorption area $A_w = \lambda^2 \times G/(4\pi)$ [morel]:

$$P_{\text{BACK}} = \frac{P \cdot G \cdot \sigma \cdot A_W}{(4\pi)^2 R^4} = \frac{P \cdot G^2 \cdot \lambda^2 \cdot \sigma}{(4\pi)^3 R^4} \tag{4.61}$$

Equation 4.61 also demonstrates that the *read range* of such a backscatter RFID system is proportional to the fourth root of the transmission power of the reader. In other words: all other things being equal, if we wish to double the range, we must multiply the transmission power by sixteen!

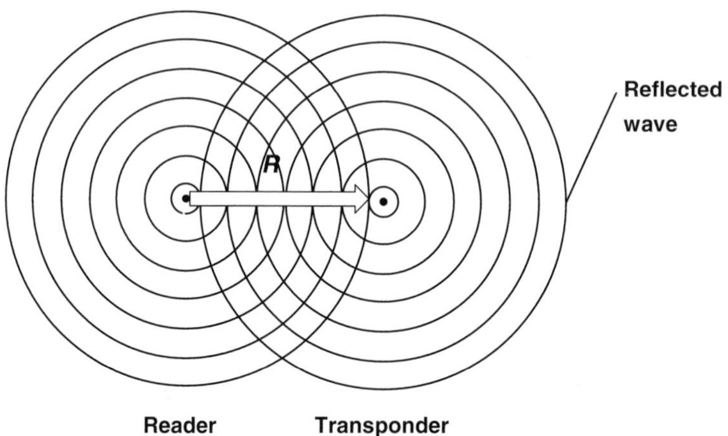

Figure 4.50: Propagation of waves transmitted and reflected by the transponder

4.2.3 *Radar cross section of an antenna*

The radar cross section can only be precisely calculated for simple surfaces such as balls, flat surfaces and the like (for example see [Baur]). The material also has a significant influence. For example, *metal surfaces* reflect much better than plastic or compound materials. Because of the significance of the dependence of the radar cross section σ on wavelength, objects are divided into three categories:

- Rayleigh range: The wavelength is large compared with the object dimensions. For objects smaller than around half the wavelength, σ shows a λ^{-4} dependency, so that the reflective properties of objects smaller than 0.1 λ can be completely disregarded in practice.
- Resonance range: The wavelength is comparable with the object dimensions. Varying the wavelength around the geometric value u_i causes σ to vary by a few dB. Objects with sharp resonance, such as sharp edges, slits, and points may, at certain wavelengths, demonstrate resonance step-up of σ. Under certain circumstances this is particularly true for antennas that are being radiated at their resonant wavelength (resonant frequency).
- Optical range: The wavelength is small compared to the object dimensions. In this case, only the geometry and position (angle of incidence of the electromagnetic wave) of the object influence the radar cross section.

Backscatter RFID systems employ antennas with different construction formats as reflection areas. Reflections at transponders occur exclusively in the resonance range. In order to make calculations about these systems we need to know the radar cross section σ of a resonant antenna, and we will therefore derive this in the following section:

The maximum received power that can be drawn from an antenna, given optimal alignment and correct polarisation, is proportional to the power density S (see equation 4.58) of an incoming wave. The proportionality factor has the dimension of an area and is thus denoted as the effective area A_e. The following applies:

$$P_e = A_e \cdot S \tag{4.62}$$

We can imagine A_e as an area at right angles to the direction of propagation, through which the power P_e travels at a given irradiance S [meinke].

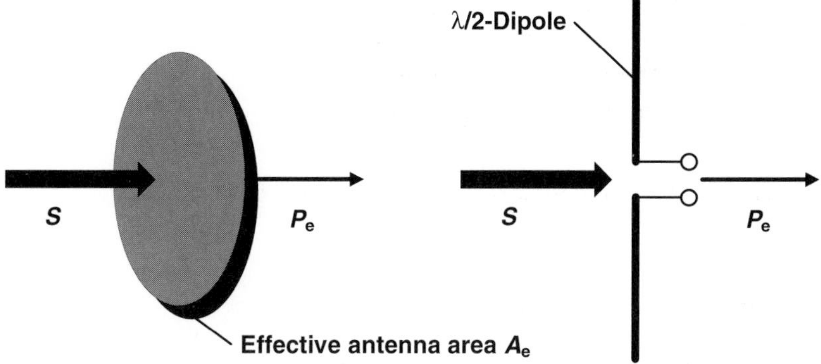

Figure 4.51: Relationship between the irradiance S and the received power P of an antenna

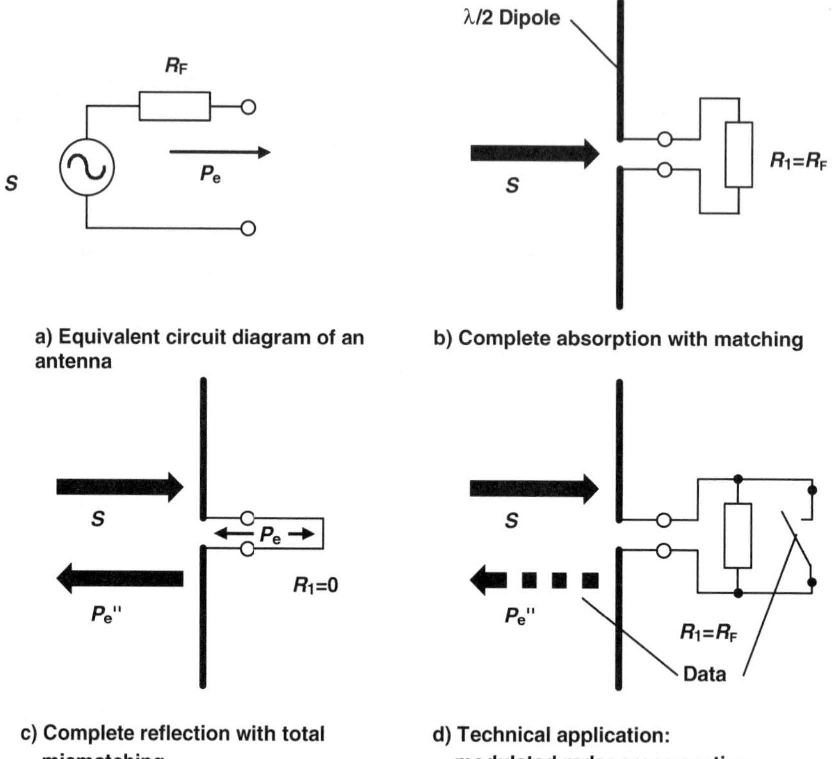

Figure 4.52 a to d: Step-by-step explanation of the modulated radar cross section as switching between absorption and reflection

The *effective area* A_e of an antenna is proportional to its gain G [meinke]. Because the gain is known for most antenna formats, the effective area is simple to calculate. The following applies:

$$A_e = \frac{\lambda_0^2}{4\pi} \cdot G \tag{4.63}$$

It follows from equation 4.63 that:

$$P_e = A_e \cdot S = \frac{\lambda_0^2}{4\pi} \cdot G \cdot S \tag{4.64}$$

The *received power* P_e can be measured at the terminals of the antenna (the base point). The antenna functions here as a voltage generator with the internal resistance R_F, the *low end* or *radiation resistance* of the antenna. An equivalent circuit diagram of an antenna is shown in Figure 4.52a.

If the terminals of the antenna are closed by a resistor R_l, the value of which corresponds with the low end resistance of the antenna (Figure 4.52b), then there is power matching between antenna and resistor R_l. All of the received power P_e is converted into heat in R_l, the layout incorporating antenna and $R_l = R_F$ thus completely absorbs the energy drawn from the field via A_e.

If, on the other hand, the antenna terminals are short circuited (or left open), then the received power P_e is completely reflected at the short circuit point and radiated back into space by the antenna (Figure 4.52c). The antenna thus reflects all of the energy arriving at A_e and acts as a reflective surface with the radar cross section $\sigma = A_e$ [knott].

For other values of R_l, σ can take on any value between 0 and σ_{max}. The following thus applies for the radar cross section of an antenna:

$$\sigma_{max} = A_e = \frac{\lambda_0^2}{4\pi} \cdot G \Big| \begin{matrix} R_1 \to 0 \\ R_1 \to \infty \end{matrix} \tag{4.65}$$

$$\sigma_{min} = 0 \big| R_1 \to R_F$$

4.2.4 Modulated radar cross section

The variation of the radar cross section in relation to the value of the load resistor R_l can be exploited for the data transfer from transponder to reader. To achieve this, the moving load R_l in the transponder is short circuited in time with a data stream to be transmitted. The radar cross section, and thus the reflected power, is hereby modulated to a certain degree by the data stream. This method is therefore termed *modulated radar cross section* (*modulated backscatter*, σ modulation) in the literature.

Because the signal reflected by the transponder is sampled in its amplitude between two limit values, the effect corresponds with ASK modulation.

4.2.5 Effective length

The equivalent circuit diagram of an antenna was shown in Figure 4.52. The open circuit voltage U_0 that occurs at the terminals is proportional to electric field strength E of the incoming wave. The proportionality factor has the dimension of a length and is therefore called the *effective length* l_0 [meinke]. The following applies:

$$U_0 = l_0 \cdot E = l_0 \cdot \sqrt{S \cdot Z_F} \tag{4.66}$$

Z_F is the field characteristic impedance from equation 4.55.

4.2.6 *Antenna construction formats for microwave transponders*

Transponders that use σ modulation operate almost exclusively in the microwave range, that is at frequencies of 2.45 GHz or 5.8 GHz. The small wavelengths of these frequency ranges (12 cm or 5.1 cm) facilitate the use of antennas with mechanical dimensions in the order of magnitude of the wavelengths. These are primarily dipoles, slot antennas and antenna formats derived from these.

4.2.6.1 Slot antennas

If we cut a strip out of the middle of a large metal surface, the length of which is λ/2, then the slot can be used as an emitter [rothammel]. The width of the slot must be small in relation to its length. The low point of the emitter is located in the middle of its longitudinal side.

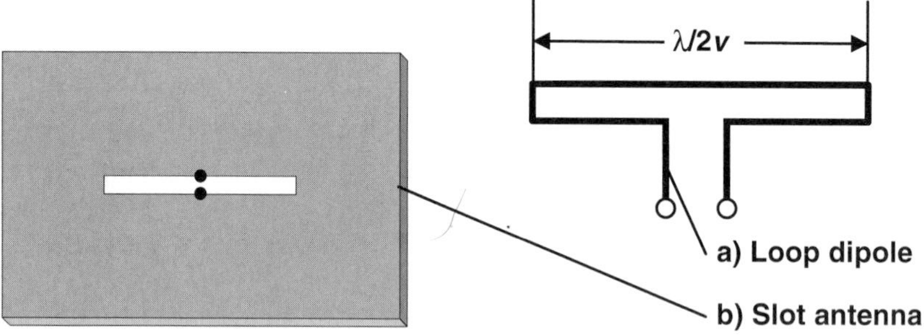

Figure 4.53: Two antenna types used in readers. Left: slot antenna, right: loop dipole

4.2.6.2 Planar antennas

This antenna format is commonly used in readers. A description can be found in Chapter 11, "Readers".

4.2.6.3. Overview – antenna parameters

The following table includes some basic antenna values, including A_e and l_e (see below for explanation).

Table 4.5: Various antenna parameters

Format:	Gain (lin.)	Effective area A_e	A_e | 2.45 GHz	Effective length l_e	Radiation resistance R_S
λ/4 Antenna upon conductive plane	3.28	0.065 λ^2	~ $1\cdot10^{-3}$ m^2	$\lambda/2\pi = 0.16\lambda$	40 Ω
λ/2 Dipole	1.64	0.13 λ^2	~ $2\cdot10^{-3}$ m^2	$\lambda/\pi = 0.32\lambda$	73 Ω
λ/2 Loop dipole	1,64	0,13 λ^2	~ $2\cdot10^{-3}$ m^2	$2\lambda/\pi = 0.64\lambda$	290 Ω
Slot antenna	3,28	0,26 λ^2	~ $4\cdot10^{-3}$ m^2	-	~ 500 Ω

5

Frequency Ranges and Radio Licensing Regulations

5.1 Frequency Ranges Used

Because RFID systems generate and radiate electromagnetic waves, they are justifiably classified as *radio systems*. The function of other *radio services* must under no circumstances be disrupted or impaired by the operation of RFID systems. It is particularly important to ensure that RFID systems do not interfere with nearby radio and television, mobile radio services (police, security services, industry), marine and aeronautical radio services and mobile telephones.

The need to exercise care with regard to other radio services significantly restricts the range of suitable operating frequencies available to an RFID system. For this reason, it is usually only possible to use frequency ranges that have been reserved specifically for industrial, scientific or medical applications. These are the frequencies classified worldwide as *ISM frequency ranges* (Industrial-Scientific-Medical), and they can also be used for RFID applications.

In addition to ISM frequencies, the entire *frequency range* below 135 kHz (in North and South America and Japan: < 400 kHz) is also suitable, because it is possible to work with high magnetic field strengths in this range, particularly when operating inductively coupled RFID systems.

The most important frequency ranges for RFID systems are therefore 0 – 135 kHz, and the ISM frequencies around 6.78 (not yet available in large parts of Europe), 13.56 MHz, 27.125 MHz, 40.68 MHz, 433.92 MHz, 869.0 MHz, 915.0 MHz (not in Europe), 2.45 GHz, 5.8 GHz, and 24.125 GHz.

5.1.1 Frequency range 9 – 135 kHz

The range below 135 kHz is heavily used by other radio services because it has not been reserved as an ISM frequency range. The propagation conditions in this *long wave* frequency range permit the radio services that occupy this range to reach areas within a radius of over 1000 km continuously at a low technical cost. Typical radio services in this frequency range are aeronautical and marine navigational radio services (LORAN C, OMEGA, DECCA), time signal services, and standard frequency services, plus military radio services. Thus, in central Europe the time signal transmitter DCF 77 in Mainflingen can be found at around the frequency 77.5 kHz. An RFID system operating at this frequency would therefore cause the failure of all radio clocks within a radius of several hundred metres around a reader.

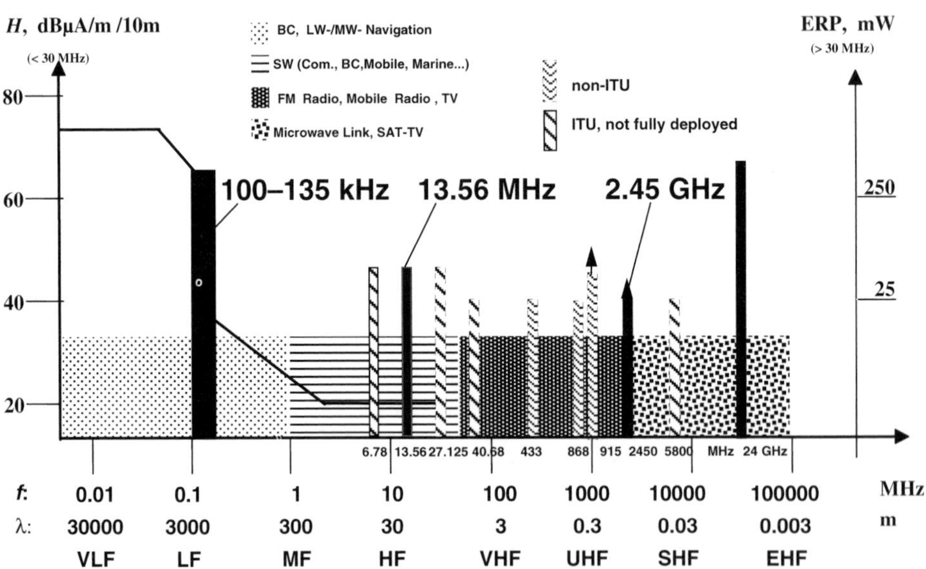

Figure 5.1: The frequency ranges used for RFID systems range from the myriametric range below 135 kHz, through short wave and ultrashort wave to the microwave range, with the highest frequency being 24 GHz. In the frequency range above 135 kHz the ISM bands available worldwide are preferred

In order to prevent such collisions, the future Licensing Act for Inductive Radio Systems in Europe, 220 ZV 122, will define a protected zone of between 70 and 119 kHz, which will no longer be allocated to RFID systems.

The radio services permitted to operate within this frequency range in Germany (source: BAPT 1997) are shown in Table 5.1.

Table 5.1: German radio services in the frequency range 9 – 135 kHz. The actual occupation of frequencies, particularly within the range 119 – 135 kHz has fallen sharply. For example, the German weather service (DWD) changed the frequency of its weather fax transmissions to 134.2 kHz as early as mid 1996

f/kHz	Class	Location	Call
16.4	FX	Mainflingen	DMA
18.5	FX	Burlage	DHO35
23.4	FX	Mainflingen	DMB
28.0	FC	Burlage	DHO36
36.0	FC	Burlage	DHO37
46.2	FX	Mainflingen	DCF46
47.4	FC	Cuxhafen	DHJ54
53.0	FX	Mainflingen	DCF53
55.2	FX	Mainflingen	DCF55
69.7	FX	Königswusterhausen	DKQ
71.4	AL	Coburg	-
74.5	FX	Königswusterhausen	DKQ2
77.5	Time	Mainflingen	DCF77
85.7	AL	Brilon	-
87.3	FX	Bonn	DEA
87.6	FX	Mainflingen	DCF87
94.5	FX	Königswusterhausen	DKQ3
97.1	FX	Mainflingen	DCF97
99.7	FX	Königswusterhausen	DIU
100.0	NL	Westerland	-
103.4	FX	Mainflingen	DCF23
105.0	FX	Königswusterhausen	DKQ4
106.2	FX	Mainflingen	DCF26
110.5	FX	Bad Vilbel	DCF30
114.3	AL	Stadtkyll	-
117.4	FX	Mainflingen	DCF37
117.5	FX	Königswusterhausen	DKQ5
122.5	DGPS	Mainflingen	DCF42
125.0	FX	Mainflingen	DCF45
126.7	AL	Portens, LORAN-C, coastal navigation	-
128.6	AL	Zeven, DECCA, coastal navigation	-
129.1	FX	Mainflingen, EVU remote control transmitter	DCF49
131.0	FC	Kiel (military)	DHJ57
131.4	FX	Kiel (military	DHJ57

Abbreviations: AL: Air navigation radio service, FC: Mobile marine radio service, FX: Fixed aeronautical radio service, MS: Mobile marine radio service, NL: Marine navigation radio service, DGPS: Differential Global Positioning System (correction data), Time: Time signal transmitter for "radio clocks"

Wire-bound carrier systems also operate at the frequencies 100 kHz, 115 kHz and 130 kHz. These include, for example, intercom systems that use the 220V supply main as a transmission medium.

5.1.2 Frequency range 6.78 MHz

The range 6.765 – 6.795 MHz belongs to the *short wave frequencies*. The propagation conditions in this frequency range only permit short ranges of up to a few 100 km in the daytime. During the nighttime hours, transcontinental propagation is possible. This frequency range is used by a wide range of radio services, for example broadcasting, weather and aeronautical radio services and press agencies.

This range has not yet been passed as an ISM range in Germany, but has been designated an ISM band by the international ITU and is being used to an increasing degree by RFID systems (in France, amongst other countries). CEPT/ERC and ETSI designate this range as a harmonised frequency in the CEPT/ERC 70-03 regulation (see Section 5.2 "International Licensing Regulations").

5.1.3 Frequency range 13.56 MHz

The range 13.553 – 13.567 MHz is located in the middle of the short wavelength range. The propagation conditions in this frequency range permit transcontinental connections throughout the day. This frequency range is used by a wide variety of radio services [siebel], for example press agencies and telecommunications (PTP).

Other ISM applications that operate in this frequency range, in addition to inductive radio systems (RFID), are remote control systems, remote controlled models, demonstration radio equipment and pagers.

5.1.4 Frequency range 27.125 MHz

The frequency range 26.565 – 27.405 is allocated to CB radio across the entire European continent as well as in the USA and Canada. Unregistered and nonchargeable radio systems with transmit power up to 4 Watts permit radio communication between private participants over distances of up to 30 km.

The ISM range between 26.957 and 27.283 MHz is located approximately in the middle of the CB radio range. In addition to inductive radio systems (RFID), ISM applications operating in this frequency range include diathermic apparatus (medical application), high frequency welding equipment (industrial application), remote controlled models and pagers.

When installing 27 MHz RFID systems for industrial applications, particular attention should be given to any high frequency welding equipment that may be located in the vicinity. HF welding equipment generates high field strengths, which may interfere with the operation of RFID systems operating at the same frequency in the vicinity. When planning

27 MHz RFID systems for hospitals (e.g. access systems), consideration should be given to any diathermic apparatus that may be present.

5.1.5 Frequency range 40.680 MHz

The range 40.660 – 40.700 MHz is located at the lower end of the *VHF range*. The propagation of waves is limited to the ground wave, so damping due to buildings and other obstacles is less marked. The frequency ranges adjoining this ISM range are occupied by mobile commercial radio systems (forestry, motorway management) and by television broadcasting (VHF range I).

The main ISM applications that are operated in this range are telemetry (transmission of measuring data) and remote control applications. The author knows of no RFID systems operating in this range, which can be attributed to the unsuitability of this frequency range for this type of systems. The ranges that can be achieved with inductive coupling in this range are significantly lower than those that can be achieved at all the lower frequency ranges that are available, whereas the wavelengths of 7.5 m in this range are unsuitable for the construction of small and cheap backscatter transponders.

5.1.6 Frequency range 433.920 MHz

The frequency range 430.000 – 440.000 MHz is allocated to amateur radio services worldwide. Radio amateurs use this range for voice and data transmission and for communication via relay radio stations or home-built space satellites.

The propagation of waves in this *UHF frequency range* is approximately optical. A strong damping and reflection of incoming electromagnetic waves occurs when buildings and other obstacles are encountered.

Depending upon the operating method and transmission power, systems used by radio amateurs achieve distances between 30 and 300 km. Worldwide connections are also possible using space satellites.

The ISM range 433.050 – 434.790 MHz is located approximately in the middle of the amateur radio band and is extremely heavily occupied by a wide range of ISM applications. In addition to backscatter (RFID) systems, baby intercoms, telemetry transmitters (including those for domestic applications, i.e. wireless external thermometer), cordless headphones, unregistered LPD walkie-talkies for short range radio, keyless entry systems (handheld transmitters for vehicle central locking) and many other applications are crammed into this frequency range. Unfortunately, mutual interference between the wide range of ISM applications is not uncommon in this frequency range.

5.1.7 Frequency range 869.0 MHz

The frequency range 868 – 870 MHz was passed for Short Range Devices (SRDs) in Europe at the end of 1997 and is thus available for RFID applications in the 43 member states of CEPT.

A few Far Eastern countries are also considering passing this frequency range for SRDs.

5.1.8 Frequency range 915.0 MHz

This frequency range is not available for ISM applications in Europe. Outside Europe (USA and Australia) the frequency ranges 888 – 889 MHz and 902 – 928 MHz are available and are used by backscatter (RFID) systems.

Neighbouring frequency ranges are occupied primarily by D-net telephones and cordless telephones as described in the CT1+ and CT2 standards.

5.1.9 Frequency range 2.45 GHz

The ISM range 2.400 – 2.4835 GHz partially overlaps with the frequency ranges used by amateur radio and radiolocation services. The propagation conditions for this UHF frequency range and the higher frequency SHF range is quasi-optical. Buildings and other obstacles behave as good reflectors and damp an electromagnetic wave very strongly at transmission (passage).

In addition to the *backscatter* (RFID) systems, typical ISM applications that can be found in this frequency range are telemetry transmitters and PC LAN systems for the wireless networking of PCs.

5.1.10 Frequency range 5.8 GHz

The ISM range 5.725 – 5.875 GHz partially overlaps with the frequency ranges used by amateur radio and radiolocation services.

Typical ISM applications for this frequency range are movement sensors, which can be used as door openers (in shops and department stores), or contactless toilet flushing, plus backscatter (RFID) systems.

5.1.11 Frequency range 24.125 GHz

The ISM range 24.00 – 24.25 GHz overlaps partially with the frequency ranges used by amateur radio and radiolocation services plus earth resources services via satellite.

This frequency range is used primarily by movement sensors, but also directional radio systems for data transmission. The author knows of no RFID systems operating in this frequency range.

5.1.12 Selection of a suitable frequency for inductively coupled RFID systems

The characteristics of the few available frequency ranges should be taken into account when selecting a frequency for an *inductively coupled* RFID systems. The usable field strength in the operating range of the planned system exerts a decisive influence on system parameters. This variable therefore deserves further consideration. In addition, the *bandwidth*, (mechanical) dimensions of the antenna coil and the availability of the frequency band should also be considered.

The path of field strength of a magnetic field in the *near* and *far field* was described in detail in Chapter 4, section 4.2. We learned that the reduction in field strength with increasing distance from the antenna was 60 dB/decade initially, but that this falls to 20 dB/decade after the transition to the far field at a distance of $\lambda/2\pi$. This behaviour exerts a strong influence on the usable field strengths in the system's operating range. Regardless of the operating frequency used, the regulation *EN 300330* specifies the maximum magnetic field strength at a distance of 10 m from a reader.

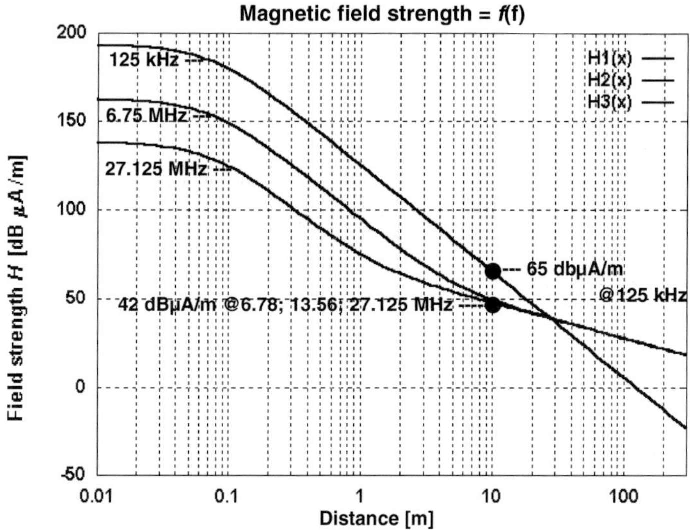

Figure 5.2: Different permissible field strengths for inductively coupled systems measured at a distance of 10 m (the distance specified for licensing procedures) and the difference in the distance at which the reduction occurs at the transition between near and far field lead to marked differences in field strength at a distance of 1 m from the antenna of the reader. For the field strength path at a distance under 10 cm, we have assumed that the antenna radius is the same for all antennas

If we move from this point in the direction of the reader, then, depending upon the wavelength, the field strength increases initially at 20 dB/decade. At an operating frequency of 6.78 MHz the field strength begins to increase by 60 dB/decade at a distance of 7.1 m – the transition into the near field. However, at an operating frequency of 27.125 MHz this steep increase does not begin until a distance of 1.7 m is reached.

It is not difficult to work out that, given the same field strength at a distance of 10 m, higher usable field strengths can be achieved in the operating range of the reader (e.g. 0 – 10 cm) in a lower frequency ISM band than would be the case in a higher frequency band. At < 135 kHz the relationships are even more favourable, firstly because the permissible field strength limit is much higher than it is for ISM bands above 1 MHz, and secondly because the 60 dB increase takes effect immediately, because the near field in this frequency range extends to at least 350 m.

If we measure the range of an inductively coupled system with the same magnetic field strength H at different frequencies we find that the range is maximised in the frequency range around 10 MHz. This is because of the proportionality $U_{ind} \sim \omega$. At higher frequencies around 10 MHz the efficiency of power transmission is significantly greater than at frequencies below 135 kHz.

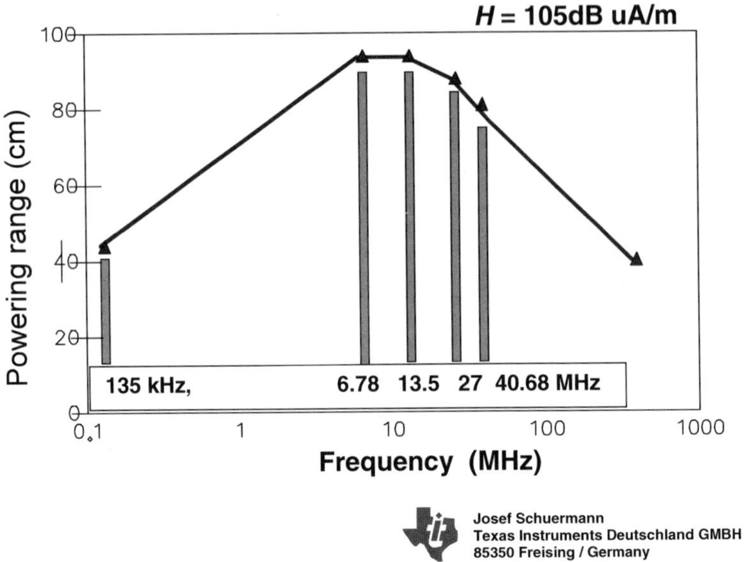

Figure 5.3: Transponder range at the same field strength. The induced voltage at a transponder is measured with the antenna area and magnetic field strength of the reader antenna held constant. (Reproduced by permission of Texas Instruments)

However, this effect is compensated by the higher permissible field strength at 135 kHz, and therefore in practice the range of RFID systems is roughly the same for both frequency ranges. At frequencies above 10 MHz the L/C relationship of the transponder resonant circuit becomes increasingly unfavourable, so the range in this frequency range starts to decrease.

Overall, the following preferences exist for the various frequency ranges:

< 135 kHz Preferred for large *ranges* and *low cost transponders*.

- High level of power available to the transponder.
- The transponder has a low power consumption due to its lower clock frequency.
- Miniaturised transponder formats are possible (animal ID) due to the use of ferrite coils in the transponder.
- Low absorption rate or high *penetration depth* in non-metallic materials and water (the high penetration depth is exploited in animal identification by the use of the bolus, a transponder placed in the rumen).

6.78 MHz Can be used for low cost and medium speed transponders.

- Worldwide ISM frequency according to ITU frequency plan, however this is not used in some countries (i.e. licence may not be used worldwide).
- Available power is a little greater than that for 13.56 MHz.
- Only half the clock frequency of that for 13.56 MHz.

13.56 MHz Can be used for high speed / high end and medium speed / low end applications.

- Available worldwide as an ISM frequency.
- Fast data transmission (typically 106 kbits/s).
- High clock frequency, so cryptological functions or a microprocessor can be realised.
- Parallel capacitors for transponder coil (resonance matching) can be realised on-chip.

27.125 MHz Only for special applications (e.g. Eurobalise)

- Not a worldwide ISM frequency.
- Large bandwidth, thus very fast data transmission (typically 424 kbits/s)
- High clock frequency, thus cryptological functions or a microprocessor can be realised.
- Parallel capacitors for transponder coil (resonance matching) can be realised on-chip.
- Available power somewhat lower than for 13.56 MHz.
- Only suitable for small ranges.

5.2 International Licensing Regulations

5.2.1 CEPT/ERC 70-03

A new CEPT harmonisation document has been available since October 1997 that serves as the basis for new national regulations. The old national regulations for Short Range Devices (SRDs) are thus being successively withdrawn. This document also refers to the ETSI standards EN 300330, EN 300220 and EN 300440 that are relevant to RFID systems.

5.2.2 EN 300330: 9 kHz – 25 MHz

The standards drawn up by the ETSI (European Telecommunications Standards Institute) serve to provide the national telecommunications authorities with a basis for the creation of national regulations for the administration of radio and telecommunications.

The basis for European *licensing regulations for inductive radio systems*, i.e. inductively coupled RFID systems, is provided by EN 300330 entitled "*Radio Equipment and Systems (RES); Short Range Devices (SRDs), Technical characteristics and test methods for radio equipment in the frequency range 9 kHz to 25 MHz and inductive loop systems in the frequency range 9 kHz to 30 MHz*", which serves as a standard. In addition to inductive radio systems, *EN 300330* also relates to anti-theft devices (for shops), alarm systems, telemetry systems and short range remote control systems, which are treated under the collective title of "Short Range Devices (SRDs)". Despite the provisional status of EN 300330, this regulation has already been adopted as a national regulation in unmodified form by the majority of European member states, and this regulation is also used for the licensing of RFID systems by many Asiatic and American states.

EN 300330 defines frequency ranges and maximum permissible field strengths for inductive radio systems, and it specifies the measuring procedures that may be used to confirm whether the specified limit values have been adhered to.

Inductive radio systems (inductive loop coil transmitters) in accordance with EN 300330 are characterised by the fact that the antenna is formed by a wire loop made up of one or several windings.

We differentiate between four classes of product type:

Class 1 Transmitter with inductive *loop antenna*, whereby the antenna is integrated into the device. Enclosed antenna area < 30 m^2.

Class 2 Transmitter with inductive loop antenna, whereby the antenna is manufactured to the customer's requirements. Devices belonging to class 2 can be tested without an antenna. The enclosed antenna area must be less than 30 m^2.

Class 3 Transmitter with large inductive loop antenna, > 30 m^2 antenna area. Class 3 devices are tested without an antenna.

Class 4 E field transmitter. These devices are tested with an antenna.

All the inductively coupled RFID systems in the frequency range 9 kHz – 30 MHz described in EN 300330 belong to the class 1 and class 2 types. The class 3 and class 4 types are therefore not considered further in this book.

5.2.2.1 Carrier power – limit values for class 1 transmitters

In class 1 inductive radio systems (integral antenna) the *H field* of the radio system is measured in the direction in which the field strength reaches a maximum. The measurement should be performed in free space, whereby the distance from the measuring antenna to the measurement antenna is 10 m. The transmitter is not modulated during the field strength measurement. Table 5.2 shows the stipulated limit values.

Table 5.2: These values only apply unreservedly for conductor loops with an enclosed area A > 0.16 m^2. Lower limit values are defined in EN 300330 in the frequency range up to 135 kHz for smaller antennas

Frequency range / MHz	Max. H field in dBμA/m @ 10 m
0.009 – 0.030	72
0.030 – 0.070 0.119 – 0.135	72 at 0.030 MHz falling by -3 dB/oct
0.070 – 0.119	42 (LW protective zone)
0.135 – 1.000	37.7 at 0.135 MHz falling by -3 dB/oct
1.000 – 10.00	29 at 1.0 MHz falling by -9 dB/oct
10.00 – 30.00	-1
6.78 ± 150 kHz (ISM) 7.4 – 8.8 (EAS) 13.56 ± 150 kHz (ISM) 27.12 ± 500 kHz (ISM)	9
6.765 – 6.795 (ISM) 13.553 – 13.567 (ISM) 26.957 – 27.283 (ISM)	42

5.2.2.2 Carrier power – limit values for class 2 transmitters

In class 2 devices the H field strength is obtained by calculation rather than measurement. An imitation antenna is constructed for this purpose. This antenna consists of a small cylindrical coil, the inductance and quality of which correspond with the inductance and quality of the intended loop antenna. The antenna current can be determined with the transmitter connected using low ohm, low inductance measurement resistors connected in series with the cylindrical coil.

The resulting H field is calculated as follows:

$$H = \frac{N \cdot I \cdot A}{2\pi d^3} \tag{5.1}$$

where N is the number of windings of the loop antenna, I is the HF current of the loop antenna, A is the enclosed area, and d is the distance from the loop antenna to the measurement antenna.

This relationship is valid when the following conditions are fulfilled. Conductor length of the loop antenna:

$$l < \frac{\lambda}{2\pi} \tag{5.2}$$

Distance between loop and measuring antenna:

$$d < \frac{\lambda}{2\pi} \tag{5.3}$$

Because EN 300330 specifies that measurements should be performed with a distance d of 10 m between the loop antenna and the measurement antenna, the frequency at which the above-mentioned conditions are only just fulfilled works out at 4.775 MHz. Class 2 measurements are therefore only permissible for transmission frequencies < 4.775 MHz.

5.2.2.3 Modulation bandwidth

The limit values for modulation sidebands that are specified in Chapter 5, Section 5.2.2.1 must also be adhered to.

5.2.2.4 Spurious emissions

Spurious emissions are emissions that cannot be attributed to the carrier frequency or the modulation sidebands, for example harmonic waves and parasitic compounds. Spurious emissions must be minimised.

The limit values specified in Chapter 5, Section 5.2.1 must be adhered to for spurious emissions in the frequency range 0 – 30 MHz. For the frequency range 30 – 1000 MHz the values specified in Table 5.3 must be adhered to, giving particular consideration to the frequency range of public radio and television that is susceptible to interference.

Table 5.3: Limit values for spurious emissions in the frequency range 30 – 1000 MHz

System state	47 – 74 MHz 87.5 – 118 MHz 174 – 230 MHz 470 – 862 MHz	Other frequencies in the range 30 – 1000 MHz
Operating state	4 nW	250 nW
Standby	2 nW	2 nW

5.2.3 EN 300220-1, EN 300220-2

The standard *EN 300220*, entitled *"Radio Equipment and Systems (RES); Short range devices, Technical characteristics and test methods for radio equipment to be used in the 25 MHz to 1000 MHz frequency range with power levels ranging up to 500 mW"*, provides the basis for national European licensing regulations for low power radio systems and comprises two sections: EN 300220-1 for transmitters and their power characteristics and EN 300220-2 in which the characteristics for the receiver are defined.

EN 300220 classifies devices according to four types – class I to IV – which are not defined in more detail. This standard covers low power radio systems, both within the ISM bands and throughout the entire frequency range (e.g. estate radio and pagers on 466.5 MHz). Typical ISM applications in these ranges are telemetry, alarm and remote control radio systems plus LPD radio telephony systems (10 mW at 433.920 MHz).

RFID systems are not mentioned explicitly, however the frequency range below 30 MHz (27.125 MHz) is covered by EN 300330 and the frequency ranges 40.680 MHz and 433.920 MHz are less typical for RFID applications.

Unlike EN 300330, which defines a maximum permissible field strength at a distance of 10m from the measurement object, EN 300220 specifies a maximum permissible *transmitter output power* at 50 Ω:

Table 5.4: Device classes within and outside the ISM bands. This standard also defines testing methods and limit values for spurious emissions, which we will not go into in more detail here

		Permissible transmission power		
Class	Range / MHz	ISM – 27 MHz	ISM – 40 MHz	ISM – 433 MHz
I	25 – 1000	10 mW	10 mW	10 mW
II	300 – 1000	-	-	25 mW
III	25 – 300	100 mW	100 mW	-
IV	300 – 1000	-	-	100 mW

5.2.4 EN 300440

The standard *EN 300440*, entitled *"Radio Equipment and Systems (RES); Short range devices, Technical characteristics and test methods for radio equipment to be used in the 1 GHz to 25 GHz frequency range with power levels ranging up to 500 mW"* provides the basis for national European licensing regulations for low power radio systems. EN 300440 classifies devices according to three types – class I to III.

RFID systems with *backscatter transponders* are classified as class II systems. Further details are governed by the CEPT recommendation *T/R 60-01* "Low power radiolocation equipment for detecting movement and for alert" (EAS) and T/R 22-04 "Harmonisation of frequency bands for Road Transport Information Systems (RTI)" (toll systems, freight identification).

Various ISM and short range applications are classified as class I and II systems. Typical applications in these classes are movement sensors (for alarm systems, door openers and similar applications), data transmission systems (wireless LANs for PC), remote control systems and telemetry systems.

EN 300440 defines the following maximum values for effective isotropic radiated power EIRP (EIRP represents the power that a fictional isotropic source ($G = 0$ dB) would have to emit in order to generate the same power flux density at the reception location as at the device under test):

Table 5.5: Permissible transmission power in accordance with EN 300440

	Frequency (GHz)		
Class	1.0 – 5.0	5.0 – 20	>20
I	10 mW	25 mW	100 mW
II*	500 mW	500 mW	500 mW
III	500 mW	2 W	2 W

* Reflective transponder systems

The following frequency ranges are reserved for ISM applications:

2.400 – 2.4835 MHz
5.725 – 5.875 GHz
24.00 – 24.25 Ghz

This standard also defines test methods and limit values for spurious emissions, which we will not describe in more detail here.

5.3 National Licencing Regulations – U.S.A

In the U.S.A. RFID systems must be licenced in accordance with licencing regulation FCC Part 15. This regulation covers the frequency range from 9 kHz to above 64 GHz and deals with the intentional generation of electromagnetic fields by low and minimum power transmitters (intentional radiators) plus the unintentional generation of electromagnetic fields (spurious radiation) by electronic devices such as radio and television receivers or computer systems. The category of low power transmitters covers a wide variety of applications, for example cordless telephones, biometry and telemetry transmitters, on-campus radio stations, toy remote controls or door openers for cars. Inductively coupled or backscatter RFID sytems are not explicitly mentioned in the FCC regulation, but they automatically fall under its scope due to their transmission frequencies, which are typically in the ISM bands, and their low transmission power.

Table 5.6: Permissible field strengths for RFID systems in accordance with FCC Part 15

Frequency range / MHz	max. E-Feld	Measuring distance	Section
1.705 ... 10.000	100 μV/m	30 m	15.223
13.553 ... 13.567	10 mV/m	30 m	15.225
26.960 ... 27.280	10 mV/m	30 m	15.227
40.660 ... 40.700	1 mV/m	3 m	15.229
49.820 ... 49.900	10 mV/m	3 m	15.235
902.0 ... 928.0	50 mV/m	3 m	15.249
2435 ... 2465	50 mV/m	3 m	15.249
5785 ... 5815	50 mV/m	3 m	15.249
24075 ... 24175	250 mV/m	3 m	15.249

Table 5.6 lists the frequency ranges that are important for RFID systems. In all other frequency ranges the permissible limit values for spurious radiation given in Table 5.7 apply. It should be noted here that, unlike the European licencing regulation ETS 300 330, the maximum permissible field strength of a reader is principally defined by the electrical field strength E. The measuring distance is selected such that a measurement is made in the far field of the generated field. This also applies for inductively coupled RFID systems in the frequency range below 30 MHz, which usually generate a magnetic high frequency.

Table 5.7: Permissible interference field strength in all other frequency ranges in accordance with FCC Part 15, Section 15.209

Frequency range / MHz	Maximum *E* field	Measuring distance
0,009 ... 0,490	2400/f μV/m	300 m
0,490 ... 1,705	24/f mV/m	30 m
1,705 ... 30,00	30 μV/m	30 m
30,00 ... 88,00	100 μV/m	3 m
88,00 ... 216	150 μV/m	3 m
216 ... 960	200 μV/m	3 m
> 960	500 μV/m	3 m

6

Coding and Modulation

The block diagram in Figure 6.1 describes a digital *communication system*. Similarly, *data transfer* between reader and transponder in an RFID system requires three main functional blocks. From the reader to the transponder – the direction of data transfer – these are: *signal coding*, (*signal processing*) and the *modulator* (*carrier circuit*) in the *reader* (*transmitter*), the *transmission medium* (*channel*), and the *demodulator* (*carrier circuit*) and *signal decoding* (*signal processing*) in the *transponder* (*receiver*).

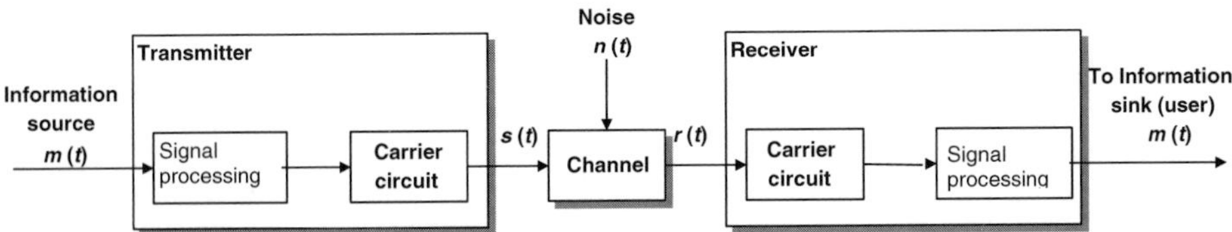

Figure 6.1: Signal and data flow in a digital communications system [couch]

A signal coding system takes the message to be transmitted and its *signal representation* and matches it optimally to the characteristics of the *transmission channel*. This process involves providing the message with some degree of protection against interference or collision and against intentional modification of certain signal characteristics [herter]. Signal coding should not be confused with modulation, and therefore it is referred to as *coding in the baseband*.

Modulation is the process of altering the signal parameters of a high frequency carrier, i.e. its amplitude, frequency or phase, in relation to a modulated signal, the baseband signal.

The transmission medium transmits the message over a predetermined distance. The only transmission media used in RFID systems are magnetic fields (inductive coupling) and electromagnetic waves (microwaves).

Demodulation is an additional modulation procedure to reclaim the signal in the baseband. As there is often an *information source* (input) in both the transponder and the reader, and information is thus transmitted alternately in both directions, these components contain both a *modulator* and a *demodulator*. This is therefore known as a *modem* (**Mo**dulator – **Dem**odulator), a term that describes the normal configuration [herter].

The task of signal decoding is to reconstruct the original message from the baseband coded *received signal* and to recognise any *transmission errors* and flag them as such.

6.1 Coding in the Baseband

Binary ones and zeros can be represented in various *line codes*. RFID systems normally use one of the following coding procedures: NRZ, Manchester, Unipolar RZ, DBP (differential bi-phase), Miller, differential coding and PP coding.

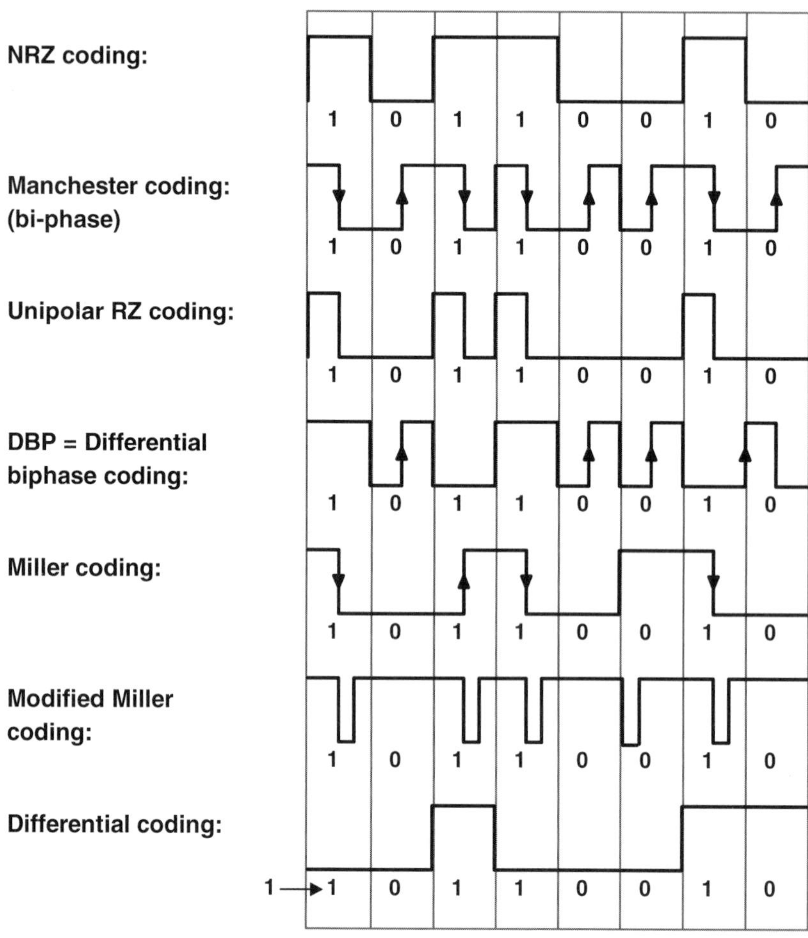

Figure 6.2: Signal coding by frequently changing line codes in RFID systems

NRZ code A binary 1 is represented by a "high" signal and a binary 0 is represented by a "low" signal. The NRZ code is used almost exclusively with FSK or PSK modulation.

Manchester code A binary 1 is represented by a negative transition in the half bit period and a binary 0 is represented by a positive transition. The Manchester code is therefore also known as *split-phase coding* [couch].

The Manchester code is often used for data transmission from the transponder to the reader based upon load modulation using a subcarrier.

Unipolar RZ code A binary 1 is represented by a "high" signal during the first half bit period, a binary 0 is represented by a "low" signal lasting for the entire duration of the bit.

DBP code A binary 0 is coded by a transition of either type in the half bit period, a binary 1 is coded by the lack of a transition. Furthermore, the level is inverted at the start of every bit period, so that the bit pulse can be more easily reconstructed in the receiver (if necessary).

Miller code A binary 1 is represented by a transition of either type in the half bit period, a binary 0 is represented by the continuance of the 1 level over the next bit period. A sequence of zeros creates a transition at the start of a bit period, so that the bit pulse can be more easily reconstructed in the receiver (if necessary).

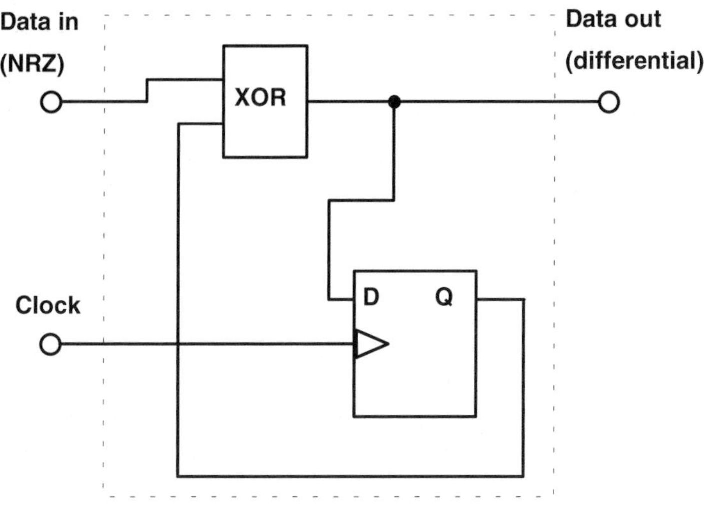

Figure 6.3: Generating differential coding from NRZ coding

Modified Miller code In this variant of the Miller code each transition is replaced by a "negative" pulse. The modified Miller code is highly suitable for use in inductively coupled RFID systems for data transfer from the reader to the transponder.

Due to the very short pulse durations ($t_{pulse} << T_{bit}$) it is possible to ensure a continuous power supply to the transponder from the HF field of the reader even during data transfer.

Differential coding In "differential coding" every binary 1 to be transmitted causes a change (toggle) in the signal level, whereas the signal level remains unchanged for a binary zero. Differential coding can be generated very simply from an NRZ signal by using an XOR gate and a D flip-flop. Figure 6.3 shows a circuit to achieve this.

Pulse-pause coding In pulse-pause coding (PPC) a binary 1 is represented by a pause duration t before the next pulse, a binary 0 is represented by a pause duration $2t$ before the next pulse. This coding procedure is popular in inductively coupled RFID systems for data transfer from the reader to the transponder. Due to the very short pulse durations ($t_{pulse} << T_{bit}$) it is possible to ensure a continuous power supply to the transponder from the HF field of the reader even during data transfer.

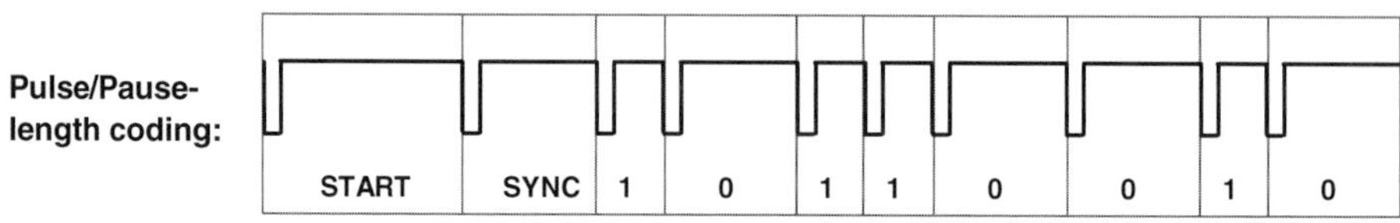

Figure 6.4: Possible signal path in pulse-pause coding

Various boundary conditions should be taken into consideration when selecting a suitable signal coding system for an RFID system. The most important consideration is the signal spectrum after modulation (see [couch], [mäusl]) and susceptibility to transmission errors. Furthermore, in the case of passive transponders (the transponder's power supply is drawn from the HF field of the reader) the power supply must not be interrupted by an inappropriate combination of signal coding and modulation procedures.

6.2 Digital Modulation Procedures

Energy is radiated from an antenna into the surrounding area in the form of electromagnetic waves. By carefully influencing of one of three signal parameters – power, frequency, phase position – of an electromagnetic wave, messages can be coded and transmitted to any point within the area. The procedure of influencing an electromagnetic wave by messages (data) is called *modulation*, an unmodulated electromagnetic wave is called a *carrier*.

By analysing the characteristics of an electromagnetic wave at any point in the area, we can reconstruct the message by measuring the change in reception power, frequency or phase position of the wave. This procedure is known as *demodulation*.

Classical radio technology is largely concerned with analogue modulation procedures. We can differentiate between *amplitude modulation*, *frequency modulation* and *phase modulation*, these being the three main variables of an electromagnetic wave. All other modulation procedures are derived from one of these three types. The procedures used in RFID systems are the digital modulation procedures *ASK* (amplitude shift keying), *FSK* (frequency shift keying) and *PSK* (phase shift keying).

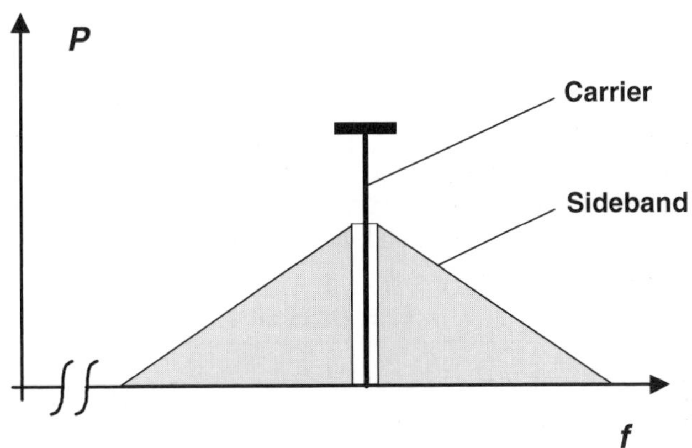

Figure 6.5: Each modulation of a sinusoidal signal – the carrier – generates so-called (modulation) sidebands

In every modulation procedure symmetric *modulation products* – so-called *sidebands* – are generated around the carrier. The spectrum and amplitude of the sidebands are influenced by the spectrum of the code signal in the baseband and by the modulation procedure. We differentiate between the upper and lower sideband.

6.2.1 Amplitude shift keying (ASK)

In *amplitude shift keying* the amplitude of a *carrier oscillation* is switched between two states u_0 and u_1 (keying) by a binary code signal. U_1 can take on values between u_0 and 0. The ratio of u_0 to u_1 is known as the *duty factor m*.

To find the duty factor m we calculate the arithmetic mean of the keyed and unkeyed amplitude of the carrier signal.

$$\hat{u}_{\mathrm{m}} = \frac{\hat{u}_0 + \hat{u}_1}{2} \tag{6.1}$$

The duty factor is now calculated from the ratio of amplitude change $\hat{u}_0$-$\hat{u}_{\mathrm{m}}$ to the mean value $\hat{u}_{\mathrm{m}}$.

$$m = \frac{\Delta\hat{u}_{\mathrm{m}}}{\hat{u}_{\mathrm{m}}} = \frac{\hat{u}_0 - \hat{u}_{\mathrm{m}}}{\hat{u}_{\mathrm{m}}} = \frac{\hat{u}_0 - \hat{u}_1}{\hat{u}_0 + \hat{u}_1} \tag{6.2}$$

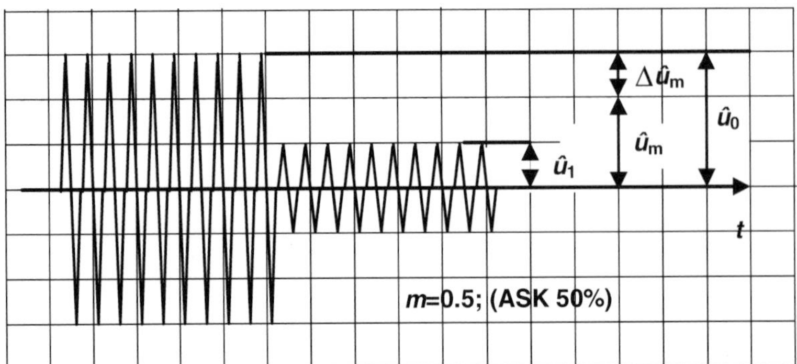

Figure 6.6: In ASK modulation the amplitude of the carrier is switched between two states by a binary code signal

In 100% ASK the amplitude of the carrier oscillation is switched between the carrier amplitude values $2\hat{u}_m$ and 0 (*On-Off keying*). In amplitude modulation using an analogue signal (sinusoidal oscillation) this would also correspond with a modulation factor of m=1 (or 100%) [mäusl].

The procedure described for calculating the duty factor is thus the same as that for the calculation of the modulation factor for amplitude modulation using analogue signals (sinusoidal oscillation). However, there is one significant difference between keying and analogue modulation. In keying, a carrier takes on the amplitude $\hat{u}_0$ in the unmodulated state, whereas in analogue modulation the carrier signal takes on the amplitude $\hat{u}_m$ in the unmodulated state.

In the literature the duty factor is sometimes referred to as the percentage carrier reduction m' during keying:

$$m' = 1 - \frac{\hat{u}_1}{\hat{u}_0} \tag{6.3}$$

For the example in Figure 6.7 the duty factor would be m'=0.66 (=66%). In the case of duty factors <15% and duty factors > 85% the differences between the two calculation methods can be disregarded.

The binary code signal consists of a sequence of 1 and 0 states, with a period duration T and a bit duration τ. From a mathematical point of view, ASK modulation is achieved by multiplying this code signal $u_{code}(t)$ by the carrier oscillation $u_{Cr}(t)$. For duty factors m<1 we introduce an additional constant (1-m), so for this case we can still multiply $u_{HF}(t)$ by 1 in the unkeyed state.

$$U_{ASK}(t) = (m \bullet u_{code}(t) + 1 - m) \bullet u_{HF}(t) \tag{6.4}$$

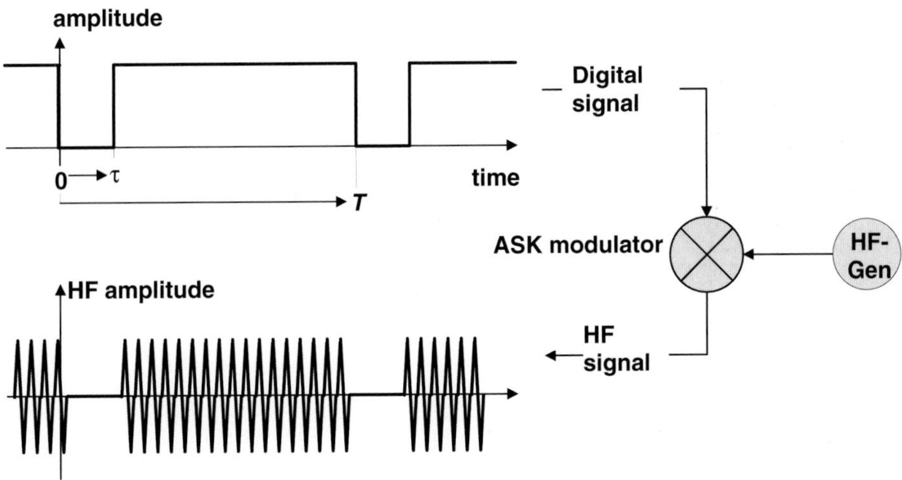

Figure 6.7: The generation of 100% ASK modulation by the keying of the sinusoidal carrier signal from a HF generator into an ASK modulator using a binary code signal

The spectrum of ASK signals is therefore found by the convolution of the code signal spectrum with the carrier frequency f_{Cr} or by multiplication of the Fourier expansion of the code signal by the carrier oscillation. It contains the spectrum of the code signal in the upper and lower sideband, symmetric to the carrier [mäusl].

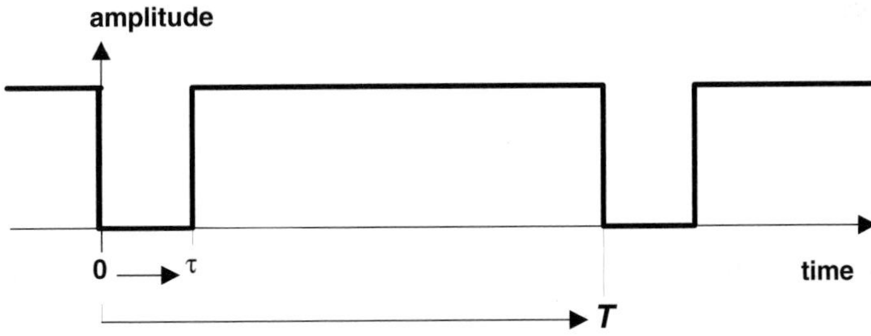

Figure 6.8: Representation of the period duration T and the bit duration τ of a binary code signal

A regular, pulse-shaped signal of period duration T and bit duration τ yields the following spectrum:

Table 6.1: Spectral lines for a pulse-shaped modulated carrier oscillation

Designation:	Frequency	Amplitude:
Carrier oscillation	f_{CR}	$u_{HF} \cdot (1-m) \cdot (T-\tau)/T$
1st spectral line	$f_{CR} \pm 1/T$	$u_{HF} \cdot m \cdot \sin(\pi \cdot \tau/T)$
2nd spectral line	$f_{CR} \pm 2/T$	$u_{HF} \cdot m \cdot \sin(2\pi \cdot \tau/T)$
3rd spectral line	$f_{CR} \pm 3/T$	$u_{HF} \cdot m \cdot \sin(3\pi \cdot \tau/T)$
*n*th spectral line	$f_{CR} \pm n/T$	$u_{HF} \cdot m \cdot \sin(n\pi \cdot \tau/T)$

6.2.2 2 FSK

In *2 frequency shift keying* the frequency of a carrier oscillation is switched between two frequencies f_1 and f_2 by a binary code signal.

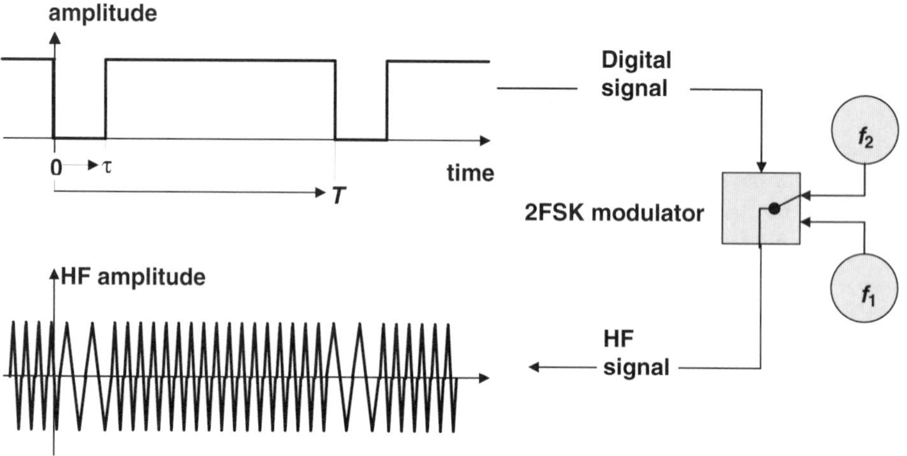

Figure 6.9: The generation of 2 FSK modulation by switching between two frequencies f_1 and f_2 in time with a binary code signal

The carrier frequency f_{CR} is defined as the arithmetic mean of the two characteristic frequencies f_1 and f_2. The difference between the carrier frequency and the characteristic frequencies is termed the frequency deviation Δf_{CR}.

$$f_{CR} = \frac{f_1 + f_2}{2} \qquad \Delta f_{CR} = \frac{|f_1 + f_2|}{2} \tag{6.5}$$

From the point of view of the time function, the *2 FSK* signal can be considered as the composition of two amplitude shift keyed signals of frequencies f_1 and f_2. The spectrum of a 2 FSK signal is therefore obtained by superimposing the spectra of the two amplitude shift keyed oscillations. The baseband coding used in RFID systems produces an asymmetric frequency shift keying:

$$\tau \neq \frac{T}{2} \tag{6.6}$$

In these cases there is also an asymmetric distribution of spectra in relation to the mid frequency Δf_{CR} [mäusl].

6.2.3 2 PSK

In *phase shift keying* the binary states "0" and "1" of a code signal are converted into corresponding phase states of the carrier oscillation, in relation to a reference phase. In 2 PSK (2 phase shift keying) the signal is switched between the phase states 0° and 180°.

Mathematically speaking, the shift keying of the phase position between 0° and 180° corresponds with the multiplication of the carrier oscillation by 1 and -1.

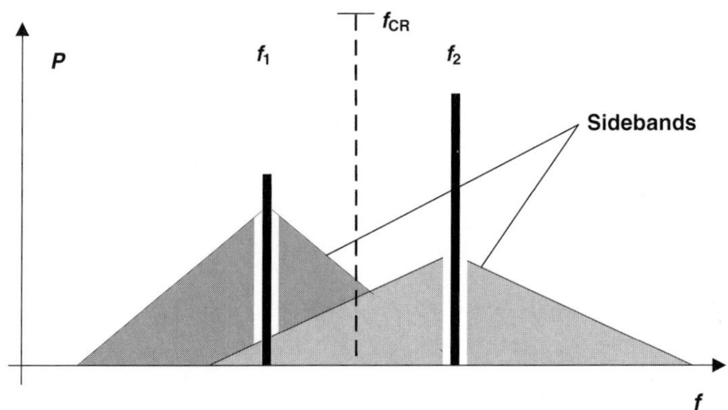

Figure 6.10: The spectrum of a 2 FSK modulation is obtained by the addition of the individual spectra of two amplitude shift keyed oscillations of frequencies f_1 and f_2

The power spectrum of a 2 PSK can be calculated as follows for a mark-space ratio τ/T of 50%,[mansukhani]:

$$P(f) = \left(\frac{P \cdot T_S}{2}\right) \cdot \left[\operatorname{sin}c^2 \pi (f - f_0) T_S + \operatorname{sin}c^2 \pi (f + f_0) T_S\right] \tag{6.6}$$

P = Transmitter power
T_s = Bit duration = τ
f_0 = Centre frequency

$$\operatorname{sin}c\,(x) = \frac{\sin\,(x)}{x}$$

The envelope of the two sidebands around the carrier frequency f_0 follows the function $(\sin(x)/x)^2$. This yields zero positions at the frequencies $f_0 \pm 1/T_S$, $f_0 \pm 2/T_S$, $f_0 \pm n/T_S$. In the frequency range $f_0 \pm 1/T_S$, 90% of the transmitter power is transmitted.

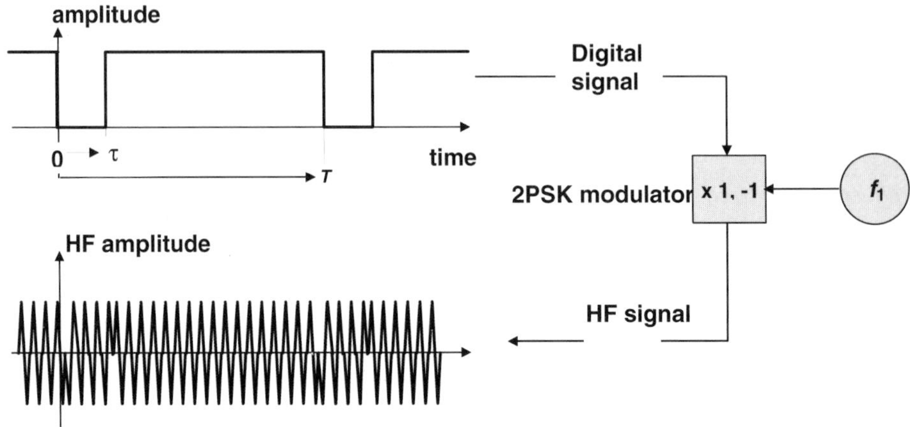

Figure 6.11: Generation of the 2 PSK modulation by the inversion of a sinusoidal carrier signal in time with a binary code signal

6.2.4 *Modulation procedures with subcarrier*

The use of a modulated *subcarrier* is widespread in radio technology. In VHF broadcasting, a stereo subcarrier with a frequency of 38 kHz is transmitted along with the baseband tone channel. The baseband contains only the mono tone signal. The differential "L-R" signal required to obtain the "L" and "R" tone channels can be transmitted "silently" by the modulation of the stereo subcarrier. The use of a subcarrier therefore represents a *multilevel modulation*. Thus, in our example, the subcarrier is first modulated with the differential signal, in order to finally modulate the VHF transmitter once again with the modulated subcarrier signal.

In RFID systems, modulation procedures using a subcarrier are primarily used in inductively coupled systems in the frequency ranges 6.78 MHz, 13.56 MHz or 27.125 MHz and in load modulation for data transfer from the transponder to the reader. The load modulation of an inductively coupled RFID system has a similar effect to ASK modulation of HF voltage at the antenna of the reader. Instead of switching the *load resistance* on and off in time with a baseband coded signal, a low frequency subcarrier is first modulated by the baseband coded data signal. ASK, FSK or PSK modulation may be selected as the modulation procedure for the subcarrier. The *subcarrier frequency* itself is normally obtained by the binary division of the operating frequency. For 13.56 MHz systems, the subcarrier frequencies 847 kHz (13.56 MHz ÷ 16), 424 kHz (13.56 Mhz ÷ 32) or 212 kHz

(13.56 MHz ÷ 64) are usually used. The modulated subcarrier signal is now used to switch the load resistor on and off.

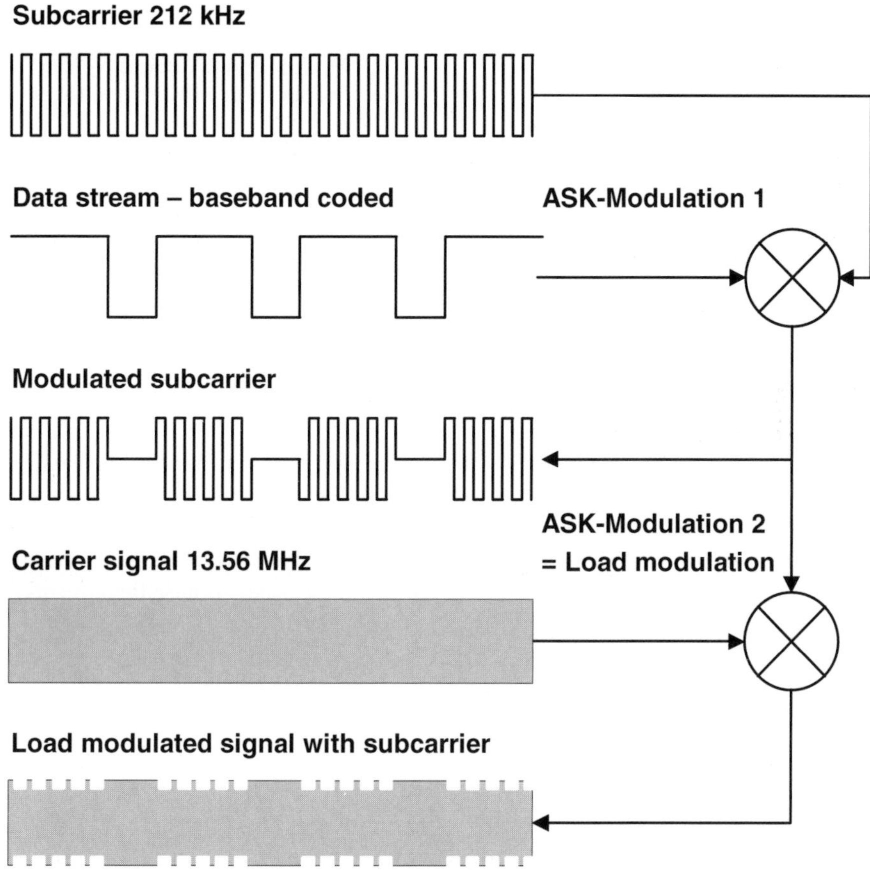

Figure 6.12: Step-by-step generation of a multiple modulation, by load modulation with ASK modulated subcarrier

The great advantage of using a subcarrier only becomes clear when we consider the frequency spectrum generated:

Load modulation with a subcarrier initially generates two spectral lines at a distance ± the subcarrier frequency f_H around the operating frequency. The actual information is now transmitted in the sidebands of the two subcarrier lines, depending upon the modulation of the subcarrier with the baseband coded data stream. If load modulation in the baseband were used, on the other hand, the sidebands of the data stream would lie directly next to the carrier signal at the operating frequency.

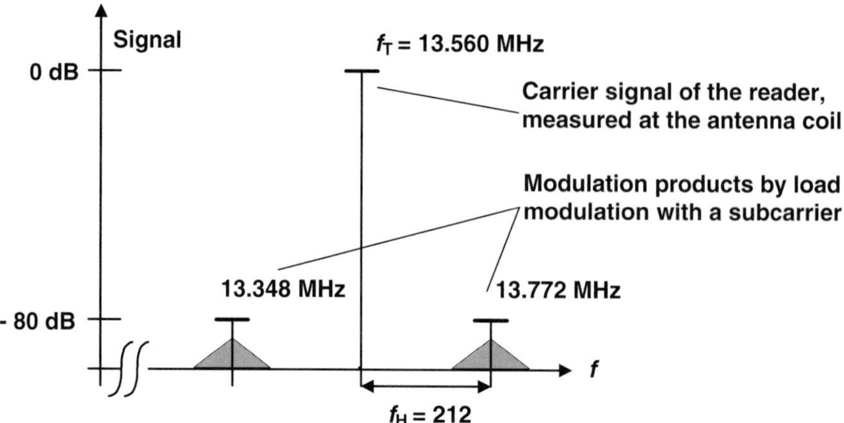

Figure 6.13: Modulation products using load modulation with a subcarrier

In very loosely coupled transponder systems the difference between the carrier signal of the reader f_T and the received modulation sidebands of the load modulation varies within the range 80 – 90 dB. One of the two subcarrier modulation products can be filtered out and demodulated by shifting the frequency of the modulation sidebands of the data stream. It is irrelevant here whether the frequencies $f_T + f_H$, or $f_T - f_H$ are used, because the information is contained in all sidebands.

7

Data Integrity

7.1 The Checksum Procedure

When transmitting data using contactless technology it is very likely that interference will be encountered, causing undesired changes to the transmitted data and thus leading to transmission errors.

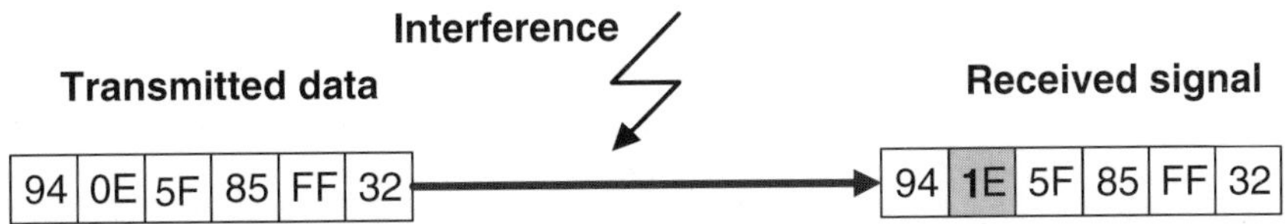

Figure 7.1: Interference during transmission can lead to errors in the data

A *checksum* can be used to recognise transmission errors and initiate corrective measures, for example the retransmission of the erroneous data blocks. The most common checksum procedures are parity checks, XOR sum and CRC.

7.1.1 Parity checking

The *parity check* is a very simple and therefore a very popular checksum procedure. In this procedure a *parity bit* is incorporated into each byte and transmitted with it with the result that 9 bits are sent for every byte. Before data transfer takes place a decision needs to be made as to whether to check for odd or even parity, to ensure that the sender and receiver both check according to the same method.

The value of the parity bit is set such that if odd parity is used an odd number of the nine bits have the value 1 and if even parity is used an even number of bits have the value 1. The even parity bit can also be interpreted as the horizontal checksum (modulo 2) of the data bit. This horizontal checksum also permits the calculation of the exclusive OR logic gating (XOR logic gating) of the data bits.

However, the simplicity of this method is balanced by its poor error recognition [pein]. An odd number of inverted bits (1, 3, 5, ...) will always be detected, but if there is an even number of inverted bits (2, 4, 6, ...) the errors cancel each other out and the parity bit will appear to be correct.

Example:
Using odd parity the number E5h has the binary representation 1110 0101 p=0.

A parity generator for even parity can be realised by the XOR logic gating of all the data bits in a byte [tietze]. The order in which the XOR operations take place is irrelevant. In the case of odd parity, the parity generator output is inverted.

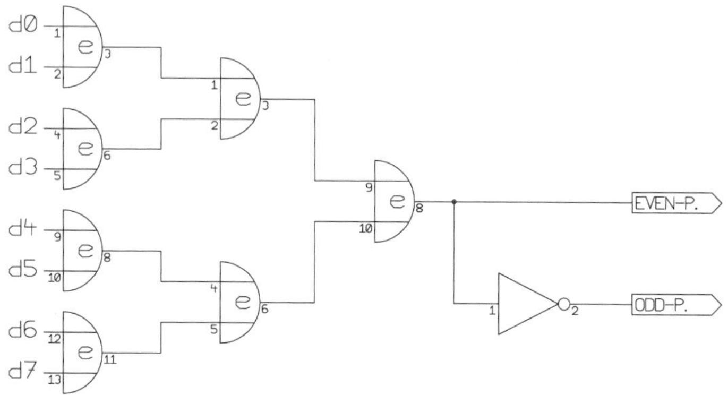

Figure 7.2: The parity of a byte can be determined by performing multiple exclusive-OR logic gating operations on the individual bits

7.1.2 LRC procedure

The XOR checksum known as the *longitudinal redundancy check* (*LRC*) can be calculated very simply and quickly.

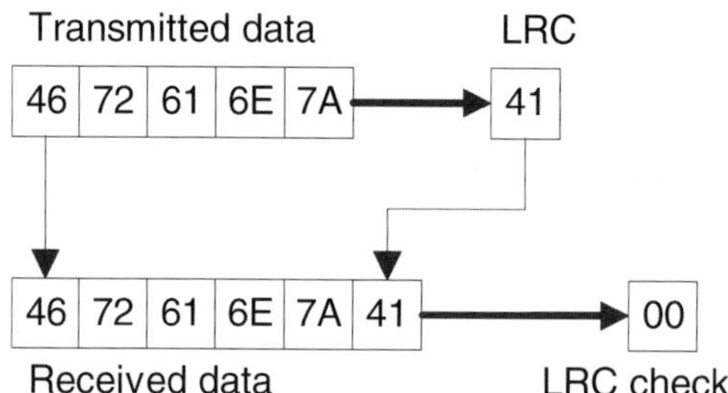

Figure 7.3: If the LCR is appended to the transmitted data, then a new LRC calculation incorporating all received data yields the checksum 00h. This permits a rapid verification of data integrity without the necessity of knowing the actual LRC sum

The XOR checksum is generated by the recursive XOR gating of all the data bytes in a data block. Byte 1 is XOR gated with byte 2, the outcome of this gating is XOR gated with byte 3, and so on. If the LRC value is appended to a data block and transmitted with it, then a simple check for transmission errors can be performed in the receiver by generating an LRC from the data block + LRC byte. The result of this operation must always be zero, any other result indicates that transmission errors have occurred.

Due to the simplicity of the algorithm, LRCs can be calculated very simply and quickly. However LRCs are not very reliable because it is possible for multiple errors to cancel each other out, and the check cannot detect whether bytes have been transposed within a data block [rankl]. LRCs are primarily used for the rapid checking of very small data blocks (e.g. 32 byte).

7.1.3 CRC procedure

The *CRC procedure* (cyclic redundancy check) was originally used in disk drives, and can generate a checksum that is reliable enough even for large data quantities. However, it is also excellently suited for error recognition in data transfer via wire-bound (telephone) or wireless interfaces (radio, RFID). The CRC procedure represents a highly reliable method of recognising transmission errors, although it cannot correct errors.

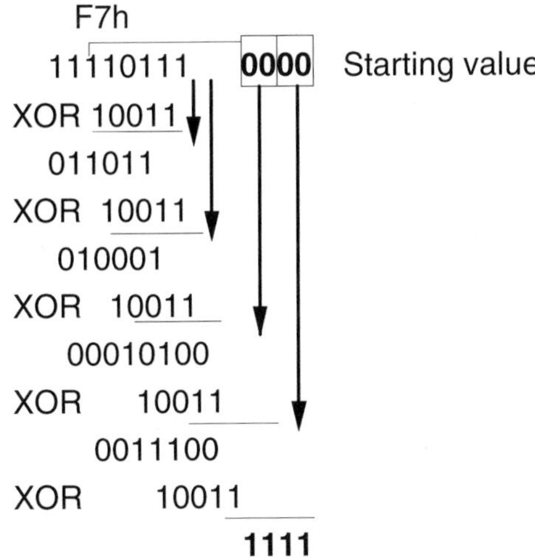

Figure 7.4: Step-by-step calculation of a CRC checksum

As the name suggests, the calculation of the CRC is a cyclic procedure. Thus the calculation of a CRC value incorporates the CRC value of the data byte to be calculated plus the CRC values of all previous data bytes. Each individual byte in a data block is checked to obtain the CRC value for the data block as a whole.

Mathematically speaking, a CRC checksum is calculated by the division of a polynomial using a so-called *generator polynomial.* The CRC value is the remainder obtained from this division. To illustrate this operation we have calculated a 4 bit CRC sum for a data block (Figure 7.4). The first byte of the data block is 7Fh, the generator polynomial is $x^4 + x + 1 =$ 10011.

To calculate a four bit CRC, we first shift the data byte four positions to the left (eight positions for CRC 8, etc.). The four positions that become free are occupied by the starting value of the CRC calculation, in the example this is 00h. The generator polynomial is now gated with the data byte by a repeated XOR operation in accordance with the following rule: "The highest value bit of the data byte is XOR logic gating with the generator polynomial. The initial zeros of the intermediate result are deleted and filled from the right with positions from the data byte or starting value, in order to carry out a new XOR gating with the generator polynomial. This operation is repeated until a 4 position remainder is left. This remainder is the CRC value for the data byte." To calculate the CRC value for the entire data block, the CRC value from the preceding data byte is used as the starting value for the subsequent data byte.

If the CRC value that has just been calculated is appended to the end of the data block and a new CRC calculation performed, then the new CRC value obtained is zero. This particular feature of the CRC algorithm is exploited to detect errors in serial data transmission.

When a data block is transmitted, the CRC value of the data is calculated within the transmitter and this value is appended to the end of the data block and transmitted with it. The CRC value of the received data, including the appended CRC byte, is calculated in the receiver. The result is always zero, unless there are transmission errors in the received block. Checking for zero is a very easy method of analysing the CRC checksum and avoids the costly process of comparing checksums. However, it is necessary to ensure that both CRC calculations start from the same initial value.

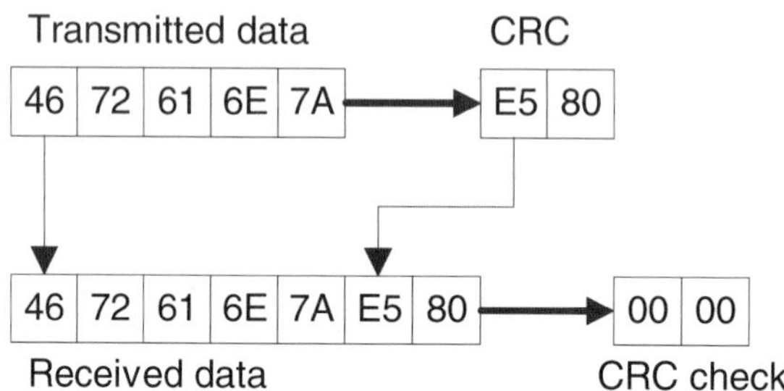

Figure 7.5: If the CRC is appended to the transmitted data a repeated CRC calculation of all received data yields the checksum 0000h. This facilitates the rapid checking of data integrity without knowing the CRC total

The great advantage of CRCs is the reliability of error recognition that is achieved in a small number of operations even where multiple errors are present [rankl]. A 16 bit CRC is suitable for checking the data integrity of data blocks up to 4 kBytes in length – above this size performance falls dramatically. The data blocks transmitted in RFID systems are considerably shorter than 4 kBytes, which means that 12 and 8 bit CRCs can also be used in addition to 16 bit CRCs.

Examples of different generator polynomials:

CRC-8 generator polynomial: $x^8 + x^4 + x^3 + x^2 + 1$

CRC-16 / disk controller generator polynomial: $x^{16} + x^{15} + x^2 + 1$

CRC-16 / CCITT generator polynomial: $x^{16} + x^{12} + x^5 + 1$

When CRC algorithms were first developed for disk controllers, priority was given to the realisation of a simple CRC processor in the form of a hardware circuit. This gave rise to a CRC processor made up of backcoupled *shift registers* and XOR gates that is very simple to realise.

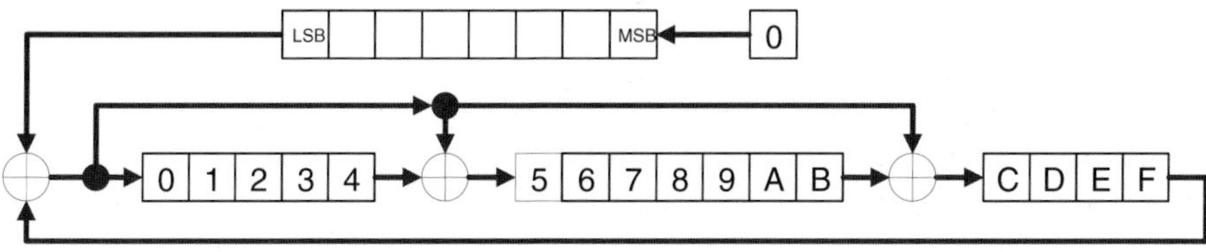

Figure 7.6: Operating principle for the generation of a CRC-16/CCITT by shift registers

When calculating CRC 16 using shift registers, the 16 bit shift register is first set to its starting value. The calculation is then initiated by shifting the data bits, starting with the lowest in value, into the backcoupled shift register one after the other. The backcoupling or polynomial division is based upon the XOR logic gating of the CRC bits. When all the bits have been shifted through the register, the calculation is complete and the content of the 16 bit CRC register represents the desired CRC [rankl].

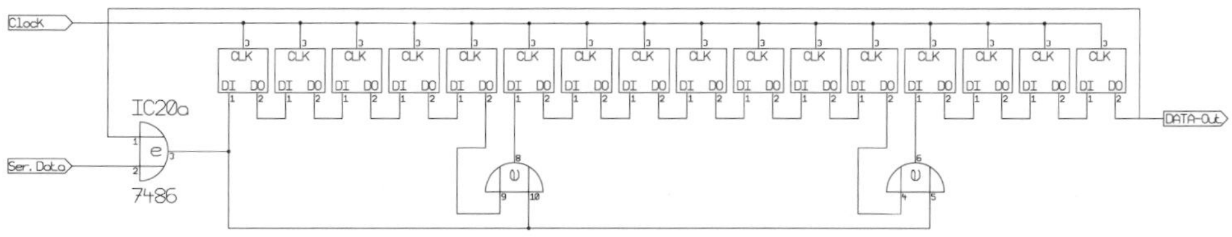

Figure 7.7: The circuit for the shift register configuration outlined above for the calculation of a CRC 16 / CCITTT

7.2 Anticollision

7.2.1 How collision arises

When RFID systems are operated we can never rule out the possibility of there being more than one transponder within the range of a single reader. When the reader sends out a command, this is processed by all the transponders within the range of the reader. It is logical to assume that all transponders will simultaneously attempt to send a reply to the transmitted command back to the reader. In the vast majority of cases, simultaneous data transmission by several transponders will lead to mutual interference, and therefore to data loss. Data loss caused by multiple access to a transmission channel is known as *collision*.

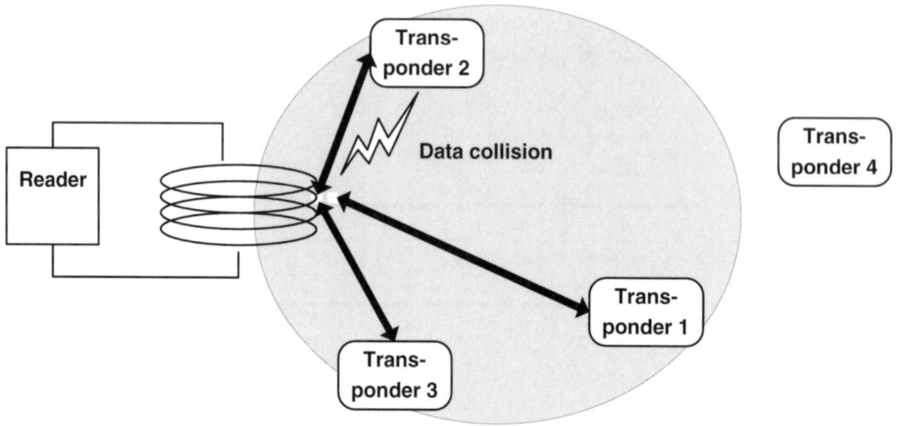

Figure 7.8: Several transponders in the interrogation zone of a single reader cause a data collision

7.2.2 Anticollision procedures

For competitive reasons, system manufacturers are generally not prepared to publish the anticollision procedures that they use. For this reason, little information is available on this subject in the technical literature. Therefore, it is impossible to deal comprehensively with this subject at this point.

An introductory overview and a concrete example of an anticollision algorithm should provide the reader with at least the basic principles of the *anticollision procedure*.

According to [hawkes-97], the procedures used in practice can be classified into three groups.

7.2.2.1 Spatial domain anticollision procedures

The basic principle underlying "spatial procedures" is the restriction of the reader's interrogation zone.

In the case of *microwave systems*, it is possible to fit the reader with a tightly bundled *directional antenna*, which scans the area around the reader – in a similar manner to a torch in a dark room – until a transponder is found by the "searchlight" of the reader.

Another option is to significantly reduce the *range* of a single reader, but to compensate for this by covering the area with a large number of readers. Such procedures have been successfully implemented in large-scale marathon events to calculate the times of runners fitted with transponders (see Chapter 13, Section 13.9 "Sporting Events").

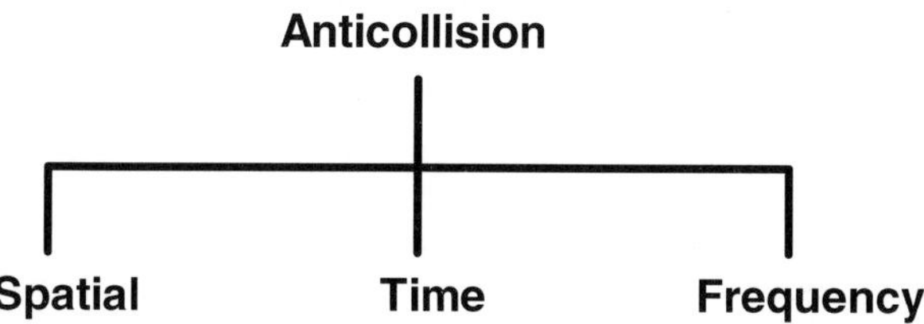

Figure 7.9: Classification of anticollision procedures into three groups

7.2.2.2 Frequency domain anticollision procedures

In this type of procedure, *frequency multiplex* or spread spectrum technologies are used for data transmission from the transponder to the reader [hawkes-97]. This type of anticollision procedure is quite rare for reasons of cost.

The author is only aware of one such system that is used for the identification of luggage tags in airports (baggage identification). The transponders contain a multi-digit number, identical to the barcode on the suitcase tag. In practice, the suitcases are passed through a tunnel reader by a conveyor belt, whereby several suitcases may be within the reader's interrogation zone at the same time. The power supply to the transponder is supplied by inductive coupling at 125 kHz, and data transmission from the transponder to the reader takes place on one of several frequencies in the range 2.7 – 4.2 MHz. The reader emits a continuous data stream, in which the frequencies currently occupied by a transponder are flagged. A transponder entering the reader's interrogation zone decodes this data stream, and sends its identification number to the reader on an unoccupied frequency.

7.2.2.3 Time domain anticollision procedures

Time domain anticollision procedures represent the largest group of anticollision procedures by some margin. This group is subdivided into procedures that are controlled by the reader (interrogator driven) and those that are controlled by the transponder (transponder driven).

Transponder driven procedures operate asynchronously, because they are not controlled by a reader. One example of such a procedure would be a group of transponders that transmit their identification numbers to a reader cyclically. The data transmission time is a fraction of the repetition time, so there are relatively long pauses between transmissions. In addition, the repetition times differ slightly for individual transponders. Thus it is highly likely that two transponders will send their identification at different times and the data transmission will not collide. A differentiation is made between "switched off" and "non switched" procedures, depending upon whether a transponder is switched off by the reader after successful data transmission.

Transponder driven procedures are naturally very slow and inflexible. Therefore, most applications use procedures that are controlled by the reader (interrogator driven). These procedures can be considered to be synchronous, because all the transponders are driven and controlled by the reader simultaneously. Interrogator driven procedures are subdivided

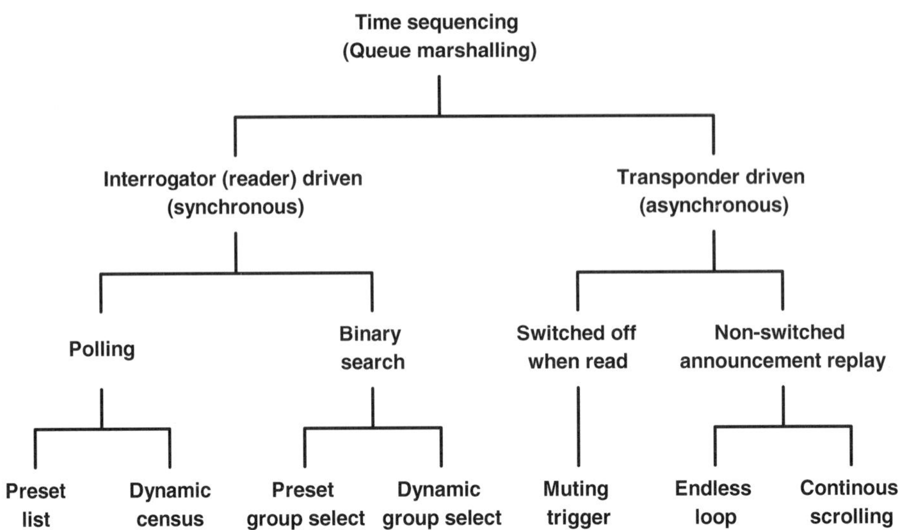

Figure 7.10: Classification of time domain anticollision procedures according to Hawkes [hawkes-97]

into *"polling"* and *"binary search" procedures.* Both procedures are based upon the identification of transponders using a unique serial number.

The "polling" procedure requires a list of all the transponder serial numbers that could be used. All of the serial numbers are requested by the reader one after the other, until a transponder with an identical serial number responds. However, this procedure may be very slow, depending upon the number of possible transponders, and is therefore only suitable for applications where there are a few known transponders in the field.

Binary search procedures are the most flexible, and therefore the most common, procedures. In a binary search procedure, a transponder is selected from a group by intentionally causing a data collision in the transmission of transponder serial numbers to the reader following a *request command* from the reader. If this procedure is to succeed it is crucial that the reader is capable of determining the precise bit position of a collision using a suitable signal coding system. Furthermore, the system must be able to preselect the *group* of transponders addressed when a request command is transmitted. This request command is repeated, each time reducing the selected group of numbers, until only one transponder responds. To clarify this process, the "binary search" procedure is described in detail in the following application example.

7.2.3 Application example – binary search algorithm

We will now illustrate the popular "binary search" procedure using a sample application. The selected transponder system is an inductively coupled system with load modulation by an ASK modulated subcarrier. A “1” level in the baseband coding switches the subcarrier on, a “0” level switches it off.

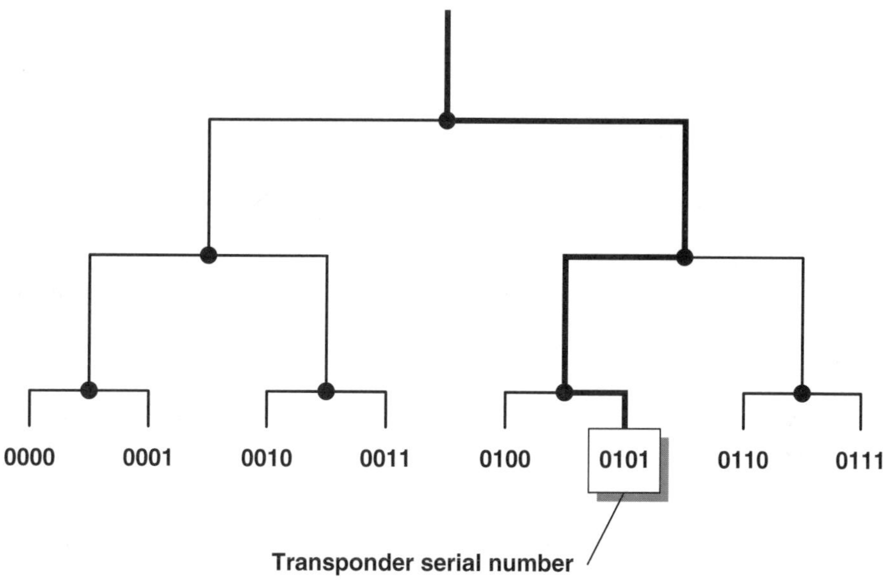

Figure 7.11: The binary search tree is the fundamental principle underlying all “binary search” procedures

A “binary search” algorithm can only be used if the reader can detect the precise bit position of a data collision. This means that a suitable *bit coding system* is needed, and for this reason we will now conduct a comparison between the *NRZ* and *Manchester* coding systems:

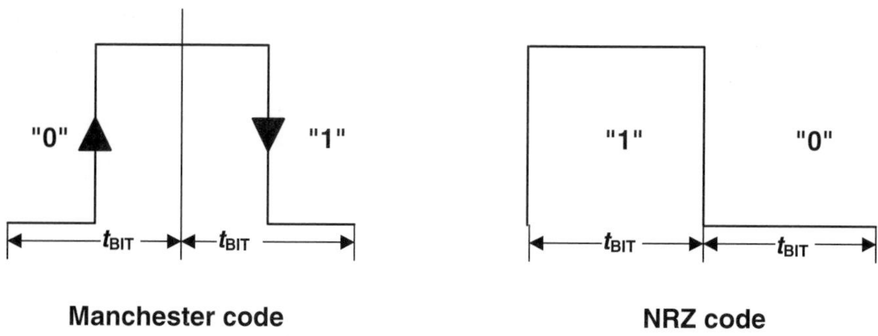

Figure 7.12: Bit coding using Manchester and NRZ code

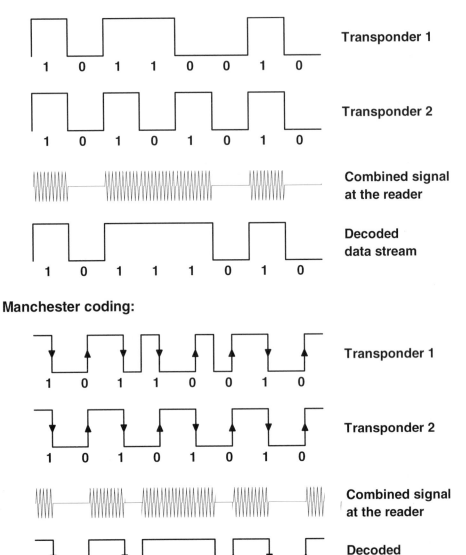

Figure 7.13: Collision behaviour in the NRZ and Manchester code. The Manchester code makes it possible to trace a collision to an individual bit

NRZ Code (Non return to zero code) The value of a bit is defined by the static level of the transmission channel within a bit window (t_{BIT}). In this example a logic "1" is coded by a static "high" level, a logic "0" is coded by a static "low" level.

If at least one of the two transponders sends a subcarrier signal, then this is interpreted by the reader as a "high" level and in our example is assigned the logic value "1". The reader cannot detect whether the sequence of bits it is receiving can be traced back to the superposition of transmissions from several transponders or the signal from a single transponder. The use of a block checksum (parity, CRC) can only detect a transmission error "somewhere" in the data block.

Manchester Code The value of a bit is defined by the change in level (negative or positive transition) within a bit window (t_{BIT}). A logic "1" is coded by a negative transition. The "no transition" state is not permissible during data transmission and is recognised as an error.

If two (or more) transponders simultaneously transmit bits of different values then the positive and negative transitions of the received bits cancel each other out, so that a subcarrier signal is received for the duration of an entire bit. This state is not permissible in the Manchester coding system and therefore leads to an error. It is thus possible to trace a collision to an individual bit.

We will therefore use Manchester coding for our "binary search" algorithm. Let us now turn our attention to the algorithm itself:

A "binary search" algorithm consists of a predefined sequence (specification) of interactions (command and response) between a reader and several transponders with the objective of being able to select any desired transponder from a large group.

For the practical realisation of the algorithm we require a set of commands that can be processed by the transponder. In addition, each transponder has a unique (binary) *transponder address* (e.g. a unique *serial number*). In our example we are using an 8 bit address, so if we are to guarantee the uniqueness of the addresses (serial numbers) a maximum of 256 transponders can be issued:

GET_TADR<(XX) This command sends a transponder address to the transponder as a parameter. If the transponder's own address is less than (or equal to) the received address, then the transponder sends its own serial number back to the reader. The group of transponders addressed can thus be preselected and reduced, as stipulated in the preceding section.

SELECT_(XX) Sends a (predetermined) transponder address (XX) to the transponder. The transponder with the identical transponder address will become available for the processing of other commands (e.g. reading and writing data). This transponder is thus selected. Transponders with different addresses will thereafter only respond to a GET_TADR command.

GET_DATA The selected transponder sends stored data to the reader. (In a real system there are also commands for authentication or writing, debiting, crediting, ...).

UNSELECT The selection of a previously selected transponder is cancelled and the transponder is "muted". In this state, the transponder is completely inactive and does not even respond to a GET_TADR command. To reactivate the transponder, it must be temporarily removed from the interrogation zone of the reader (= no power supply) so that a reset operation is performed.

We will now illustrate the use of the commands defined above in a "binary search" algorithm by describing the behaviour of two transponders in the interrogation zone of the reader.

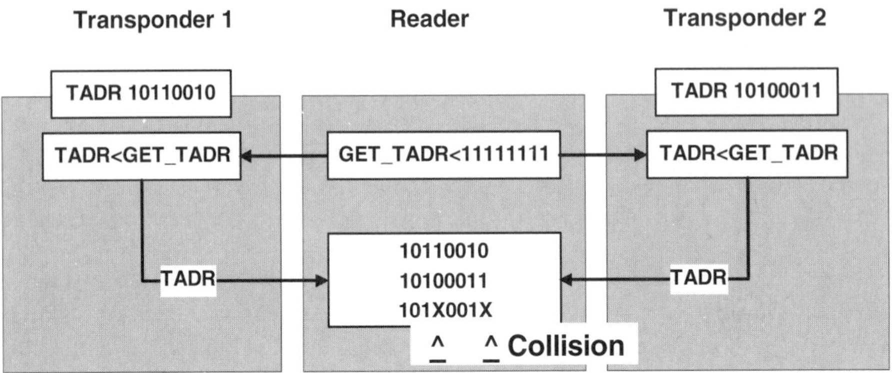

Figure 7.14: The different addresses (serial numbers) of the transponders lead to a data collision

The algorithm begins with the transmission of the command **GET_TADR<11111111**. The transponder address 11111111b is the highest possible address of this system. The addresses of all transponders in the interrogation zone of the reader must therefore be less than 11111111b, so all transponders will respond to this command by sending their own transponder address.

The superposition of the different bit sequences of the two transponders in bit 0 and bit 4 of the received transponder addresses leads to a collision (X). The two collisions indicate that there are at least two transponders in the interrogation zone of the reader. To be precise, the received sequence of bits 101**X**001**X** can be traced back to four possible transponder addresses: 101**1**001**0**, 101**1**001**1**, 101**0**001**0**, 101**0**001**1**.

We now set bit 4 to "0" (101**0**001**X**), which excludes half of the options. All lower value bits are set to "1" – the result in bit 0 is now disregarded.

The reader now sends the command **GET_TADR<10101111** to the transponder (only transponders in the address range 101**0**0000 to 101**0**1111 will respond). If there is another collision, then the procedure is repeated (highest value $X = 0$, all lower bits = 1). This procedure is repeated until only one transponder responds to the GET_TADR command. In our example this is transponder 2, which will be recognised in the second pass.

Using a **SELECT** command, transponder 2 is selected and can now be written to or read from by the reader without interruption. All other transponders remain silent, as only a selected transponder may respond to a write/read command – **GET DATA**.

After the completion of the write/read operation, transponder 2 is deactivated altogether by an **UNSELECT** command, and therefore no longer responds even to a **GET_TADR** command. Therefore, if the anticollision algorithm were repeated in our example it would automatically lead to the selection of the previously unprocessed transponder 1.

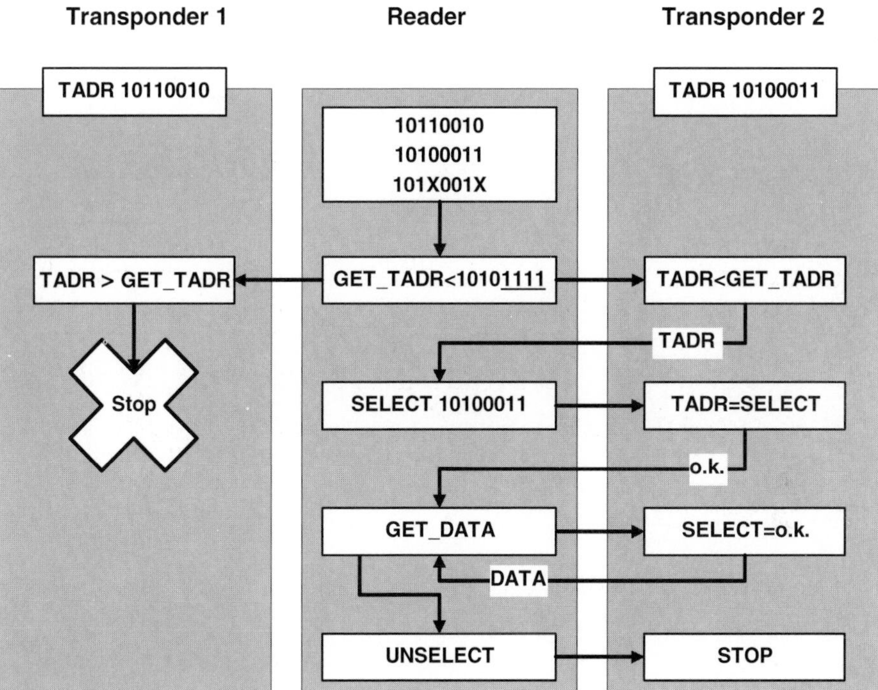

Figure 7.15: A single transponder can be selected by the successive reduction of the number range

8

Data Security

RFID systems are increasingly being used in high security applications, such as access systems and systems for making payments or issuing tickets. However, the use of RFID systems in these applications necessitates the use of security measures to protect against *attempted attacks*, in which people try to trick the RFID system in order to gain unauthorised access to buildings or avail themselves of services (tickets) without paying. This is nothing new, we only have to look to myths and fairy stories to find examples of attempts outsmart *security systems*. For example, *Ali Baba* was able to gain access to the supposedly secure hideout of the 40 thieves by discovering the secret password.

Modern *authentication protocols* also work by checking knowledge of a secret (i.e. a cryptographic key). However, suitable algorithms can be employed to prevent the secret key being cracked. High security RFID systems must have a defence against the following individual attacks:

- Unauthorised reading of a data carrier in order to duplicate and/or modify data.
- The placing of a foreign data carrier within the interrogation zone of a reader with the intention of gaining unauthorised access to a building or receiving services without payment.
- Eavesdropping into radio communications and replaying the data, in order to imitate a genuine data carrier ("replay and fraud").

When selecting a suitable RFID system, consideration should be given to cryptological functions. Applications that do not require a security function (e.g. industrial automation, tool recognition) would be made unnecessarily expensive by the incorporation of cryptological procedures. On the other hand, in high security applications (e.g. ticketing, payment systems) the omission of cryptological procedures can be a very expensive oversight if manipulated transponders are used to gain access to services without authorisation.

8.1 Mutual Symmetrical Authentication

Mutual authentication between reader and transponder is based upon the principle of "Three Pass Mutual Authentication" in accordance with *ISO 9798-2*, in which both participants in the communication check the other party's knowledge of a secret (secret cryptological key).

In this procedure, all the transponders and receivers that form part of an application are in possession of the same secret *cryptological key K* (→ symmetrical procedure). When a transponder first enters the interrogation zone of a reader it cannot be assumed that the two participants in the communication belong to the same application. From the point of view of the reader, there is a need to protect the application from *manipulation* using falsified data. Likewise, on the part of the transponder there is a need to protect the stored data from unauthorised reading or overwriting.

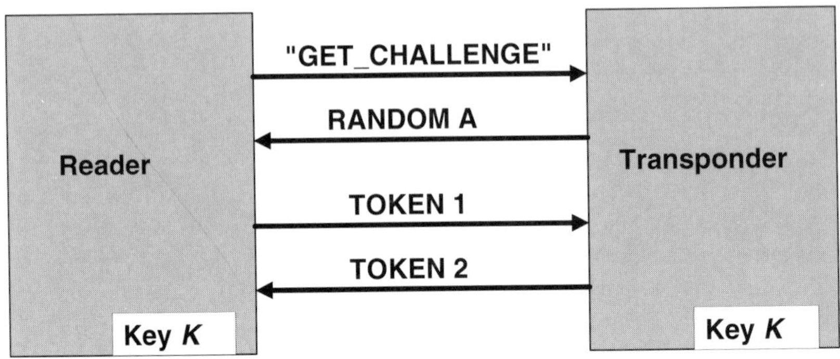

Figure 8.1: Mutual authentication procedure between transponder and reader

The mutual authentication procedure begins with the reader sending a "GET_CHALLENGE" command to the transponder. A *random number* R_A is then generated in the transponder and sent back to the reader (response challenge-response procedure). The reader now generates a random number R_B. Using the common secret key K and a common key algorithm e_k, the reader calculates an encrypted data block (token 1), which contains both random numbers and additional control data, and sends this data block to the transponder.

$$\text{Token 1} = e_K (R_B \| R_A \| ID_A \| \text{Text1})$$

The received token 1 is decrypted in the transponder and the random number R_A' contained in the plain text is compared to the previously transmitted R_A. If the two figures correspond, then the transponder has confirmed that the two common keys correspond. Another random number R_{A2} is generated in the transponder and this is used to calculate an encrypted data block (token 2), which also contains R_B and control data. Token 2 is sent from the transponder to the reader.

$$\text{Token 2} = e_K (R_{A2} \| R_B \| \text{Text2})$$

The reader decrypts token 2 and checks whether R_B, which was sent previously, corresponds with R_B', which has just been received. If the two figures correspond, then the reader is satisfied that the common key has been proven. Transponder and reader have thus ascertained that they belong to the same system and further communication between the two parties is thus legitimised.

To sum up, the mutual authentication procedure has the following advantages:

- The secret keys are never transmitted over the airwaves, only encrypted random numbers are transmitted.
- Two random numbers are always encrypted simultaneously. This rules out the possibility of performing an inverse transformation using R_A to obtain token 1, with the aim of calculating the secret key.
- The token can be encrypted using any algorithm.
- The strict use of random numbers from two independent sources (transponder, reader) means that recording an authentication sequence for playback at a later date (replay attack) would fail.
- A random key (session key) can be calculated from the random numbers generated, in order to cryptologically secure the subsequent data transmission.

8.2 Authentication Using Derived Keys

One disadvantage of the authentication procedure described in Section 8.1 is that all transponders belonging to an application are secured using an identical cryptological key *K*. For applications that involve vast quantities of transponders (e.g. the ticketing system for the public transport network, which uses several millions of transponders) this represents a potential source of danger. Because such transponders are accessible to everyone in uncontrolled numbers, the small probability that the key for a transponder will be discovered must taken into account. If this occurred, the procedure described above would be totally open to manipulation.

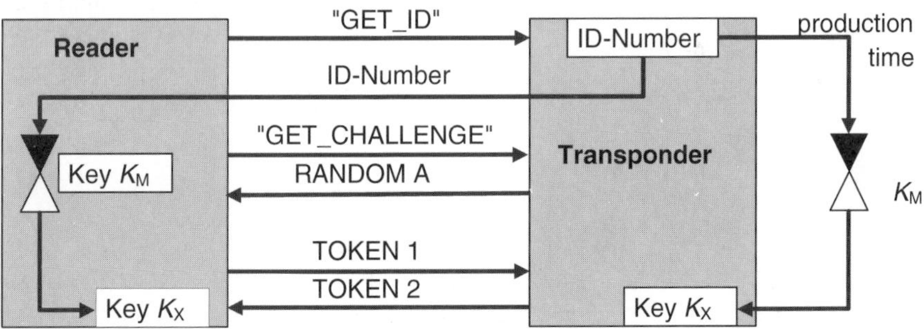

Figure 8.2: In an authentication procedure based upon derived keys, a key unique to the transponder is first calculated in the reader from the serial number (ID number) of the transponder. This key must then be used for authentication

A significant improvement on the authentication procedure described can be achieved by securing each transponder with a different cryptological key. To achieve this, the serial number of each transponder is read out during its production. A key K_X is calculated (derived) using a cryptological algorithm and a *master key* K_M, and the transponder is thus initialised. Each transponder thus receives a key linked to its own ID number and the master key K_M.

The mutual authentication begins by the reader requesting the ID number of the transponder. In a special security module in the reader, the *SAM* (security authentication module), the transponder's specific key is calculated using the master key K_M, so that this can be used to initiate the authentication procedure. The SAM normally takes the form of a smart card with contacts incorporating a cryptoprocessor, which means that the stored master key can never be read.

8.3 Encrypted Data Transfer

Chapter 7 described methods of dealing with interference caused by physical effects during data transmission. Let us now extend this model to a potential attacker. We can differentiate between two basic types of attack: Attacker 1 behaves passively and tries to eavesdrop into the transmission to discover confidential information for wrongful purposes. Attacker 2, on the other hand, behaves actively to manipulate the transmitted data and alter it to his benefit.

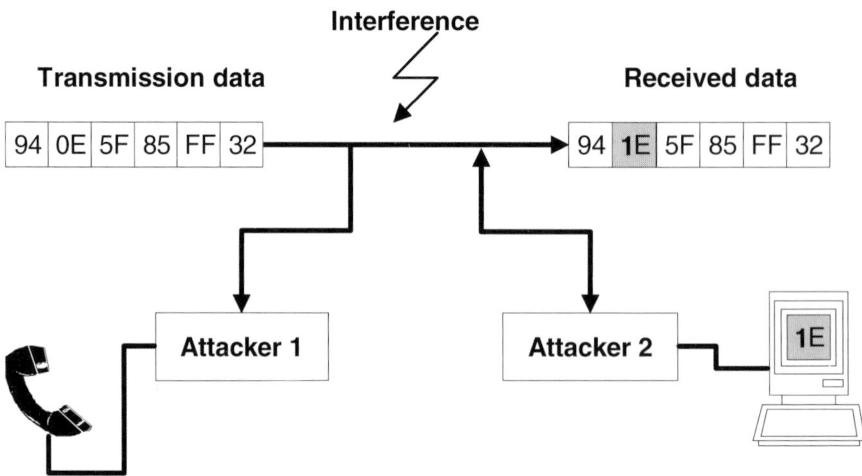

Figure 8.3: Attempted attacks on a data transmission. Attacker 1 attempts to eavesdrop, whereas attacker 2 maliciously alters the data

Cryptological procedures are used to protect against both passive and active attacks. To achieve this, the transmitted data (plain text) can be altered (encrypted) prior to transmission so that a potential attacker can no longer draw conclusions about the actual content of the message (plain text).

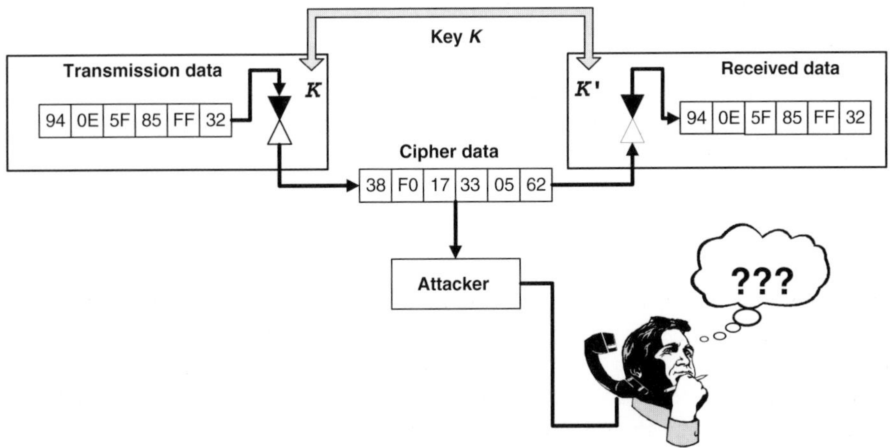

Figure 8.4: By encrypting the data to be transmitted, this data can be effectively protected from eavesdropping or modification

Encrypted data transmission always takes place according to the same pattern: The transmission data (plain text) is transformed into cipher data (cipher text) (*encryption, ciphering*) using a secret *key* K and a secret algorithm. Without knowing the encryption algorithm and the secret key K a potential attacker is unable to interpret the recorded data. It is not possible to recreate the transmission data from the cipher data.

The cipher data is transformed back to its original form in the receiver using the secret key K' and the secret algorithm (*decryption, deciphering*).

If the keys K for ciphering and K' for deciphering are identical ($K=K'$) or in a direct relationship to each other, the procedure is a *symmetrical key procedure*. If knowledge of the key K is irrelevant to the deciphering process, the procedure is an *asymmetrical key procedure*. RFID systems have for a long time used only symmetrical procedures, therefore we will not describe other procedures in further detail here.

If each character is individually encrypted prior to transmission, the procedure is known as *sequential ciphering* (or *stream ciphering*). If, on the other hand, several characters are incorporated into a block then we talk of a block cipher. Because block ciphers are generally very calculation intensive, they play a less important role in RFID systems. Therefore the emphasis is placed on sequential ciphers in what follows.

A fundamental problem of all cryptological procedures is the secure distribution of the secret key K, which must be known by the authorised communication participants prior to the start of the data transfer procedure.

8.3.1 Stream cipher

Sequential ciphers or stream ciphers are encryption algorithms in which the sequence of plain text characters is encrypted sequentially using a different function for every step

[fumy]. The ideal realisation of a stream cipher is the so-called "*one time pad*", also known as the "*Vernam cipher*" after its discoverer [longo].

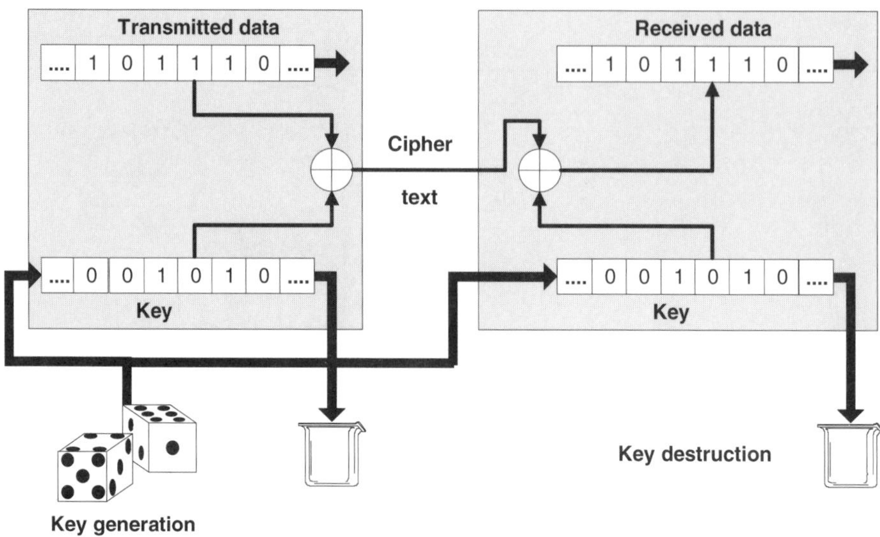

Figure 8.5: In the "one time pad" keys generated from random numbers (dice) are used only once and then destroyed (wastepaper basket). The problem here is the secure transmission of the key between sender and recipient

In this procedure a random key K is generated, for example using dice, prior to the transmission of encrypted data, and this key is made available to both parties. The key sequence is linked with the plain text sequence by the addition of characters or using XOR gating. The random sequence used as a key must be at least as long as the message to be encrypted, because periodic repetitions of a typically short key in relation to the plain text would permit cryptoanalysis and thus an attack on the transmission. Furthermore, the key may only be used once, which means that an extremely high level of security is required for the secure distribution of keys. Stream ciphering in this form is completely impractical for RFID systems.

To overcome the problem of key generation and distribution, systems have been created based upon the principle of the "one time pad" stream cipher, that use a so-called *pseudorandom sequence* instead of an actual random sequence. Pseudorandom sequences are generated using so-called pseudorandom generators.

Figure 8.6 shows the fundamental principle of a sequential cipher using a pseudorandom generator: because the encryption function of a sequential cipher can change (at random) with every character, the function must be dependent not only upon the current input character but also upon an additional feature, the internal state M. This internal state M is changed after every encryption step by the state transformation function $g(K)$. The pseudorandom generator is made up of the components M and $g(K)$. The security of the cipher depends principally upon the number of internal states M and the complexity of the transformation function $g(K)$. The study of sequential ciphers is thus primarily concerned with the analysis of pseudorandom generators.

The *encryption function* $f(K)$ itself, on the other hand, is generally very simple and can only comprise an addition or XOR logic gating [fumy] [glogau].

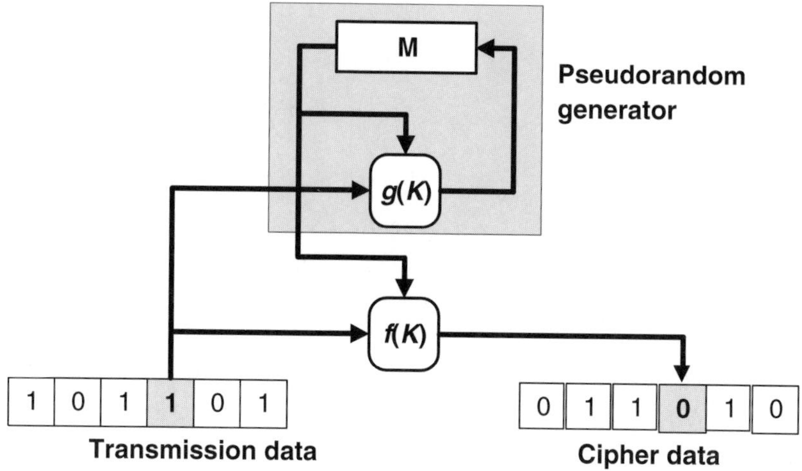

Figure 8.6: The principle underlying the generation of a secure key by a pseudorandom generator

From a circuitry point of view, pseudorandom generators are realised by state machines. These consist of binary storage cells, so-called flip-flops. If a state machine has n storage cells then it can take on 2^n different internal M states. The state transformation function $g(K)$ is represented by combinatorial logic (a more detailed explanation of the functionality of state machines can be found in the Chapter 10 "The Architecture of Electronic Data Carriers"). The implementation and development of pseudorandom generators can be greatly simplified if we restrict ourselves to the use of linear feedback shift registers.

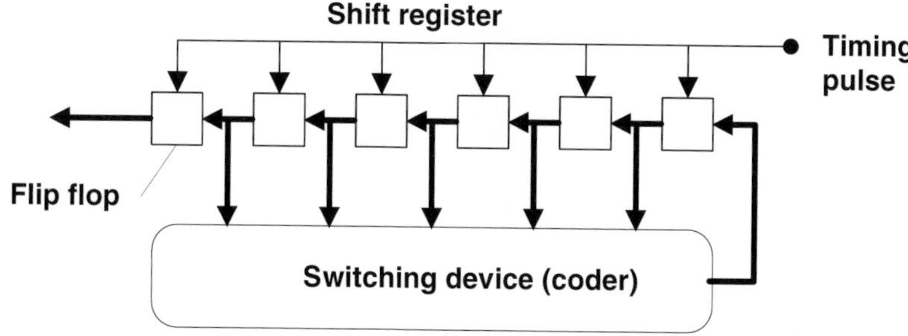

Figure 8.7: Basic circuit of a pseudorandom generator incorporating a linear feedback shift register (LFSR)

A shift register is realised by the serial connection of flip-flops ($output_n$ is connected with $input_{n+1}$) and the parallel connection of all timing inputs. The content of the flip-flop cell is shifted forwards by one position with every timing pulse. The content of the last flip-flop is output [rueppel] [golomb].

9

Standardisation

The development of standards is the responsibility of the technical committee of the ISO. The ISO is the worldwide union of national standardisation institutions, such as DIN (Germany) or ANSI (USA).

The description of standards in this chapter merely serves to aid our technical understanding of the RFID applications dealt with in this book and no attempt has been made to describe the standards mentioned in their entirety. Furthermore, standards are updated from time to time and are thus subject to change. When working with the RFID applications in question the reader should not rely on the parameters specified in this chapter. We recommend that copies of the original versions in question are procured. The necessary addresses are listed in Chapter 14, Section 14.3 "Contact Addresses, Technical Periodicals" at the end of this book.

9.1 Animal Identification

ISO standards 11784 and 11785 deal with the identification of animals using RFID systems.

- ISO 11784: "Radio-frequency identification of animals – Code structure"
- ISO 11785: "Radio-frequency identification of animals – Technical concept"

The constructional form of the transponder used is not specified in the standards and therefore the form can be designed to suit the animal in question. Small, sterile glass transponders that can be injected into the fatty tissues of the animal are normally used for the identification of cows, horses and sheep. Ear tags or collars are also possible.

9.1.1 ISO 11784 – Code structure

The identification code for animals comprises of a total of 64 bit (8 Byte). Table 9.1 shows the significance of the individual bits.

The national identification code should be managed by the individual countries. The bits 27 to 64 may also be allocated to differentiate between different animal types, breeds, regions within the country, breeders etc., but this not specified in this standard.

Table 9.1: Identification codes for animals

Bit no.:	Information:	Description:
1	animal (1) / non-animal application (0)	Specifies whether the transponder is used for animal identification or other purposes
2 – 15	reserved	Reserved for future applications
16	data block (1) follows / no data block (0)	Specifies whether additional data will be transmitted after the identification code
17 – 26	Country code as per ISO 3166	Specifies the country of use, the code 999 describes a test transponder
27 – 64	National identification code	Unique, country specific registration number

9.1.2 ISO 11785 – Technical concept

This standard defines the transmission method for the transponder data and the reader specifications for activating the data carrier (transponder). A central aim in the development of this standard as to facilitate the interrogation of transponders from an extremely wide range of manufacturers using a common reader. A reader for *animal identification* in compliance with the standard recognises and differentiates between transponders that use a full/half duplex system (load modulation) and transponders that use a sequential system.

9.1.2.1 Requirements

The standard specifies the operating frequency for the reader as 134.2 kHz ± 1.8 kHz. The emitted field provides a power supply for the transponder and is therefore termed the "activation field".

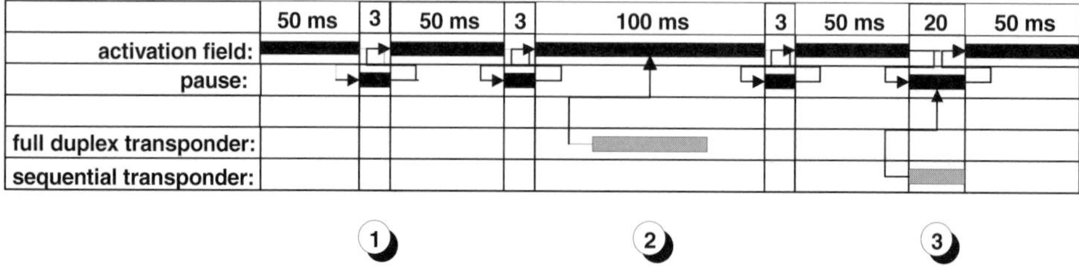

Figure 9.1: Path of the activation field of a reader over time: ① no transponder in interrogation zone ② full/half duplex (= load modulated) transponder in interrogation zone ③ sequential transponder in the interrogation zone of the reader

The activation field is periodically switched on for 50 ms at a time and then switched off for 3 ms (1). During the 50 ms period when it is switched on it waits for the response from a full/half duplex transponder – a sequential transponder in the field requires the activation field to charge up its charging capacitor.

If a full/half duplex transponder is present within the range of the activation field, then this transponder sends its data during the operating interval of the field (2). Whilst data is being received the operating interval can be extended to 100 ms if the data transfer is not completed within the first 50 ms.

A sequential transponder in the range of the activation field (3) begins to transmit data within the 3 ms pause. The duration of the pause is extended to a maximum of 20 ms to permit the complete transmission of a data record.

If portable or stationary readers are operated in the vicinity of one another, then there is a high probability that a reader will emit its activation field during the 3 ms pause of the other reader. This would result in neither of the readers being able to receive the data signal of a sequential transponder. Due to the relatively strong activation field in comparison to the field strength of a sequential transponder this effect occurs in a multiple of the reader's normal read radius. Appendix C of the standard therefore describes procedures for the *synchronisation* of several readers to circumvent this problem.

Portable and stationary readers can be tested for the presence of a second reader (B) in the vicinity by extending the pause duration to 30 ms. If the activation field of a second reader (B) is received within the 30 ms pause, then the standard stipulates that the activation field of the reader (A) should be switched on for a maximum of 50 ms as soon as the previously detected reader (B) switches its activation field on again after the next 3 ms pause. In this manner, a degree of synchronisation can be achieved between two neighbouring readers. Because data is only transmitted from the transponder to the reader (and the activation field thus always represents an unmodulated HF field), an individual transponder can be read by two portable readers simultaneously. To maintain the stability of the synchronisation, every tenth pause cycle is extended from 3 ms to 30 ms to detect any other readers that have recently entered the area.

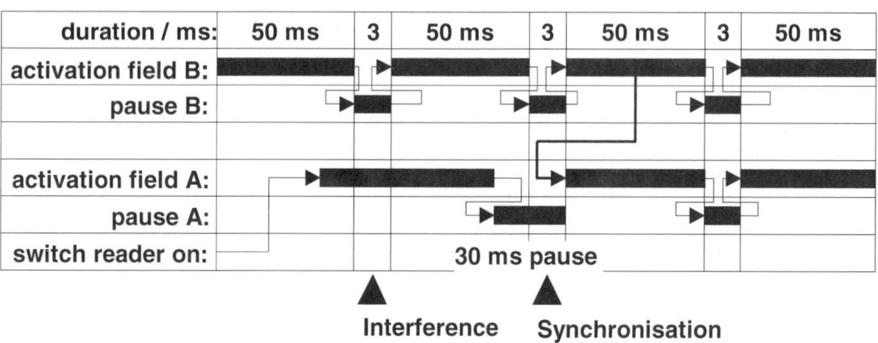

Figure 9.2: Automatic synchronisation sequence between readers A and B. Reader A inserts an extended pause of a maximum of 30 ms after the first transmission pulse following activation so that it can listen for other readers. In the diagram, the signal of reader B is detected during this pause. The reactivation of the activation field of reader B after the next 3 ms pause triggers the simultaneous start of the pulse pause cycle of reader A

Stationary readers also use a *synchronisation cable* connected to all readers in the system. The synchronisation signal at this cable is a simple logic signal with a low and high level. The resting state of the cable is a logic low level.

If one of the connected readers detects a transponder, then the synchronisation cable switches to the high level whilst data is transmitted from the transponder to the reader. All other readers extend their current phase (activation / pause).

If the detected data carrier is a full/half duplex transponder, then the synchronised readers are in the "activation field" phase. The activation period of the activation field is now extended until the synchronisation cable is once again switched to low level (but with a maximum of 100 ms).

If the signal of a sequential transponder is received, the synchronised readers are in the "pause" phase. The synchronisation signal at the cable extends the pause duration of all readers to 20 ms (fixed value).

9.1.2.2 Full/half duplex system

Full/half duplex transponders, which receive their power supply through an activation field, begin to transmit the stored identification data immediately. For this a *load modulation procedure* without a subcarrier is used, whereby the data is represented in a differential bi-phase code (DBP). The bit rate is derived by dividing the reader frequency by 32. At 134.2 kHz the transmission speed (bit rate) is 4194 bit/s.

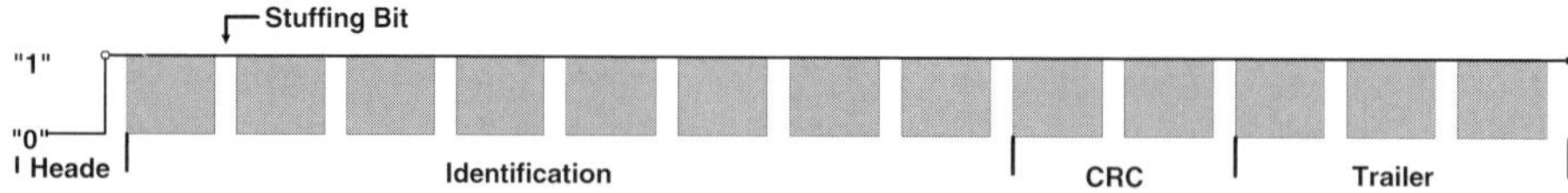

Figure 9.3: Structure of the load modulation data telegram comprising of starting sequence (header), ID code and check sum and trailer

A full/half duplex data telegram comprises of an 11 bit header, 64 bit (8 byte) useful data, 16 bit (2 byte) CRC and 24 bit (3 Byte) trailer. After every eight transmitted bits a stuffing bit with a logic "1" level is inserted to avoid the chance occurrence of the header "00000000001".

The transmission of the total of 128 bits takes around 30.5 ms at the given transmission speed.

9.1.2.3 Sequential system

After every 50 ms the activation field is switched off for 3 ms. A sequential transponder that has previously been charged with energy from the activation field begins to transmit the stored identification data approximately 1 to 2 ms after the activation field has been switched off.

The modulation method used by the transponder is frequency shift keying (2 FSK). The bit coding uses NRZ (comparable to RS232 on a PC). A logic "0" corresponds with the basic frequency 134.2 kHz, a logic "1" corresponds to the frequency 124.2 kHz.

The bit rate is derived by dividing the transmission frequency by 16. The bit rate varies between 8387 bit/s for a logic "0" and 7762 bit/s for a logic "1" depending upon the frequency shift keying.

The sequential data telegram comprises of an 8 bit header 01111110b, 64 bit (8 Byte) useful data, 16 bit (2 Byte) CRC and 24 bit (3 Byte) trailer. Stuffing bits are not inserted.

The transmission of the total of 112 bits takes a maximum of 14.5 ms at the given transmission speed ("1" sequence).

9.2 Contactless Smart Cards

9.2.1 ISO 10536 – Close coupling smart cards

The ISO standard 10536 entitled "Identification cards – contactless integrated circuit(s) cards" describes the structure and operating parameters of contactless close coupling smart cards. *ISO 10536* consists of the following four sections:

- Part 1: Physical characteristics
- Part 2: Dimensions and location of coupling areas
- Part 3: Electronic signals and reset procedures
- Part 4: Answer to reset and transmission protocols

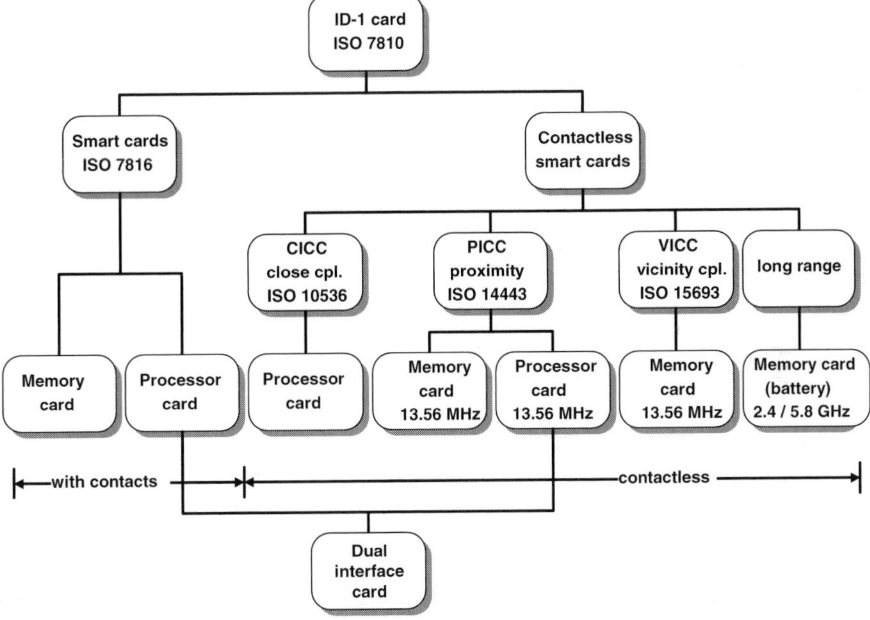

Figure 9.4: Family of smart cards (contactless and with contacts) giving the applicable standards

9.2.1.1 Part 1 – Physical characteristics

The physical characteristics of close coupling cards are defined in Part 1 of the standard. The specifications are identical to those for the mechanical dimensions of smart cards with contacts.

9.2.1.2 Part 2 – Dimensions and locations of coupling areas

Part 2 of the standard specifies the position and dimensions of the coupling elements. Both *inductive* (H1 – 4) and *capacitive coupling elements* (E1 – 4) are used. The arrangement of the coupling elements is selected so that a close coupling card can be operated in an insertion reader in all four positions.

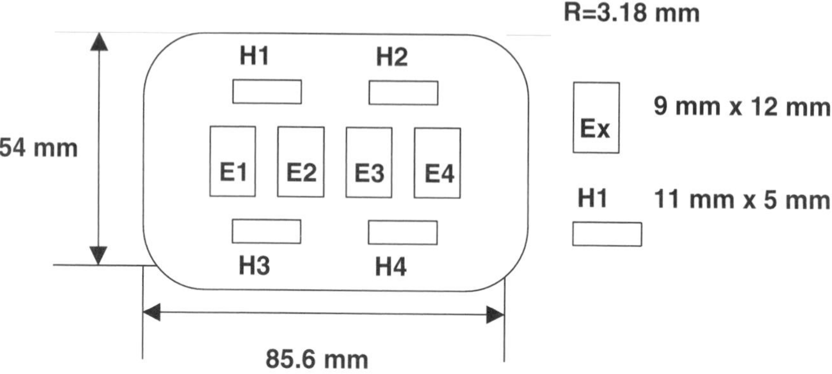

Figure 9.5: Position of capacitive (E1 – E4) and inductive coupling elements (H1 – H4) in a close coupling smart card

9.2.1.3 Part 3 – Electronic signals and reset procedures

Power supply The power supply for close coupling cards is derived from the four inductive coupling elements H1 – H4. The inductive alternating field should have a frequency of 4.9152 MHz. The coupling elements H1, H2 are designed as coils but have opposing directions of winding, so that if the coupling elements are fed at the same time there is a phase difference of 180° between the associated magnetic fields F1 and F2 (e.g. by a u-shaped core in the reader). The same applies for the coupling elements H3 and H4.

The readers must be designed such that that power of 150 mW can be provided to the contactless card from any of the magnetic fields F1 – F4. However, the card may not draw more than 200 mW via all four fields together.

Data transmission card → reader Either inductive or capacitive coupling elements may be used for data transmission between card and reader. However, it is not possible to switch between the two types of coupling during communication.

Inductive Load modulation with a subcarrier is used for the transmission of data via the coupling fields H1 – H4. The subcarrier frequency is 307.2 kHz and the subcarrier is modulated using 180° PSK. The reader is designed such that a load change of 10% of the base load at one or more of the fields F1 – F4 can be recognised as a load modulation signal. The specified minimum load change for a card is 1 mW.

Capacitive In this procedure the coupling fields E1, E2 or E3, E4 are used in pairs. In both cases the paired coupling fields are controlled by a differential signal. The voltage difference $U_{diff} = U_{E1} - U_{E2}$ should be measured such that a voltage level of at least 0.33 V is present at the reader coupling surfaces $E1'$ and E2'. Data transmission takes place using *NRZ coding* in the baseband (i.e. no subcarrier). The data rate after reset is 9600 bit/s, however a higher data rate can be used during operation.

Data transmission reader → card The standard gives preference to the inductive method for data transmission to the card. The modulation procedure is a 90° PSK of the fields F1 – F4, whereby the phase position of all fields is modulated synchronously.

Depending upon the position of the card in the insertion reader the following phase relationships are possible between the coupling fields during modulation

Table 9.2: Position 1

A	A'
ΦF1	Φ'F1 = ΦF1 – 90°
ΦF3 = ΦF1 + 90°	Φ'F3 = ΦF3 + 90°

Table 9.3: Position 2

A	A'
F1	Φ'F1 = Φ'F1 + 90°
ΦF3 = ΦF1 – 90°	Φ'F3 = Φ'F3 – 90°

(State A: unmodulated, State A': modulated)

Data transmission takes place using *NRZ coding* in the baseband (i.e. no subcarrier). The data rate after reset is 9600 bit/s, however a higher data rate can be used during operation.

9.2.1.4 Part 4 - Answer to reset and transmission protocols

This part of the ISO 10536 describes the transmission protocol between reader and card. We will not describe Part 4 here because it is still under development by the standardisation committee in question, and may therefore be subject to change.

9.2.2 ISO 14443 – Proximity coupling smart cards

This standard will provide a consistent basis for *proximity coupling smart cards* (inductive coupling). The definitive version of the standard is not expected until 2001 at the earliest. Therefore, Table 9.4 can only give a brief overview of the provisional draft.

Table 9.4: Provisional parameters from the working paper for *ISO 14443-2*. Data transmission reader (PCD) → smart card (PICC) [berger]

PCD → PICC	Type A	Type B
Modulation:	ASK 100%	ASK 10% (duty factor 8% - 12%)
Bit coding:	Modified Miller code	NRZ code
Synchronisation:	On bit level	1 start and 1 stopbit per Byte (specification in Part 3)
Baud rate:	106 kBd	106 kBd

Table 9.5: Provisional parameters from ISO 14443-2. Data transmission smart card (PICC) → reader (PCD) [berger]

PICC → PCD	Type A	Type B
Modulation:	Load modulation with subcarrier 847 kHz, ASK modulated	Load modulation with subcarrier 847 kHz, BPSK modulated
Bit coding:	Manchester code	NRZ code
Synchronisation:	1 Bit "frame synchronisation"	1 start and 1 stop bit per Byte (specification in Part 3)
Baud rate:	106 kBd	106 kBd

9.2.3 *ISO 15693 – Vicinity coupling smart cards*

This standard aims to provide a consistent basis for vicinity coupling smart cards (inductive coupling). These are smart cards with a read range of up to 1 m, like those used in hands free access control systems. A definitive version of the standard is expected in 2002. Therefore, Table 9.6 can only give a brief overview of the provisional draft.

Table 9.6: Provisional parameters from the working paper for ISO 15693-3 [berger]

Parameter:	Value:	Comment:
Power supply	13.56 MHz ± 7 kHz	Inductive coupling
Data transmission reader → card		
Modulation	10% ASK, 100% ASK	Card supports both
Bit coding	"long distance mode": "1 of 256" "fast mode": "1 of 4"	Card supports both
Baud rate	"long distance mode": 1.65 kb/s "fast mode": 26.48 kb/s	
Data transmission card → reader		
Modulation	Load modulation with subcarrier	
Bit coding	Manchester, subcarrier is ASK (423 kHz) or FSK (423/485 kHz) modulated	
Baud rate	"long distance mode": 6.62 kb/s "fast mode": 26.48 kb/s	Selected by reader

9.3 ISO 69873 – Data Carriers for Tools and Clamping Devices

This standard specifies the dimensions for contactless data carriers and their mounting space in tools and cutters. Normally the data carriers are placed in a *quick release taper shaft* in accordance with *ISO 69 871* or in a *retention knob* in accordance with *ISO 69 872*. The standard gives installation examples for this.

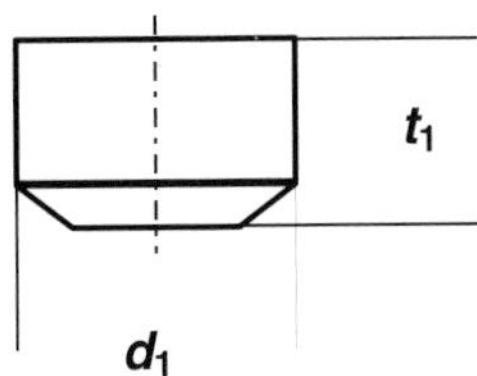

Figure 9.6: Format of a data carrier for tools and cutters

The dimensions of a data carrier are specified in *ISO 69 873* as d_1 = 10 mm and t_1 = 4.5 mm. The standard also gives the precise dimensions for the mounting space.

9.4 ISO 10374 – Container Identification

This standard describes an automatic identification system for containers based upon microwave transponders. The optical identification of containers is described in the standard *ISO 6346* and is reflected in the data record of the transponder based *container identification*.

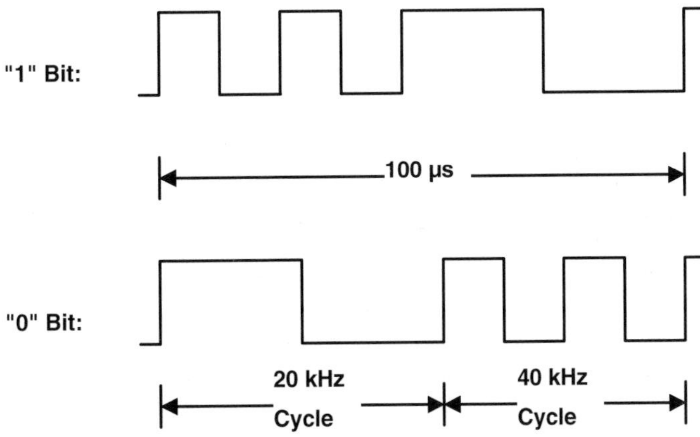

Figure 9.7: Coding of data bits using the modified FSK subcarrier procedure

Active – i.e. battery supported – microwave transponders are used. These are activated by an unmodulated carrier signal in the frequency range 850 – 950 MHz and 2400 – 2500 MHz. The sensitivity of the transponder is defined with an electric field strength E of a maximum of 150 mV/m. The transponder responds by backscatter modulation (modulated reflection cross section), using a modified FSK subcarrier procedure. The signal is modulated between the two subcarrier frequencies 40 kHz and 20 kHz.

The transmitted data sequence corresponds with the following example:

Table 9.7: Data sequence of a container transponder

Bit number:	Data	Unit	Minimum value	Maximum value
0 – 4	Object recognition	-	1	32
5 – 6	Reflector type	Type code	0	3
7 – 25	Owner code	alphabetic	AAAA	ZZZZ
26 – 45	Serial number	numeric	000000	999999
46 – 49	Check digit	numeric	0	9
50 – 59	Length	Centimetre	1	2000
60 – 61	Checksum	-	-	-
62 – 63	Structure bits	-	-	-
64	Length	-	-	-
65 – 73	Height	Centimetre	1	500
74 – 80	Width	Centimetre	200	300
81 – 87	Container format	Type code	0	127
88 – 96	Laden weight	100 kg	19	500
97 – 103	Tare weight	100 kg	0	99
104 – 105	Reserve	-	-	-
106 – 117	Security	-	-	-
118 – 123	Data format code	-	-	-
124 – 125	Check sum	-	-	-
126 – 127	Data frame end	-	-	-

9.5 VDI 4470 – Anti-theft Systems for Goods

9.5.1 Part 1 – Detection gates – inspection guidelines for customers

The VDI 4470 guideline provides a practical introduction to the inspection and testing of installed systems for electronic article surveillance (EAS systems). It describes definitions and test procedures for checking the decisive system parameters – the *false alarm rate* and *detection rate*.

The term "false alarms" is used to mean alarms that are not triggered by an active security tag, whereas the detection rate represents the ratio of alarms to the total number of active tags.

9.5.1.1 Ascertaining the false alarm rate

The number of false alarms should be ascertained immediately after the installation of the EAS system during normal business. This means that all equipment, e.g. tills and computer are in operation. During this test phase the products in the shop should not be fitted with security tags. During a monitoring period of one to three weeks an observer records all alarms and the conditions in which they occur (e.g. person in gates, cleaning, storm). Alarms that are caused by a security tag being carried through the gates by accident (e.g. a tag brought from another shop) are not counted.

9.5.1.2 Ascertaining the detection rate

The detection rate may be ascertained using either real or artificial products.

Real products In this case a number of representative products vulnerable to theft are selected and carried through the gateways by a test person in a number of typical hiding places – hood, breast pocket, shoe, carrier bag, etc. When selecting test products, remember that the material of a product (e.g. metal surfaces) may have a quite marked effect on the detection rate.

The detection rate of a system is calculated as the proportion of alarms triggered to the totality of tests carried out.

Artificial products This test uses a wooden rod with a tag in the form of a label attached to the middle. A test person carries this reference object through reference points in the gateway that are precisely defined by VDI 4470 at a constant speed.

The detection rate of a system is calculated as the proportion of alarms triggered to the totality of tests carried out.

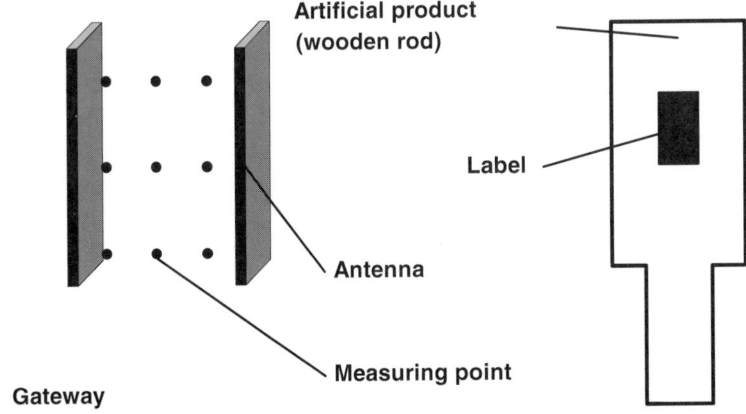

Figure 9.8: Left: Measuring points in a gateway for inspection using artificial products. Right: Artificial product

9.5.1.3 Forms in VDI 4470

In order to simplify the testing of objects and to allow tests to be performed in a consistent manner in all branches, VDI 4470 provides various forms:

- Form 1: "Test for False Alarms"
- Form 2: "Test with Real Products"
- Form 3a: "Test with Artificial Products"
- Form 3b: "Test with Artificial Products"
- Form 4a: "Test with Artificial Products"
- Form 4b: "Test with Artificial Products"

9.5.2 Part 2 – Deactivation devices, inspection guidelines for customers

As well as the option of removing hard tags (e.g. microwave systems) at the till, various tags can also be "neutralised", i.e. deactivated (e.g. RF procedure, electro-magnetic procedure).

The objective is to achieve the complete deactivation of all tags placed in a *deactivation device*, in order to avoid annoying or worrying customers by unjustified false alarms. Deactivation devices must therefore generate optical or acoustic signals, which either indicate a successful or an unsuccessful deactivation.

Deactivation devices are tested during the normal activities of the shop. A minimum of 60 protected products are required, which are checked for functionality before and after the test. The protected products are each put into/onto the deactivation device one after the other and the output from the signalling device recorded.

To ascertain the *deactivation rate* the successfully deactivated tags are divided by the total number of tags. This ratio must be 1, corresponding with a 100% deactivation rate. Otherwise, the test has not been successful.

10

The Architecture of Electronic Data Carriers

Before we describe the functionality of the data carriers used in RFID systems we must first differentiate between two fundamental operating principles: there are *electronic data carriers* based upon integrated circuits (*microchips*) and data carriers that exploit physical effects for data storage. 1 bit transponders and surface wave components belong to the latter category.

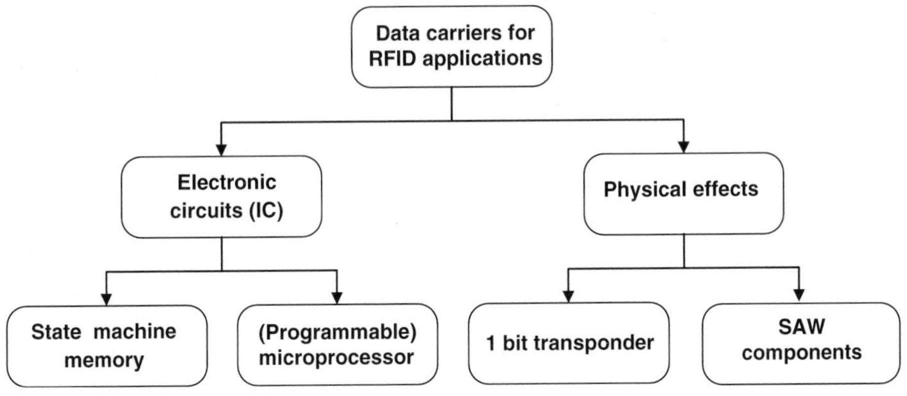

Figure 10.1: Overview of the different operating principles used in RFID data carriers

Electronic data carriers are further subdivided into data carriers with a pure memory function and those that incorporate a programmable microprocessor.

The following chapter deals exclusively with the functionality of electronic data carriers. The simple functionality of physical data carriers has already been described in Chapter 3 "Fundamental Operating Principles".

10.1 Transponder with Memory Function

Transponders with a memory function range from the simple *read only transponder* to the *high end transponder* with intelligent cryptological functions.

Transponders with a memory function contain RAM, ROM, EEPROM or FRAM and a *HF interface* to provide the *power supply* and permit communication with the reader. The main distinguishing characteristic of this family of transponders is the realisation of address and security logic on the chip using a *state machine*.

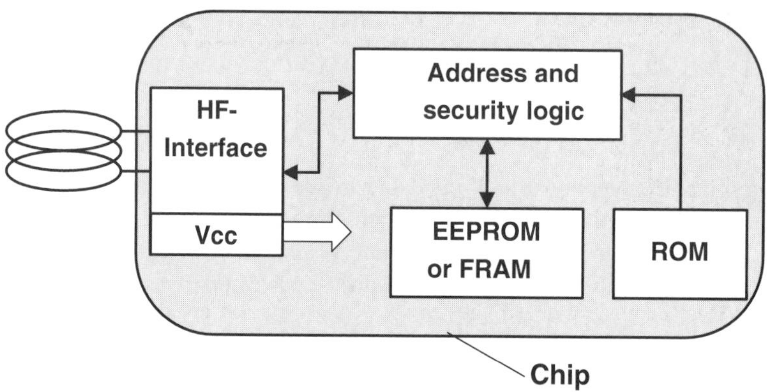

Figure 10.2: Block diagram of an RFID data carrier with a memory function

10.1.1 HF interface

The HF interface forms the interface between the analogue, high frequency transmission channel from the reader to the transponder and the digital circuitry of the transponder. The HF interface therefore performs the functions of a classical modem (Modulator – Demodulator) used for analogue data transmission via telephone lines.

The modulated HF signal from the reader is reconstructed in the HF interface by *demodulation* to create a digital serial data stream for reprocessing in the address and security logic. A clock-pulse generation circuit generates the system clock for the data carrier from the carrier frequency of the HF field.

The HF interface incorporates a *load modulator* or *backscatter modulator* (or an alternative procedure, e.g. frequency divider), controlled by the digital data being transmitted, to return data to the reader.

Passive transponders, i.e. transponders that do not have their own power supply, are supplied with energy via the HF field of the reader. To achieve this, the HF interface draws current from the transponder antenna, which is rectified and supplied to the chip as a regulated supply voltage.

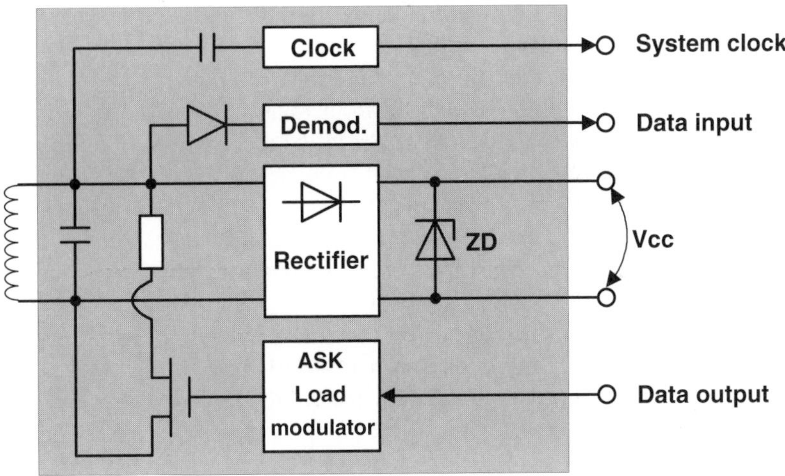

Figure 10.3: Block diagram of the HF interface of an inductively coupled transponder with a load modulator

10.1.2 Address and security logic

The *address and security logic* forms the heart of the data carrier and controls all processes on the chip.

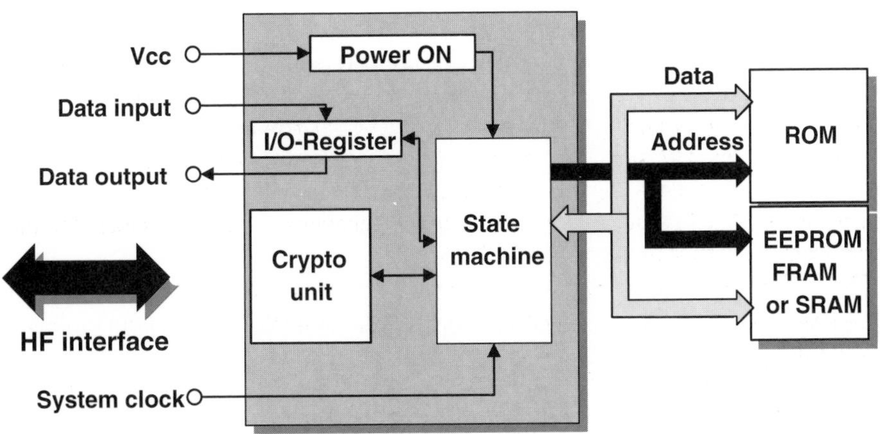

Figure 10.4: Block diagram of address and security logic module

The *power on logic* ensures that the data carrier takes on a defined state as soon as it receives an adequate power supply upon entering the HF field of a reader. Special I/O registers perform the data exchange with the reader. An optional *cryptological unit* is required for authentication, data encryption and key administration.

The data memory, which comprises a ROM for permanent data such as serial numbers, and EEPROM or FRAM is connected to the address and security logic via the address and data bus inside the chip.

The *system clock* required for sequence control and system synchronisation is derived from the HF field by the HF interface and supplied to the address and security logic module. The state-dependent control of all procedures is performed by a state machine ("hard wired software"). The complexity that can be achieved using state machines comfortably equals the performance of microprocessors (high end transponders). However the "programme sequence" of these machines is determined by the chip design. The functionality can only be changed or modified by modifying the chip design and this type of arrangement is thus only of interest for very large production runs.

10.1.2.1 State machine

A *state machine* (also switching device, Mealy machine) is an arrangement used for executing logic operations, which also has the capability of storing variable states. The output variable Y depends upon both the input variable X and what has gone before, which is represented by the switching state of flip-flops [tietze].

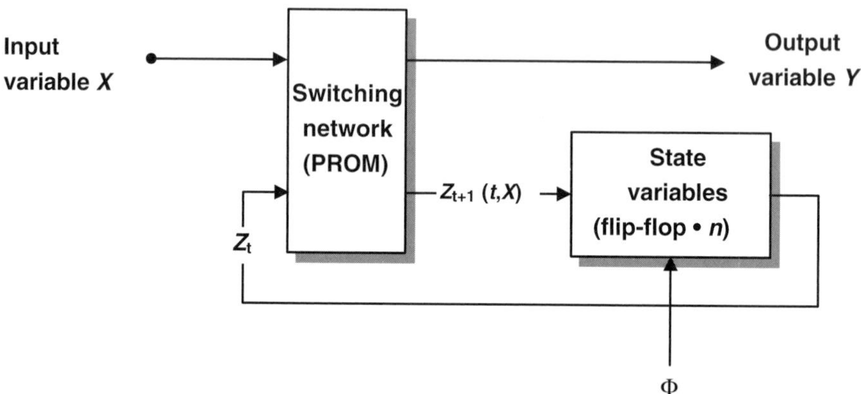

Figure 10.5: Block diagram of a state machine, consisting of the state memory and a backcoupled switching network

The state machine therefore passes through different states, which can be clearly represented in a *state diagram* (Figure 10.6). Each possible state S_Z of the system is represented by a circle. The transition from this state into another is represented by an arrow. The arrow caption indicates the conditions that the transition takes place under. An arrow with no caption indicates an unspecified transition (power on $\rightarrow S_1$). The current new state $S_Z(t+1)$ is determined primarily by the old state $S_Z(t)$ and, secondly, by the input variable x_i.

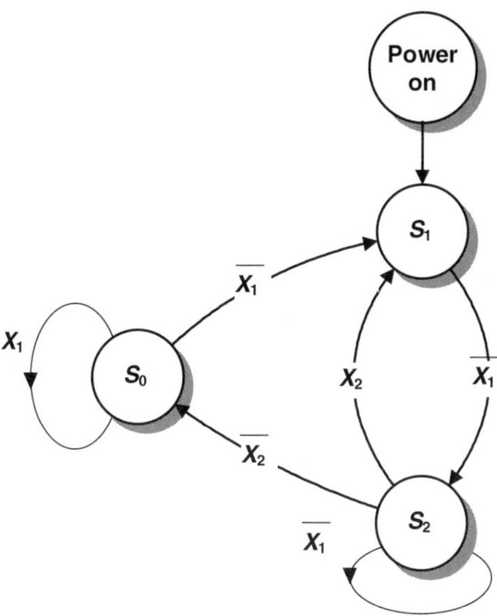

Figure 10.6: Example of a simple state diagram to describe a state machine

The order in which the states occur may be influenced by the input variable x. If the system is in state S_Z and the transition conditions that could cause it to leave this state are not fulfilled, the system remains in this state.

A switching network performs the required classification: If the state variable $Z(t)$ and the input variable are fed into its inputs, then the new state $Z(t+1)$ will occur at the output (Figure 10.5). When the next timing signal is received this state is transferred to the output of (transition triggered) flip-flops and thus becomes the new system state ($S(t+1)$ of the state machine.

10.1.3 Memory architecture

10.1.3.1 Read-only transponder

This type of transponder represents the low end, low cost segment of the range of RFID data carriers. As soon as a *read only transponder* enters the interrogation zone of a reader it begins to continuously transmit its own identification number. This identification number is normally a simple *serial number* of a few bytes with a check digit attached. Normally, the chip manufacturer guarantees that each serial number is only used once. More complex codes are also possible for special functions.

The transponder's unique identification number is incorporated into the transponder during chip manufacture. The user cannot alter this serial number, nor any data on the chip.

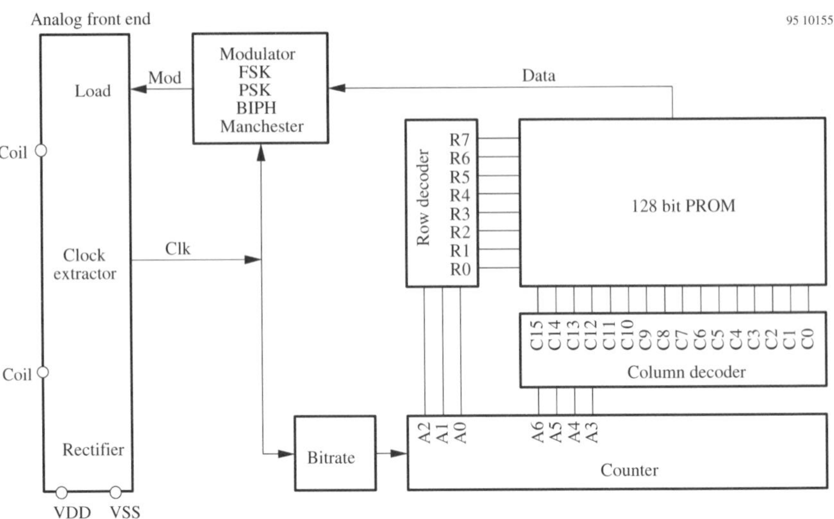

Figure 10.7: Block diagram of a read-only transponder. When the transponder enters the interrogation zone of a reader a counter begins to interrogate all addresses of the internal memory (PROM) sequentially. The data output of the memory is connected to a load modulator which is set to the baseband code of the binary code (modulator). In this manner the entire content of the memory (128 bit serial number) can be emitted cyclically as a serial data stream. (Reproduced by permission TEMIC Semiconductor GmbH, Heilbronn)

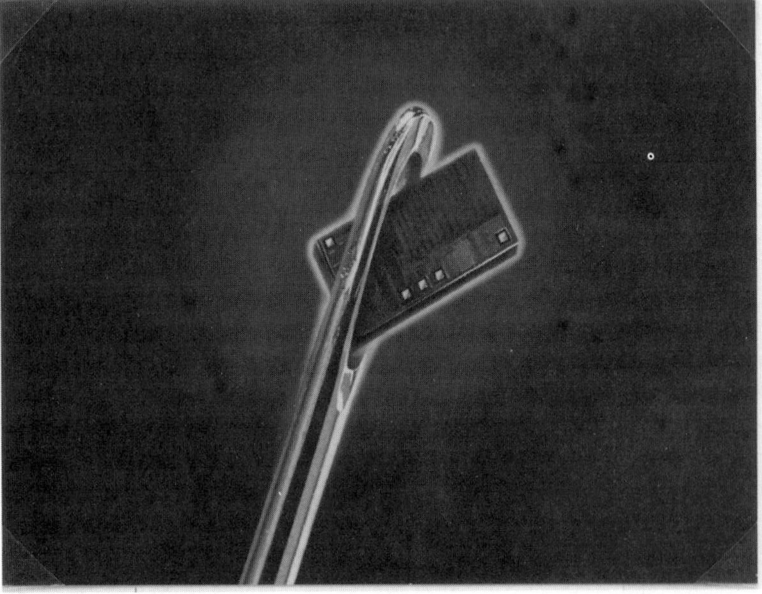

Figure 10.8: Size comparison: Low cost transponder chip in the eye of a needle. (Reproduced by permission of Philips Electronics N.V.)

Communication with the reader is unidirectional, with the transponder sending its identification number to the reader continuously. Data transmission from the reader to the transponder is not possible. However, because of the simple layout of the data carrier and reader, read-only transponders can be manufactured extremely cheaply.

Read-only transponders are used in price sensitive applications that do not require the option of storing data in the transponder. The classic fields of application are therefore animal identification, access control and industrial automation with central data management.

10.1.3.2 Writeable transponder

Transponders that can be written with data by the reader are available with memory sizes ranging from just 1 byte ("pigeon transponder") to 64 kBytes (microwave transponders with SRAM).

Write and read access to the transponder is often in blocks. Where this is the case, a block is formed by assembling a predefined number of bytes, which can then be read or written as a single unit. To change the data content of an individual block, the entire block must first be read from the transponder, whereafter the same block, including the modified bytes, can be written back to the transponder.

Current systems use block sizes of 16 bits, 4 bytes or 16 bytes. The *block structure* of the memory facilitates simple addressing in the chip and by the reader.

10.1.3.3 Transponder with cryptological function

If a writeable transponder is not protected in some way, any reader that is part of the same RFID system can read from it, or write to it. This is not always desirable, because sensitive applications may be impaired by unauthorised reading or writing of data in the transponder. Two examples of such applications are the contactless cards used as tickets in the public transport system and transponders in vehicle keys for electronic immobilisation systems.

There are various procedures for preventing unauthorised access to a transponder. One of the simplest mechanisms is read and write protection by checking a *password*. In this procedure, the card compares the transmitted password with a stored reference password and permits access to the data memory if the passwords correspond.

However, if mutual authorisation is to be sought or it is necessary to check that both components belong to the same application, then authentication procedures are used. Fundamentally, an *authentication procedure* always involves a comparison of two secret *keys*, which are not transmitted via the interface. (A detailed description of such procedures can be found in Chapter 8 "Data Security"). Cryptological authentication is usually associated with the encryption of the data stream to be transmitted. This provides an effective protection against attempts to eavesdrop into the data transmission by monitoring the wireless transponder interface using a radio receiver.

In addition to the memory area allocated to application data, transponders with cryptological functions always have an additional memory area for the storage of the secret key and a *configuration register (access register*, Acc) for selectively write protecting selected address areas. The secret key is written to the *key memory* by the manufacturer before the transponder is supplied to the user. For security reasons, the key memory can never be read.

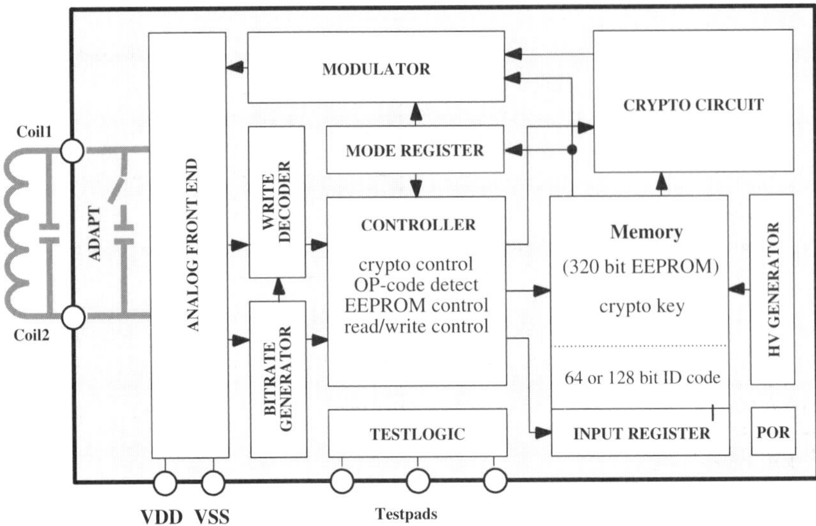

Figure 10.9: Block diagram of a writeable transponder with a cryptological function to perform authentication between transponder and reader. (Reproduced by permission of TEMIC Semiconductor GmbH, Heilbronn)

Hierarchical key concept Some systems provide the option of storing two separate keys – key A and key B – that give different access rights. The authentication between transponder and reader may take place using key A or key B. The option of allocating different *access rights* (Acc) to the two keys may therefore be exploited in order to define hierarchical security levels in an application.

Figure 10.10 illustrates this principle for clarification. The transponder incorporates two key memories, which are initialised by the two keys A and B. The access rights that the readers are allocated after successful authentication depends upon the setting that has been selected in the transponder (access register) for the key that has been used.

Reader 1 is only in possession of key A. After successful authentication, the selected settings in the access register (Acc) only permit it to read from the transponder memory. Reader 2, on the other hand, is in possession of key B. After successful authentication using key B, the settings selected in the access register (Acc) permit it to write to the transponder memory as well as reading from it.

Sample application – hierarchical key Let us now consider the system of travel passes used by a public transport network as an example of the practical use of *hierarchical keys*. We can differentiate between two groups of readers: the "devaluers" for fare payments and the "revaluers" which revalue the contactless smart cards.

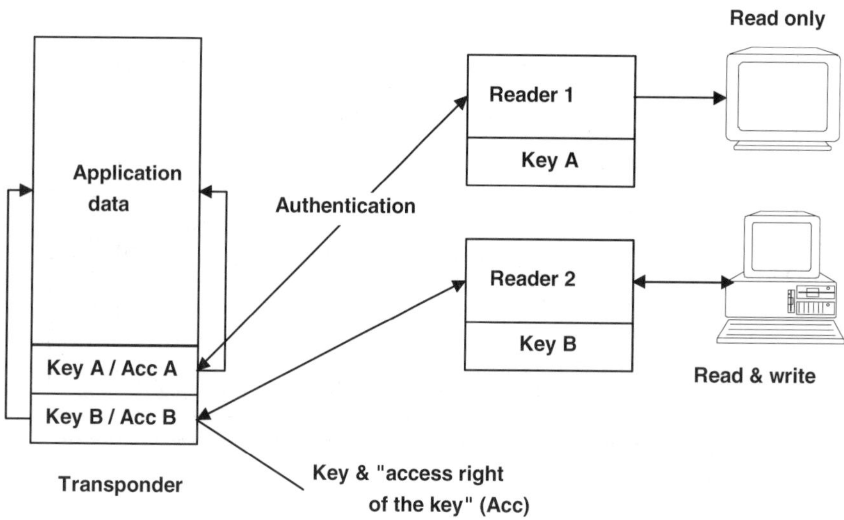

Figure 10.10: A transponder with two key memories facilitates the hierarchical allocation of access rights, in connection with the authentication keys used

The access rights to the transponder's two access registers A and B are configured such that, after successful authentication using key A, the system only permits the deduction of monetary amounts (the devaluation of a counter in the transponder). Only after authentication with key B may monetary amounts be added (the revaluation of the same counter).

In order to protect against attempted fraud, the readers in vehicles or subway entrances, i.e. devaluers, are only provided with key A. This means that a transponder can never be revalued using a devaluer, not even if the software of a stolen devaluer is manipulated. The transponder itself refuses to add to the internal counter unless the transaction has been authenticated by the correct key.

The high-security key B is only loaded into selected secure readers that are protected against theft. The transponder can only be revalued using these readers.

10.1.3.4 Segmented memory

Transponders can also be protected from access by readers that belong to other applications using authentication procedures, as we described in a previous chapter. In transponders with large memory capacities, it is possible to divide the entire memory into small units called segments, and protect each of these from unauthorised access with a separate key. A *segmented transponder* like this permits data from different applications to be stored completely separately.

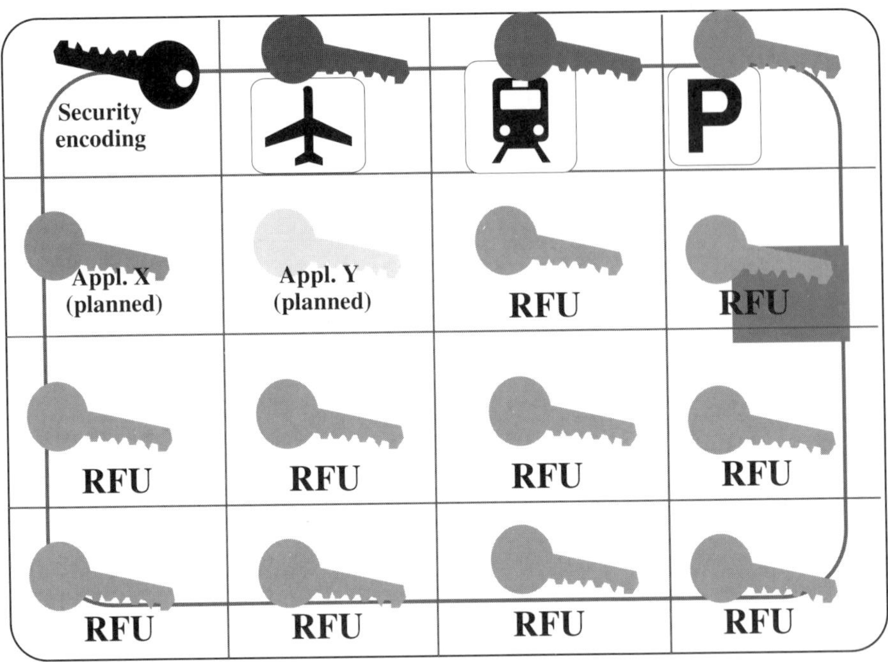

Figure 10.11: Several applications on one transponder – each protected by its own secret key

Access to an individual segment can only be gained after successful authentication with the appropriate key. Therefore, a reader belonging to one application can only gain access to its "own" segment if it only knows the *application's own key*.

The majority of segmented memory systems use fixed segment sizes. In these systems, the storage space within a segment cannot be altered by the user. A fixed segment size has the advantage that it is very simple and cheap to realise upon the transponder's microchip.

However, it is very rare for the storage space required by an application to correspond with the segment size of the transponder.

In small applications, valuable storage space is wasted on the transponder, because the segments are only partially used. Very large applications, on the other hand, need to be distributed across several segments, which means that the application specific key must be stored in each of the occupied segments. This multiple storage of an identical key also wastes valuable storage space.

A much better use of space is achieved by the use of variable length segments. Here, the memory allocated to a segment can be matched to the requirements of the application using the memory area. Because of the difficulty in realising *variable segmentation*, this variant is rare in transponders with state machines.

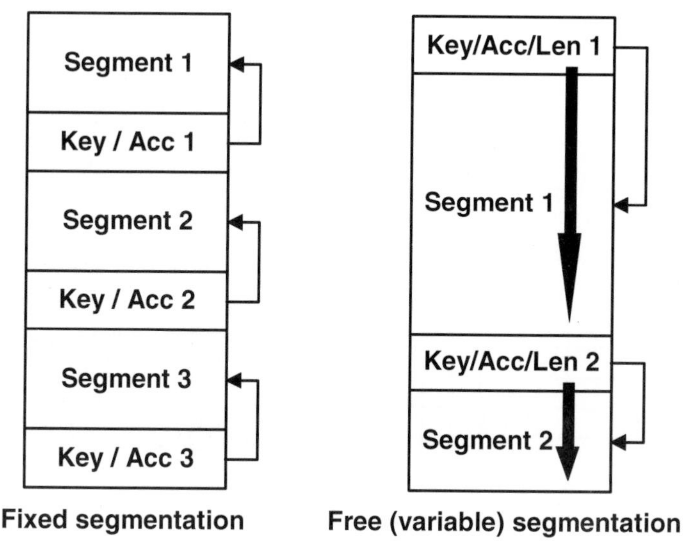

Figure 10.12: Differentiation between fixed segmenting and free segmenting

Figure 10.13 illustrates the memory configuration of a transponder with fixed segmentation. The available memory, totalling 128 bytes, is divided into 4 segments, known as "pages". Each of the 4 segments can be protected against unauthorised reading or writing by its own password. The access register of this transponder ("OTP write protection") consists of an additional memory area of 16 bits per segment. Deleting a single bit from the access register, permanently protects 16 bits of the application memory against overwriting.

10.1.3.5 MIFARE® application directory

The memory of a *MIFARE® transponder* is divided into 16 independent segments, known as sectors. Each sector is protected against unauthorised access by two different keys (hierarchical structure). Different access rights can be allocated to the two keys in its own access register (config.) [koo]. Thus, 16 independent *applications* that are protected from each other by secret keys can be loaded onto the transponder. None of the applications can be read without the secret key, not even for checking or identification. So it is not even possible to determine what applications are stored on the transponder.

Let us now assume that the city of Munich has decided to issue a contactless City-Card, which citizens can use to avail themselves of city services, and which occupies only a small part of the available memory on the card. The remaining memory units on the card could be used by other service providers for their own applications, such as local transport tickets, car rental, filling station cards, parking passes, bonus cards for restaurants and supermarket chains, and many others. However, we cannot find out which of the many possible applications are currently available on the card, because each reader belonging to an application only has access to its own sector, for which it also has the correct key.

To get around this problem, the author, in conjunction with Philips Semiconductors Gratkorn (was Mikron), has developed an *application directory* for the MIFARE® smart card. Figures 10.15 and 10.16 illustrate the data structure of this directory, the *MAD* (MIFARE® application directory):

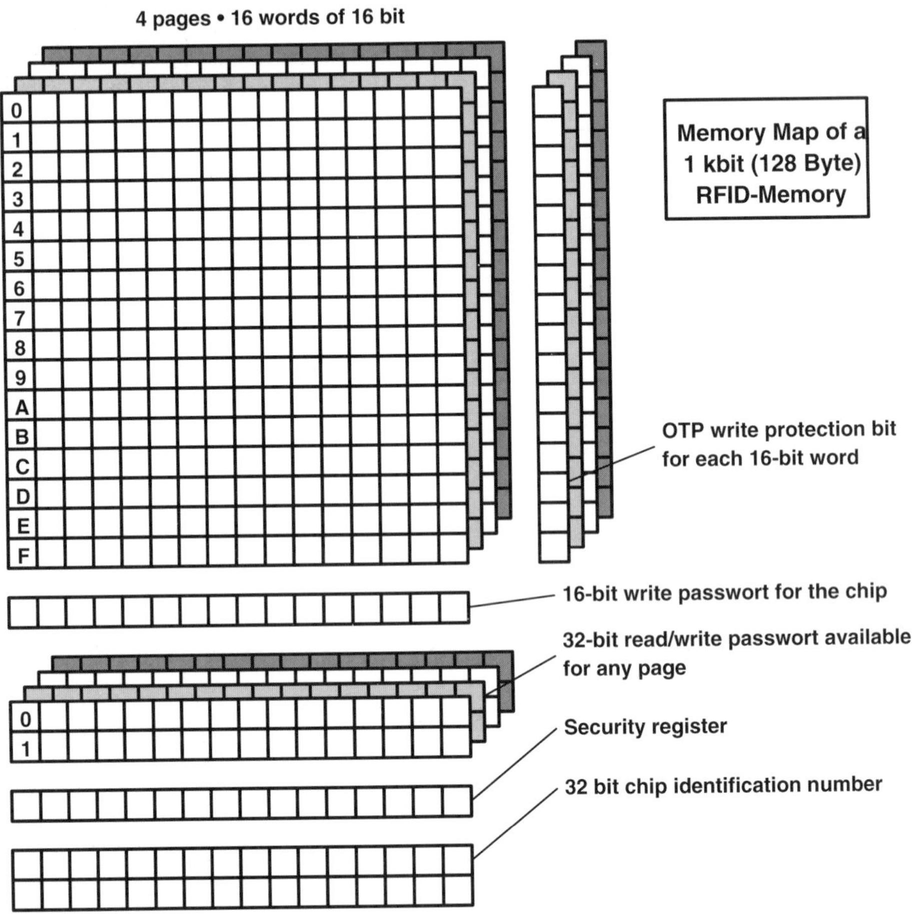

Figure 10.13: Example of a transponder with fixed segmented memory (IDESCO MICROLOG®): The four "pages" can be protected against unauthorised reading or writing by different passwords [idesco]

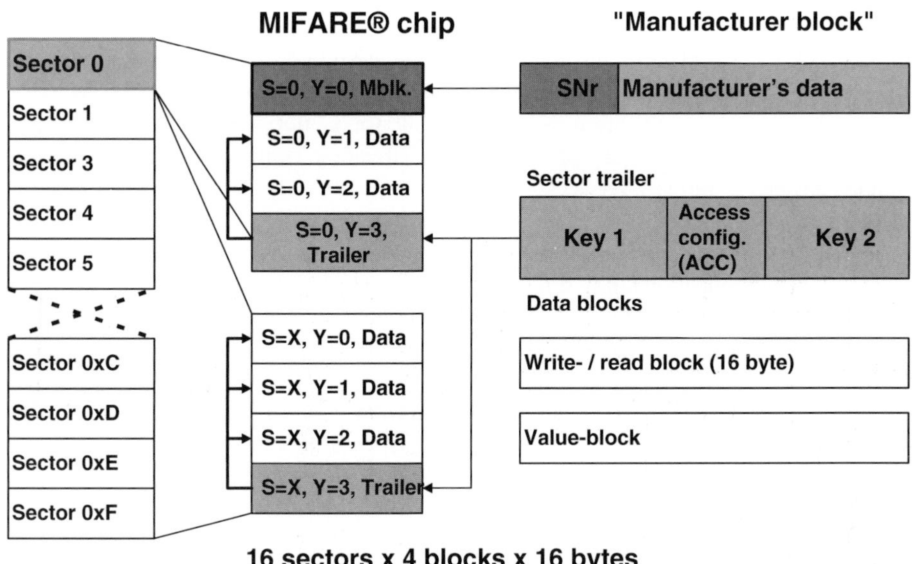

Figure 10.14: Memory configuration of a MIFARE® data carrier [koo]. The entire memory is divided into 16 independent sectors. Thus a maximum of separate 16 applications can be loaded onto a MIFARE® card

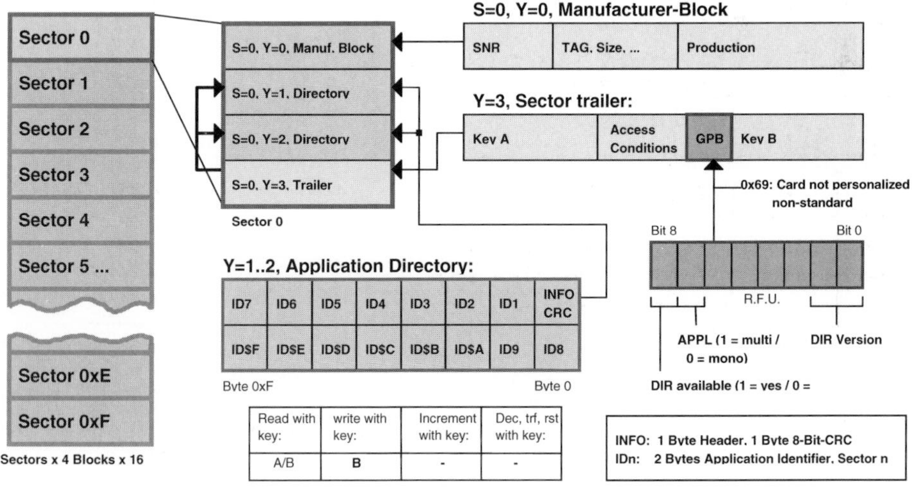

Figure 10.15: The data structure of the MIFARE® application directory consists of an arrangement of 15 pointers (ID1 to ID$F), which point to the subsequent sectors

Blocks 1 and 2 of sector 0 are reserved for the MAD, leaving 32 bytes available for the application directory. Two bytes of each make up a pointer, ID1 to ID$F, to one of the remaining 15 sectors. Reading the content of the pointer yields 2 bytes, the *function cluster* and the *application code*, which can be used to look the application up in an external database. Even if the application we are looking for is not registered in the available database, we can still gain an approximate classification from the function cluster, for example "airlines", "railway services", "bus services", "city card services", "ski ticketing", "car parking", etc.

Each application is allocated a unique identification number, made up of the function cluster code and application code. It is possible to request an identification number from the developer of MIFARE® technology, Philips Semiconductors Gratkorn (Mikron) at Graz.

If a function cluster is set at 00h, then this is an *administration code* for the management of free or reserved sectors.

Sector 0 itself does not require an ID pointer, because the MAD itself is stored in sector 0. The 2 bytes that this leaves free are used to store an 8 bit CRC, which is used to check the MAD structure for errors, and an info byte. A note can be recorded in the lowest 4 bits of the info byte, giving the sector ID of the card publisher. In our example, this would be the sector ID of one of the sectors in which the data belonging to the city of Munich is stored. This allows the reader to determine the card publisher, even if more than one application is recorded on the smart card.

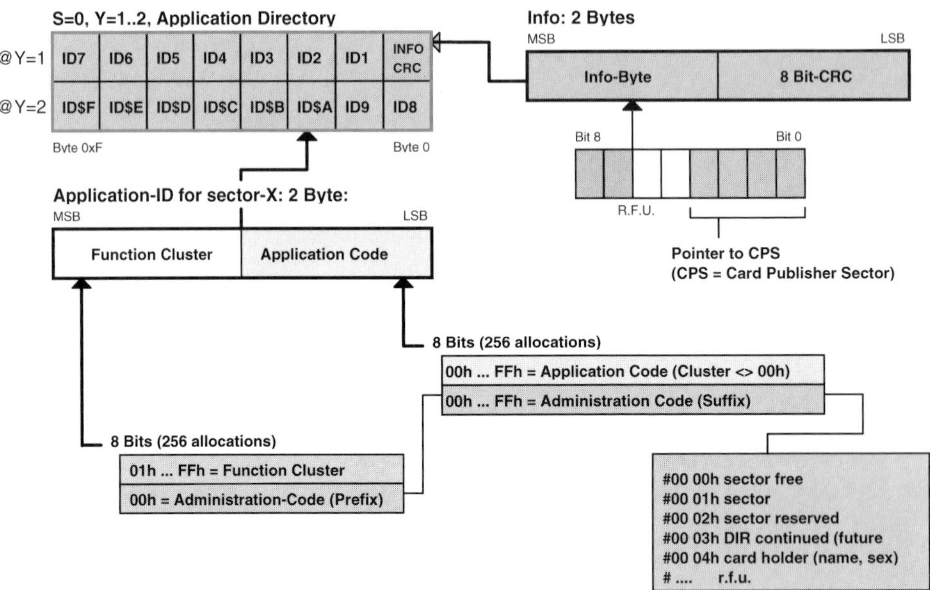

Figure 10.16: Data structure of the MIFARE® application directory: It is possible to find out what applications are located in each sector from the contents of the 15 pointers (ID1 to ID$F)

Another special feature is MAD's key management system. Whilst key A, which is required for reading the MAD, is published, key B, which is required for recording further applications, is managed by the card publisher. This means that joint use of the card by a secondary service provider is only possible after a joint use contract has been concluded and the appropriate key issued.

10.2 Microprocessors

Transponders with *microprocessors* will become increasingly common in applications using contactless smart cards in the near future. Instead of the inflexible state machine, the transponder in these cards incorporates a microprocessor.

Industry standard microprocessors, such as the familiar 8051 or 6805, are used as the microprocessor at the heart of the chip. In addition, some manufacturers are offering simple mathematical coprocessors (cryptological unit) on the same chip, which permit the rapid performance of the calculations required for encryption procedures.

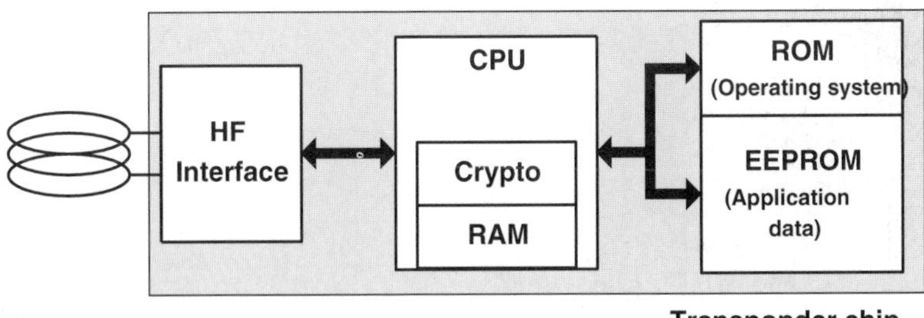

Figure 10.17: Block diagram of a transponder with a microprocessor. The microprocessor contains a coprocessor (cryptological unit) for the rapid calculation of the cryptological algorithms required for authentication or data encryption

Contactless smart cards with microprocessors incorporate their own *operating system*, as has long been the case in contact based cards. The tasks of the operating system in a contactless smart card are data transfer from and to the smart card, command sequence control, file management and the execution of cryptographic algorithms (e.g. encryption, authentication).

The programme modules are written in ROM code and are incorporated into the chip at the chip manufacturing stage by an additional exposure mask (mask programming).

The typical command processing sequence within a smart card operating system is as follows: commands sent from the reader to the contactless smart card are received by the smart card via the HF interface. Error recognition and correction mechanisms are performed by the I/O manager irrespective of higher-level procedures. An error-free command received by the secure messaging manager is decrypted or checked for integrity. After decryption the higher-level command interpreter attempts to decode the command. If this is not possible, then the return code manager is called, which generates the appropriate return code and sends it back to the reader via the I/O manager.

If a valid command is received, then the actual programme code associated with this application command is executed. If access to the application data in the EEPROM is necessary, this is performed exclusively by the file management system and the memory manager, which convert all symbolic addresses into the corresponding physical addresses of the memory area. The file manager also checks access conditions (authorisation) for the data in question.

A more detailed introduction into the procedures for the development of operating systems and smart card applications can be found in the book "The Smart Card Handbook" published by John Wiley & Sons.

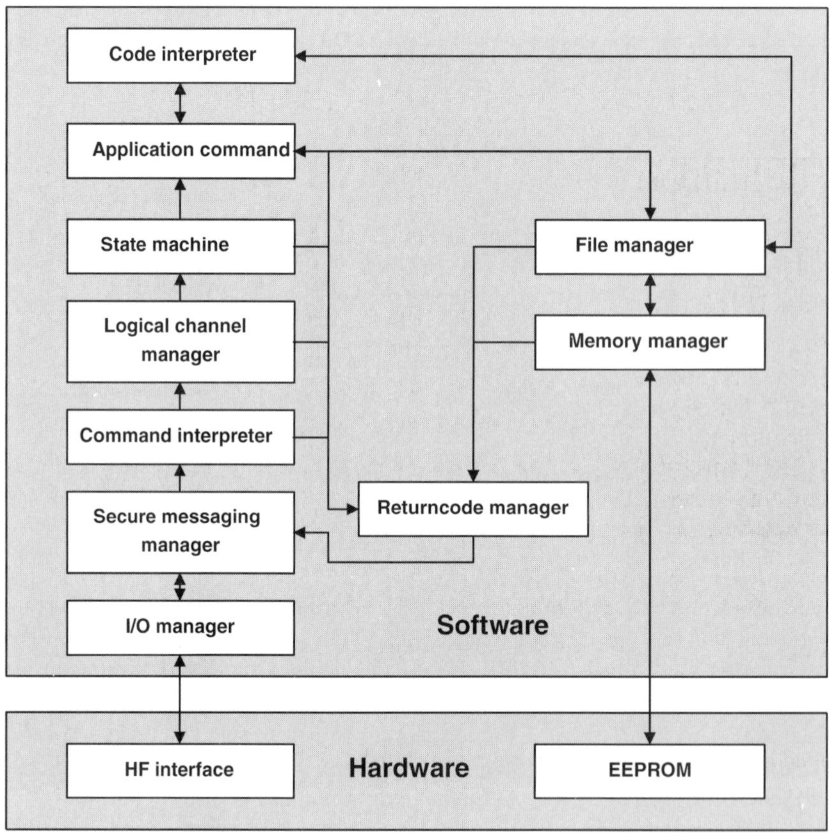

Figure 10.18: Command processing sequence within a smart card operating system [rankl]

10.2.1 Dual interface card

One very interesting development is the merging of contactless and contact smart cards into the *Dual Interface Card* (or *Combicard*). Both functions are unified on a single card. A Dual Interface Card can thus be addressed by either contactless or contact interfaces as required.

This idea is based upon the complete independence between smart card interface and smart card logic. The interface forms the lowest level of an application. This is transparent for the transmitted data, which means that from the point of view of the application software it is irrelevant what interface is used. The interface can thus be altered at will and interface and logic components can be combined as required. This provides the greatest advantage of the dual interface card for user and system operator: the option of introducing new applications using an existing infrastructure.

Smart Cards - Functional Blocks

Interface	Logic	Memory
• Contact (ISO 7816) • Contactless - close coupling (ISO 10536) - proximity (ISO 14443-draft) and/or industrial standard	• Memory Card • Password protected Memory Card • Hardwired logic (µC-functionalities) • Microcontroller • Microcontroller with Crypto-Coprocessors	• EEPROM • FRAM

Figure 10.19: Functional blocks of contactless and contact smart cards. (Reproduced by permission of Philips Electronics N.V.)

Depending upon the system, the change-over between contacts and HF interface can be performed either by the microprocessor's operating system or an additional switching matrix on the chip.

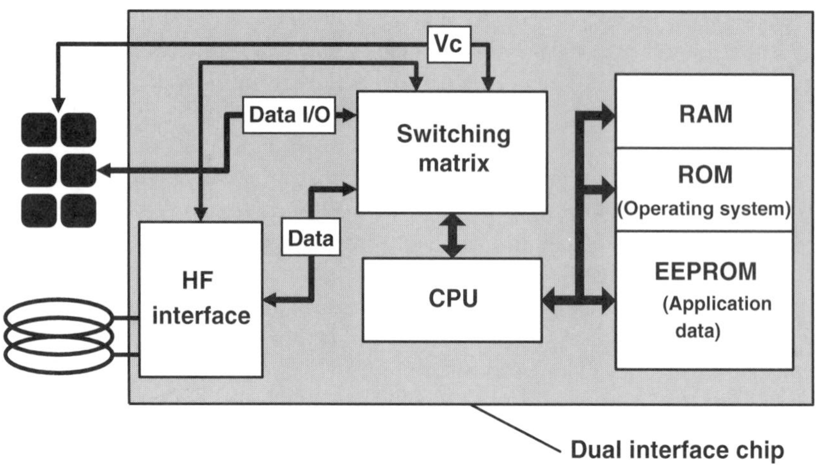

Figure 10.20: Block diagram of a dual interface card. Using a switching matrix, it is possible to communicate with the microprocessor using either a smart card reader (contact) or a contactless reader (inductive coupling)

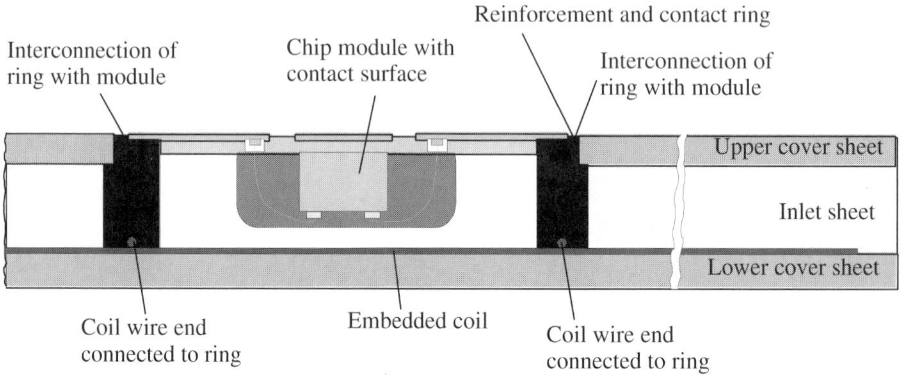

Figure 10.21: Possible layout of a dual interface smart card. The chip module has both contact surfaces (like a telephone smart card) and a transponder coil (Reproduced by permission of Amatech GmbH & Co. KG, Pfronten)

If an automatic changeover is performed by a switching matrix, the selection criterion is whether power is supplied via the HF interface or via the contacts. When the card is inserted into a contact reader or is approached by a contactless reader the interface is selected that first supplies the chip with power. The other interface is fully deactivated, so that simultaneous operation via both interfaces is impossible. Another possible method of automatic changeover is the evaluation of clock signals that are provided to the chip via the HF interface or the contacts.

If the changeover is performed by the operating system itself, the validity of data received via one of the two interfaces is used as the selection criterion.

10.2.1.1 MIFARE® plus dual interface card

This chip, which was developed jointly by the companies Philips Semiconductors Gratkorn and Siemens HL, uses a completely different principle to create a dual interface card. The heart of the chip is a 8 kByte EEPROM memory, in which the application data is stored. Unlike the functionality of a pure microprocessor card described above, access to the EEPROM memory takes place via two completely separate logic structures. As is the case for a Dual Port RAM, the EEPROM can be accessed by two separate interfaces, as desired.

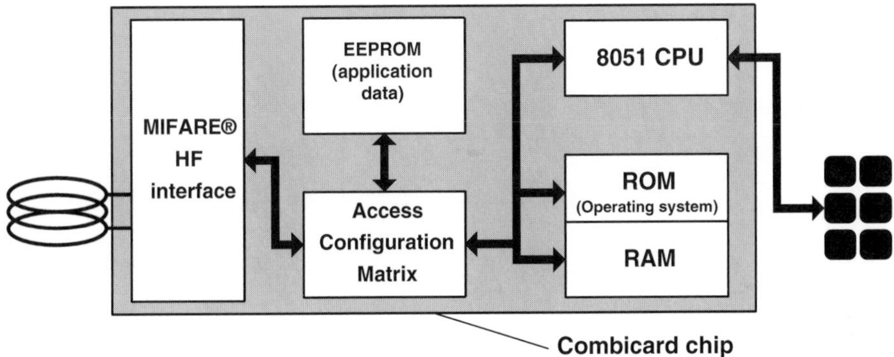

Figure 10.22: Block diagram of the MIFARE®-plus "Dual Interface Card" chip. In the contactless operating state the EEPROM is accessed via a MIFARE®-compatible state machine. When operating via contacts, a microprocessor accesses the same memory using its own operating system

The contactless interface is based upon a state machine, which fully reproduces a contactless MIFARE® memory card. Therefore, from the point of view of a contactless reader, communication with a MIFARE® plus dual interface card is no different from a conventional contactless MIFARE® memory card.

Communication via the contact interface, on the other hand, takes place via a 8051 microprocessor with its own operating system. The architecture of this interface corresponds with that of a microprocessor card, as described above.

The access rights to the EEPROM via the two interfaces is managed by an Access Configuration Matrix. The *Access Configuration Matrix* can allocate separate access rights to a selected memory area for the two interfaces. This permits the use of hierarchical security concepts like those described in the preceding chapter.

10.3 Memory Technology

After the state machine or microprocessor, the most important component of a data carrier is the memory that user data is read from or written to. Read only data is defined at the manufacturing stage by the chip mask (exposure mask) or permanently burnt into the memory by a laser. The use of a laser also makes it possible to programme *unique numbers* (*serial numbers* that are issued only once), or consecutive numbers into the data carrier.

If data is to be written to the data carrier, then RAM, EEPROM or FRAM cells are also incorporated into the chip. However, only EEPROM and FRAM cells can store the written data for long periods (typical retention periods are 10 years) without a power supply.

10.3.1 RAM

RAM is memory that can be used for the storage of temporary data. When the power supply is removed, the stored data is lost forever. In transponders, RAM is mainly used for the temporary storage of data that exists briefly during operation in the interrogation zone of a reader. In active transponders that have their own battery, RAMs with battery back-ups are sometimes used for the long-term storage of data.

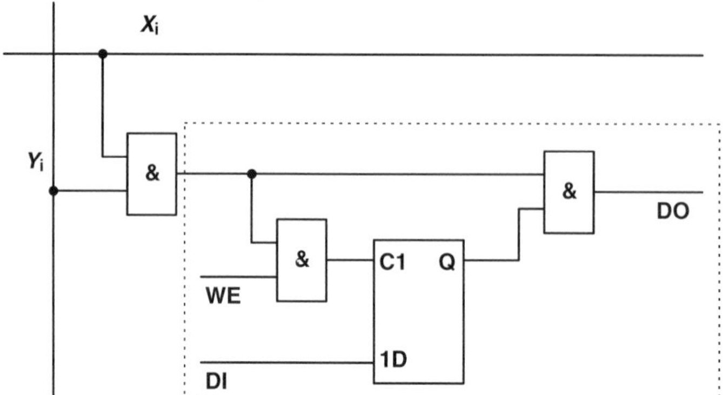

Figure 10.23: Simplified functional block diagram of a (S)RAM cell

The main component of the (S)RAM memory cell is a D-flip-flop. Figure 10.23 shows the block diagram for a single memory cell. Each memory cell has the connections DI (Data Input), WE (Write Enable) and DO (Data Out). If data is only to be read from the memory cell, it is sufficient to activate the selected cell with logic 1 levels at the allocated address connections Y_i and X_i.

To write data to the memory cell, the WE connection must also be switched to the 1 level. If there is a 1 level at C1 input data is written to the flip-flop.

10.3.2 EEPROM

The operating principle of an EEPROM cell is based upon the ability of a capacitor (condenser) to store electric charge over long periods. An EEPROM therefore represents a tiny capacitor that can be charged or discharged. A charged capacitor represents a logic "1", a discharged capacitor represents a logic "0".

In its simplest form, an EEPROM cell basically consists of a modified field effect transistor on a carrier material (substrate) made of silicon. The EEPROM cell contains an additional gate between the control gate of the field effect transistor and the substrate, which is not connected to an external power supply, and which is positioned at a very short distance (~10 nm) from the carrier material. This so-called *floating gate* can be charged or discharged via the substrate using the tunnel effect, and therefore represents a capacitor. For the tunnel effect to exist there must be a sufficiently large potential difference at the thin insulating tunnelling oxidation layer between the floating gate and the substrate.

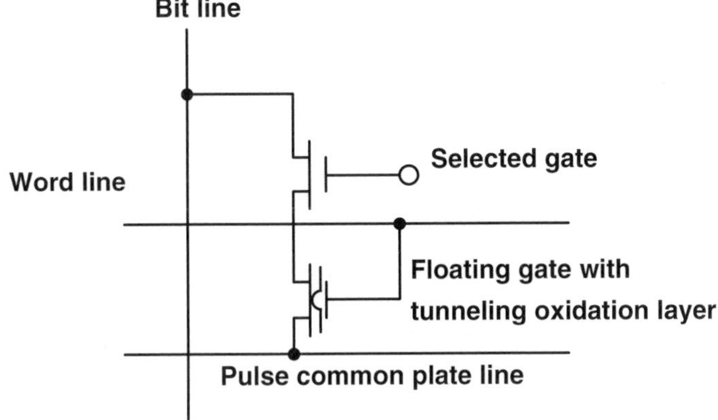

Figure 10.24: The EEPROM cell consists of a modified field effect transistor with an additional floating gate

The flow of current between source and drain can be controlled by the stored charge of the floating gate. A negatively charged floating gate gives rise to a high threshold voltage between the source and drain of the field effect transistor, meaning that this is practically blocked. The current flow through the field effect transistor of an EEPROM cell is evaluated by signal amplification of the memory chip, whereby the strength of the current clearly indicates a "0" or "1".

To write a "0" or "1" to an EEPROM cell, a high positive or negative voltage is applied to the control gate, which activates the tunnel effect. The voltage required to charge the EEPROM cell is around 17 V at the control gate which falls to 12 V at the floating gate. However, RFID data carriers are supplied with 3 V or 5 V from the HF interface (or a battery). Therefore a voltage of 25 V is generated from the low supply voltage of the chip using a cascaded charging pump integrated into the chip, which provides the required 17 V after stabilisation.

It takes between 5 and 10 ms to charge an EEPROM cell. The number of possible write cycles is limited to between 10 000 and 100 000 for EEPROM cells. This is because in every write operation electrons are captured by the tunnelling oxidation layer and these are never released. These electrons influence the threshold voltage of the field effect transistor, with the effect becoming greater with every write operation. As soon as this parasitic effect of the tunnelling oxidation layer becomes greater than the primary influence of the floating gate the EEPROM cell has reached its *lifetime* [rankl].

A charged floating gate loses its charge due to insulation losses and quantum mechanical effects. However, according to the semiconductor manufacturer's figures, EEPROMs still provide reliable data retention for 10 years. If the EEPROM cell is nearing its lifetime, then information is only stored for short periods, which are determined by the parasitic influence of the oxide layer. For this reason, a plausibility test should be carried out on stored data using checksums (e.g. CRC) in RFID data carriers with EEPROM memories.

10.3.3 FRAM

High power consumption during writing and high write times of around 5 – 10 ms have a detrimental effect on the performance of RFID systems that employ EEPROM technology. A new, non-transient memory technology, which should improve this situation, has been under development for around 20 years: the ferroelectrical RAM, or *FRAM* . At the end of the 1980s the company Ramtron was established, and they collaborated with Hitachi on the development of this technology. The first RFID systems using FRAM technology were produced by the Ramtron subsidiary Racom. However, the development of FRAMs is still associated with many problems, and so RFID systems using FRAMs are still not widespread.

The principle underlying the FRAM cell is the ferroelectric effect, i.e. the capability of a material to retain an electrical polarisation even in the absence of an electric field. The polarisation is based upon the alignment of an elementary dipole within a crystal lattice in the ferroelectric material due to the effect of an electric field that is greater than the coercive force of the material. An opposing electric field causes the opposite alignment of the internal dipole. The alignment of the internal dipole takes on one of two stable states, which are retained after the electric field has been removed.

Figure 10.25 shows a simplified model of the ferroelectric lattice. The central atom moves into one of the two stable positions, depending upon the field direction of the external electric field. Despite this, FRAM memories are completely insensitive to foreign electric interference fields and magnetic fields.

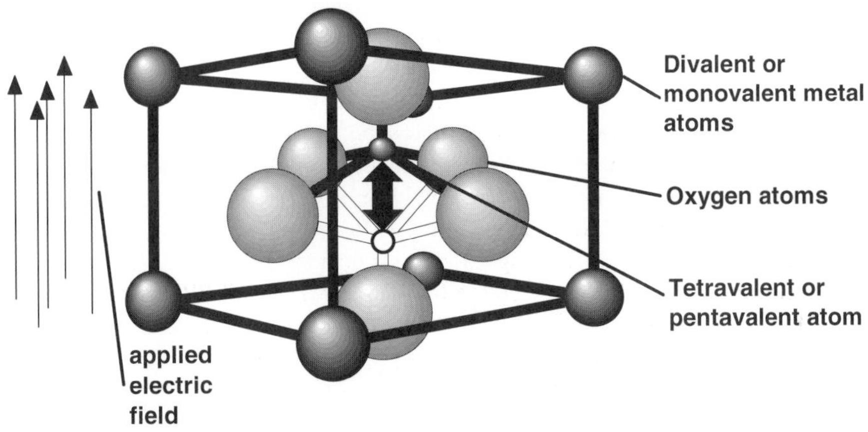

Figure 10.25: Basic configuration of a ferroelectric crystal lattice: an electric field steers the inner atom between two stable states

To read the FRAM cell (Figure 10.26), an electric field (U_{CC}) is applied to the ferroelectric capacitor via a switching transistor. If the stored information represents a logic "1" then the cell is in position "A" on the hysteresis loop. If, on the other hand, it represents a logic "0", the cell is in position "C". By the application of the voltage U_{CC} we move to point "B" on the hysteresis loop, releasing electric charge, which is captured and evaluated by the signal amplifiers on the memory chip. The magnitude of escaping charge clearly indicates a "1" or "0", because a significantly greater charge escapes in the transition from state "A" to "B" than in the transition from state "C" to "B".

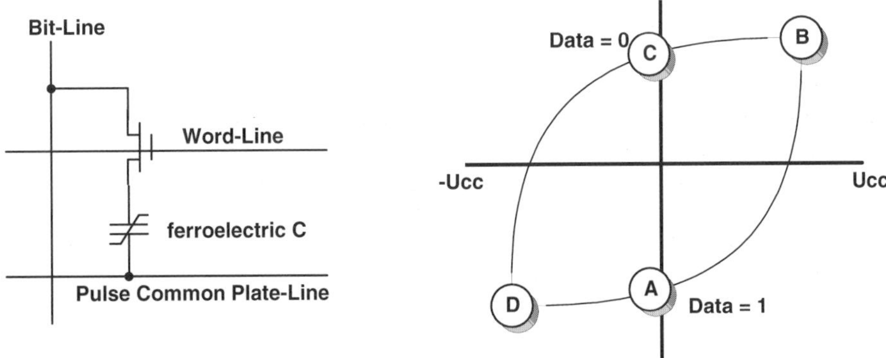

Figure 10.26: FRAM cell structure (1 bit) and hysteresis loop of the ferroelectric capacitor

After the external (read) field U_{CC} has been removed, the FRAM cell always returns to state "C", and thus a stored "1" is lost, because state "C" represents a "0". For this reason, as soon as a "1" is read, the memory chip's logic automatically performs a re-write operation. This involves applying an opposing electric field $-U_{CC}$ to the ferroelectric capacitor, which changes the state of the FRAM cell, moving it to point "D" on the hysteresis loop. After the removal of the electric field the FRAM cell falls into state "D", which recreates the originally stored state "A" [haber].

Writing a "1" or "0" to the FRAM cell is achieved simply by the application of an external voltage $-U_{CC}$ or $+U_{CC}$. After the voltage is removed the FRAM cell returns to the corresponding residual state "A" or "C".

10.3.4 Performance comparison FRAM – EEPROM

Unlike EEPROM cells, the write operation of a FRAM cell occurs at a high speed. Typical *write times* lie in the region of 0.1 μs. FRAM memories can therefore be written in "real time", i.e. in the bus cycle time of a microprocessor or the cycle time of a state machine.

FRAMs also beat EEPROMs in terms of power consumption by orders of magnitude. FRAM memory was therefore predestined for use in RFID systems. However, problems in combining CMOS processors (microprocessor) and analogue circuits (HF interface) with FRAM cells on a single chip still prevent the rapid spread of this technology.

Table 10.1: Comparison between FRAM and EEPROM [pana]

	FRAM	EEPROM
Size of memory cell	~ 80 μm^2	~130 μm^2
Lifetime in write cycles	10^{12}	10^5
Write voltage	2 V	12 V
Energy for writing	0.0001 μJ	100 μJ
Write time	0.1 μs	10 ms (10 000 μs)

10.4 Measuring Physical Variables

10.4.1 Transponder with sensor functions

Battery operated *telemetry transmitters* in the frequency range 27.125 MHz or 433 MHz are normally used for the detection of *sensor data*. The fields of application of these systems are very limited, however, and are restricted by their size and the lifetime of the battery.

Specially developed RFID transponders incorporating an additional *A/D converter* on the ASIC chip facilitate the measurement of physical variables. In principle, any sensor can be used, in which the resistance alters in proportion to physical variables. Due to the availability of miniaturised *temperature sensors* (NTC), this type of system was first developed for temperature measurement.

Temperature sensor, transponder ASIC, transponder coil and back-up capacitors are located in a glass capsule, like those used in animal identification systems (see Chapter 13,

Section 13.6 "Example Applications – Animal Identification") [ruppert]. The passive RFID technology with no battery guarantees the lifelong functioning of the transponder and is also environmentally friendly.

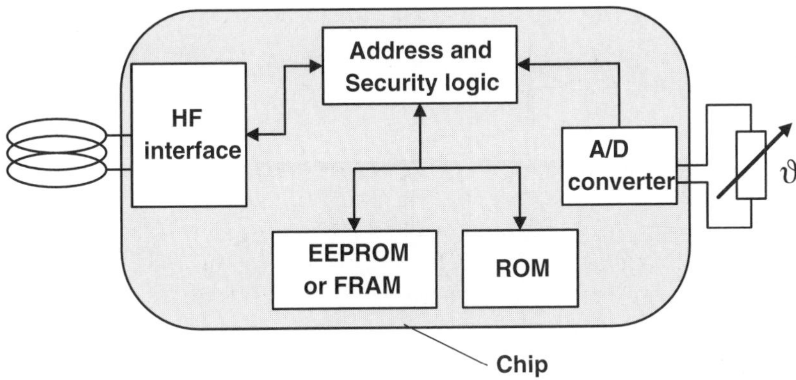

Figure 10.27: Inductively coupled transponder with additional temperature sensor

The measured value of the A/D converter can be read by a special reader command. In read-only transponders the measured value can also be appended to a periodically emitted identification number (serial number).

Nowadays, the main field of application for transponders with sensor functions is wireless temperature measurement in animal keeping. In this application the body temperatures of domestic and working animals are measured for health monitoring and breeding and birth control. The measurement can be performed automatically at feed and watering points or manually using a portable reader [ruppert].

In industrial usage, transponders with a sensor function may be used anywhere where physical variables need to be measured in rotating or moving parts where cable connections are impossible.

10.4.2 Measurements using microwave transponders

Industry standard microwave transponders can also be used to measure speed and distance by the analysis of the *Doppler effect* and *signal travelling times*.

The Doppler effect occurs in all electromagnetic waves and is particularly easy to measure in microwaves. If there is a relative movement between the transmitter and a receiver, then the receiver detects a different frequency than the one emitted by the transmitter. If the receiver moves closer to the transmitter, then the wavelength will be shortened by the distance that the receiver has covered during one oscillation. The receiver thus detects a higher frequency.

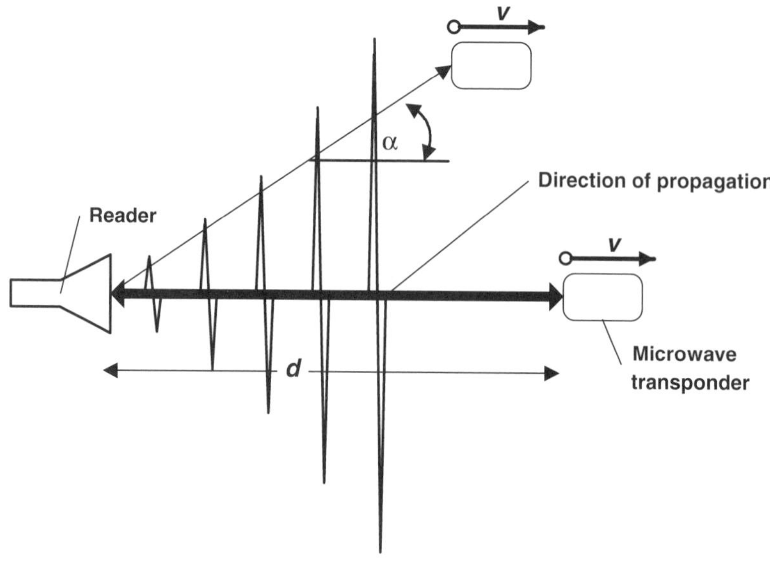

Figure 10.28: Distance and speed measurements can be performed by exploiting the Doppler effect and signal travelling times

If the electromagnetic wave is reflected back to the transmitter from an object that has moved, then the received wave contains twice the frequency shift. There is almost always an angle α between the direction of propagation of the microwaves and the direction of movement of the "target". This leads to a second, expanded Doppler equation:

$$f_d = \frac{f_{TX} \cdot 2v}{c} \cdot \cos\alpha \qquad v = \frac{f_d \cdot c}{2 f_{TX} \cdot \cos\alpha} \qquad (10.1),\ (10.2)$$

The Doppler frequency f_d is the difference between the transmitted frequency f_{TX} and the received frequency f_{RX}. The relative speed of the object is $v \cdot \cos\alpha$, c is the speed of light, $3 \cdot 10^8$m/s.

A transmission frequency of 2.45 GHz yields the Doppler frequencies shown in Table 10.2 at different speeds.

To measure the distance d of a transponder, we analyse the travelling time t_d of a microwave pulse reflected by a transponder:

$$d = \frac{1}{2} \cdot t_d \cdot c \qquad (10.3)$$

The measurement of the speed or distance of a transponder is still possible if the transponder is already a long way outside the normal interrogation zone of the reader, because this operation does not require communication between reader and transponder.

Table 10.2: Doppler frequencies at different speeds

f_d, Hz	v, m/s	v, km/h
0	0	0
10	0.612	1.123
20	1.224	4.406
50	3.061	9.183
100	6.122	18.36
200	12.24	36.72
500	30.61	110.2
1000	61.22	220.39
2000	122.4	440.6

11

Readers

11.1 Data Flow in an Application

A *software application* that is designed to read data from a contactless data carrier (transponder) or write data to a contactless data carrier, requires a contactless *reader* as an interface. From the point of view of the application software, access to the data carrier should be as transparent as possible. In other words, the read and write operations should differ as little as possible from the process of accessing comparable data carriers (smart card with contacts, serial EEPROM).

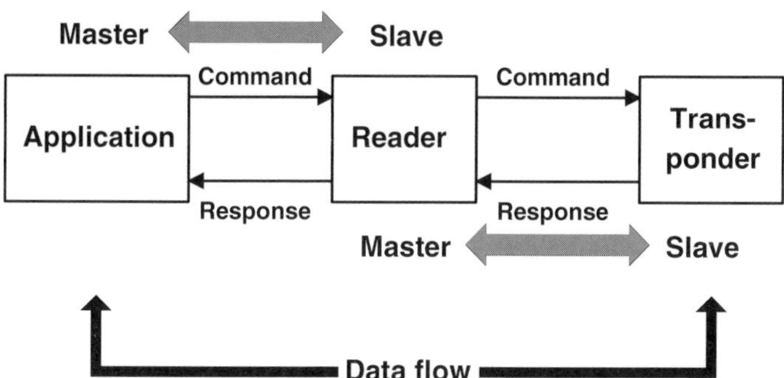

Figure 11.1: Master-slave principle between application software (application), reader and transponder

Write and read operations involving a contactless data carrier are performed on the basis of the *master-slave principle*. This means that all reader and transponder activities are initiated by the application software. In a hierarchical system structure the application software represents the master, whilst the reader, as the slave, is only activated when write / read commands are received from the application software.

To execute a command from the application software, the reader first enters into communication with a transponder. The reader now plays the role of the master in relation to the transponder. The transponder therefore only responds to commands from the reader and is never active independently (except for the simplest read-only transponders. See Chapter 10 "The Architecture of Electronic Data Carriers).

A simple read command from the application software to the reader can initiate a series of communication steps between the reader and a transponder. In the following example, a read command first leads to the activation of a transponder, followed by the execution of the authentication sequence and finally the transmission of the requested data:

Table 11.1: Example of the execution of a read command by the application software, reader and transponder

Application ↔ Reader	Reader ↔ Transponder	Comment
→ Blockread_Address[00]		Read transponder memory [address]
	→ Request	Transponder in the field?
	← ATR_SNR[4712]	Transponder operates with serial number
	→ GET_Random	Initiate authentication
	← Random[081514]	
	→ SEND_Token1	
	← GET_Token2	Authentication successfully completed
	→ Read_@[00]	Read command [address]
	← Data[9876543210]	Data from transponder
← Data[9876543210]		Data to application

The reader's main functions are therefore to activate the data carrier (transponder), structure the communication sequence with the data carrier, and transfer data between the application software and a contactless data carrier. All features of the contactless communication, i.e. making the connection, and performing anticollision and authentication procedures, are handled entirely by the reader.

11.2 Components of a Reader

A number of contactless transmission procedures have already been described in the preceding chapters. Despite the fundamental differences in the type of coupling (inductive – electromagnetic), the communication sequence (FDX, HDX, SEQ), the data transmission procedure from the transponder to the reader (load modulation, backscatter, subharmonic) and, last but not least, the frequency range, all readers are similar in their basic operating principle and thus in their design.

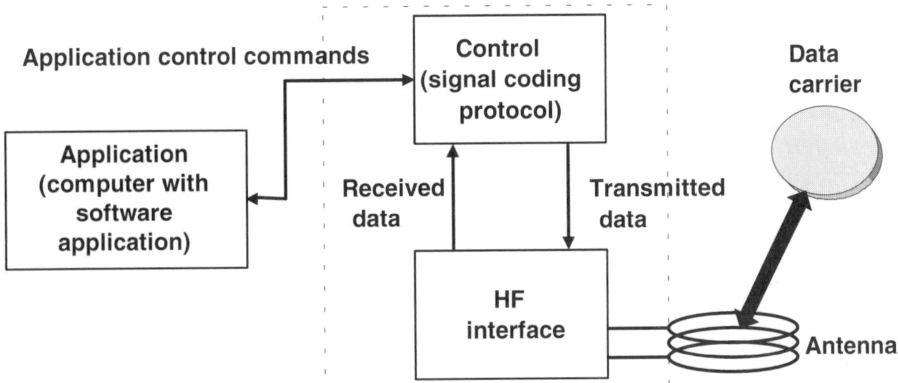

Figure 11.2: Block diagram of a reader consisting of control system and HF interface. The entire system is controlled by an external application via control commands

Readers in all systems can be reduced to two fundamental functional blocks: the control system and the *HF interface*, consisting of a transmitter and receiver. Figure 11.3 shows a reader for an inductively coupled RFID system. On the right hand side we can see the HF interface, which is shielded against undesired spurious emissions by a tinplate housing. The control system is located on the left-hand side of the reader and, in this case, it comprises an ASIC module and microcontroller. In order that it can be integrated into a software application, this reader has an RS232 interface to perform the data exchange between the reader (slave) and the external application software (master).

Figure 11.3: Example of a reader. The two functional blocks, HF interface and control system, can be clearly differentiated. (MIFARE® reader, reproduced by permission of Philips Electronics N.V.)

11.2.1 HF-interface

The reader's HF interface performs the following functions:

- Generation of high frequency transmission power to activate the transponder and supply it with power
- Modulation of the transmission signal to send data to the transponder
- Reception and demodulation of HF signals transmitted by a transponder

The HF interface contains two separate signal paths to correspond with the two directions of data flow from and to the transponder. Data transmitted to the transponder travels through the *transmitter arm*. Conversely, data received from the transponder is processed in the *receiver arm*. We will now analyse the two signal channels in more detail, giving consideration to the differences between the different systems.

11.2.1.1 Inductively coupled system, FDX/HDX

First, a signal of the required operating frequency, i.e. 135 kHz or 13.56 MHz is generated in the transmitter arm by a stable (frequency) quartz oscillator. To avoid worsening the noise ratio in relation to the extremely weak received signal from the transponder, the *oscillator* is subject to high demands regarding phase stability and sideband noise.

The oscillator signal is fed into a modulation module controlled by the baseband signal of the signal coding system. This *baseband signal* is a keyed direct voltage signal (TTL level), in which the binary data is represented using a serial code (Manchester, Miller, NRZ). Depending upon the modulator type, *ASK* or *PSK modulation* is performed on the oscillator signal.

FSK modulation is also possible, in which case the baseband signal is fed directly into the frequency synthesiser.

The modulated signal is then brought to the required level by a power output module and can then be decoupled to the antenna box.

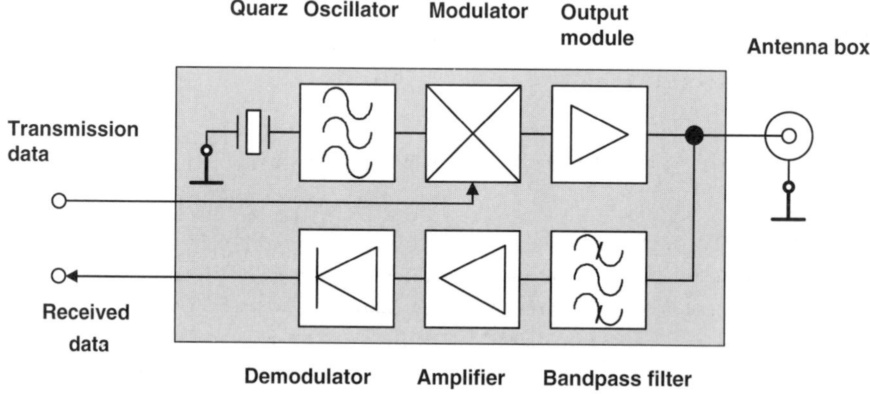

Figure 11.4: Block diagram of an HF interface for an inductively coupled RFID system

The *receiver arm* begins at the antenna box, with the first component being a steep edge bandpass filter or a notch filter. In FDX/HDX systems this filter has the task of largely blocking the strong signal from the transmission output module and filtering out just the response signal from the transponder. In subharmonic systems, this is a simple process, because transmission and reception frequencies are usually a whole octave apart. In systems with load modulation using a *subcarrier* the task of developing a suitable filter should not be underestimated because, in this case, the transmitted and received signals are only separated by the subcarrier frequency. Typical subcarrier frequencies in 13.56 MHz systems are 847 kHz or 212 kHz.

Some LF systems with load modulation and no subcarrier use a notch filter to increase the modulation depth (duty factor) – the ratio of the level to the load modulation sidebands – and thus the duty factor by reducing their own carrier signal. A different procedure is the rectification and thus demodulation of the (load) amplitude modulated voltage directly at the reader antenna. A sample circuit for this can be found in Section 11.3 "Reader IC U2270 for Electronic Immobilisation".

11.2.1.2 Microwave systems – half duplex

The main difference between *microwave systems* and low frequency inductive systems is the frequency synthesising: the operating frequency, typically 2.45 GHz, cannot be generated directly by the quartz oscillator, but is created by the multiplication (excitation of harmonics) of a lower oscillator frequency. Because the modulation is retained during frequency multiplication, modulation is performed at the lower frequency.

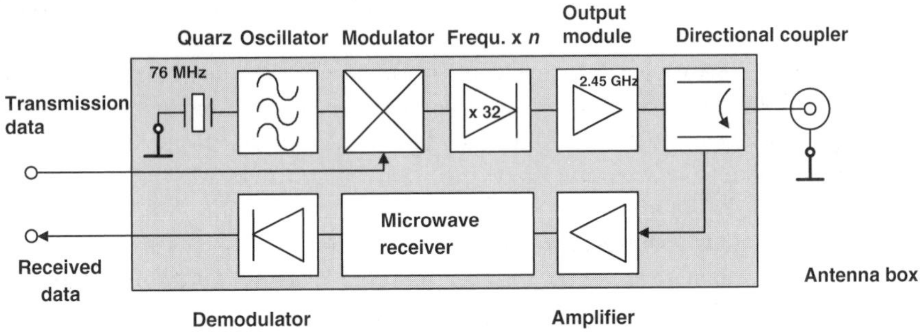

Figure 11.5: Block diagram of an HF interface for microwave systems

Some microwave systems employ a *directional coupler* to separate the system's own transmission signal from the weak backscatter signal of the transponder [isd].

A directional coupler consists of two continuously coupled homogeneous wires [meinke]. If all four ports are matched and power P_1 is supplied to port ①, then the power is divided between ports ② and ③, with no power occurring at the decoupled port ④. The same applies if power is supplied to port ③, in which case the power is divided between ports ① and ②.

A directional coupler is described by its *coupling loss*

$$a_k = -20 \cdot \ln |P②/P①| \tag{11.1}$$

and *directivity*:

$$a_D = -20 \cdot \ln | P \ /P \ | \tag{11.2}$$

Directivity is the logarithmic magnitude of the ratio of undesired overcoupled power P_4 to desired coupled power P_2.

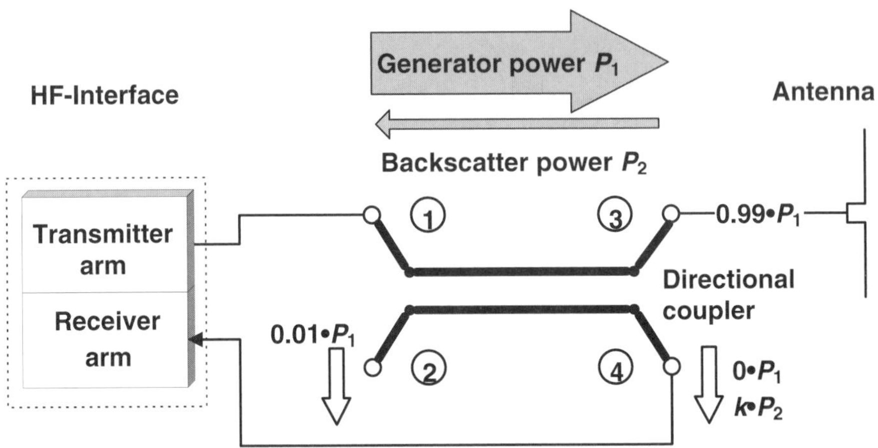

Figure 11.6: Layout and operating principle of a directional coupler for a backscatter RFID system

A directional coupler for a backscatter RFID reader should have the maximum possible directivity to minimise the decoupled signal of the transmitter arm at port . The coupling loss, on the other hand, should be low to decouple the maximum possible proportion of the reflected power P_2 from the transponder to the receiver arm at port . When a reader employing decoupling based upon a directional coupler is commissioned, it is necessary to ensure that the transmitter antenna is well (anechoically) set up. Power reflected from the antenna due to poor adjustment is decoupled at port as backwards power. If the directional coupler has a good coupling loss, even a minimal mismatching of the transmitter antenna (e.g. by environmental influences) is sufficient to increase the backwards travelling power to the magnitude of the reflected transponder power. Nevertheless, the use of a directional coupler gives a significant improvement compared to the level ratios achieved with a direct connection of transmitter output module and receiver input.

11.2.1.3 Sequential systems – SEQ

In a sequential RFID system the HF field of the reader is only ever transmitted briefly to supply the transponder with power and/or send commands to the transponder.

The transponder transmits its data to the reader whilst the reader is not transmitting. The transmitter and receiver in the reader are thus active sequentially, like a walkie-talkie, which also transmits and receives alternately.

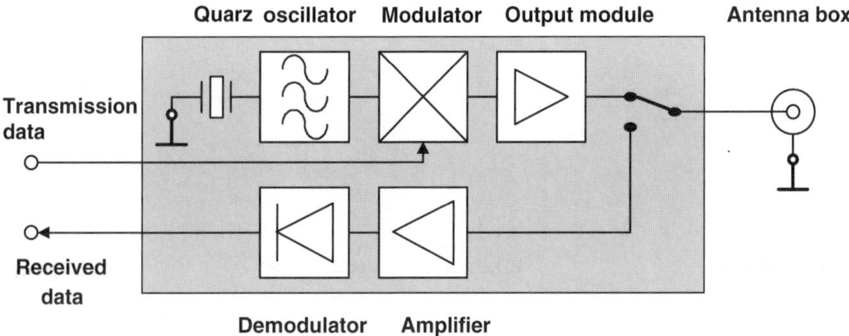

Figure 11.7: HF interface for a sequential reader system

The reader contains an instantaneous switching unit to switch between transmitter and receiver mode. This function is normally performed by PIN diodes in radio technology.

No special demands are made of the receiver in an SEQ system. Because the strong signal of the transmitter is not present to cause interference during reception, the SEQ receiver can be designed to maximise sensitivity. This means that the range of the system as a whole can be increased to correspond with the *energy range*, i.e. the distance between reader and transponder at which there is just enough energy for the operation of the transponder.

11.2.2 Control unit

The reader's control unit performs the following functions:

- Communication with the application software and the execution of commands from the application software;
- Control of the communication with a transponder (master-slave principle);
- Signal coding and decoding.

In more complex systems the following additional functions are available:

- Execution of an anticollision algorithm;
- Encryption and decryption of the data to be transferred between transponder and reader;
- Performance of authentication between transponder and reader.

The control unit is usually based upon a microprocessor to perform these complex functions. Cryptological procedures, such as stream ciphering between transponder and reader, and also signal coding, are often performed in an additional ASIC module to relieve the processor of calculation intensive processes. For performance reasons the ASIC is accessed via the microprocessor bus (register orientated).

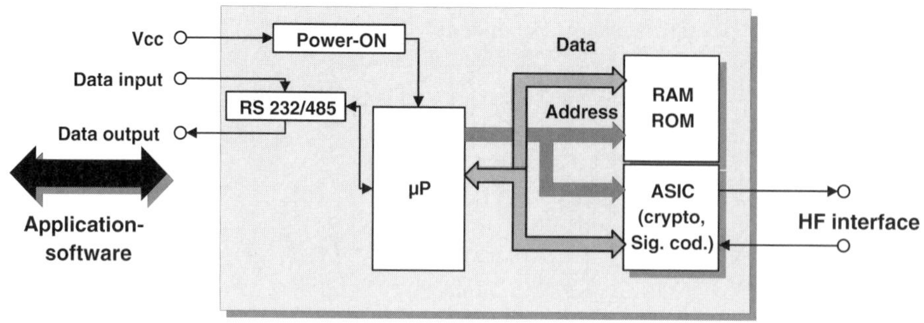

Figure 11.8: Block diagram of the control unit of a reader. There is a serial interface for communication with the higher application software

Data exchange between *application software* and the reader's control unit is performed by an RS232 or RS485 interface. As is normal in the PC world, NRZ coding (8 bit asynchronous) is used. The baud rate is normally a multiple of 1200 Bd (4800 Bd, 9600 Bd, etc.). Various, often self-defined, protocols are used for the communication protocol. Please refer to the handbook provided by your system supplier.

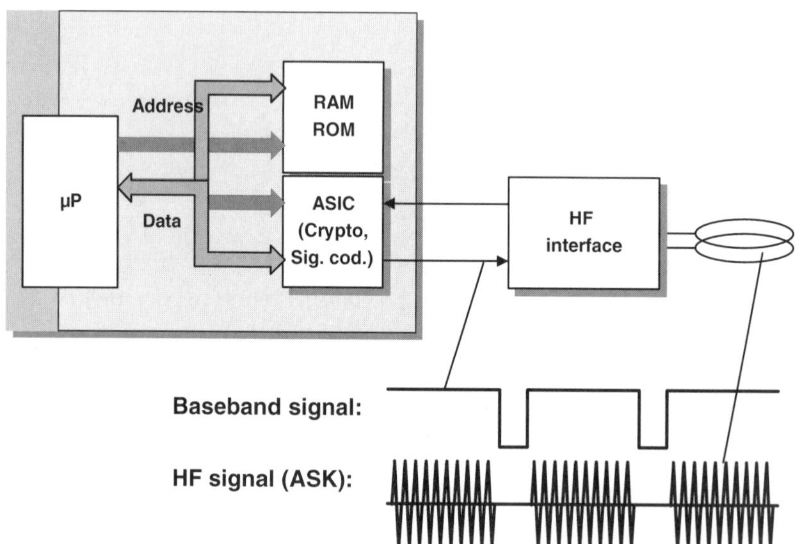

Figure 11.9: Signal coding and decoding is also performed by the control unit in the reader

The interface between the HF interface and the control unit represents the state of the HF interface as a binary number. In an ASK modulated system a logic "1" at the modulation input of the HF interface represents the state "HF signal on", a logic "0" represents the state "HF signal off" (further information in Chapter 10 "HF Interface").

11.3 Low Cost Configuration – Reader IC U2270B

It is typical of applications that use contactless identification systems that they require only a few readers, but a very large number of transponders. For example, in a public transport system, several tens of thousands of contactless smart cards are used, but only a few hundred readers are installed in vehicles. In applications such as animal identification or container identification, there is also a significant difference between the number of transponders used and the corresponding number of readers. There are also a great many different systems, because there are still no applicable standards for inductive or microwave RFID systems. As a result, readers are only ever manufactured in small batches of a few thousand.

Electronic *immobilisation* systems, on the other hand, require a vast number of readers. Because since 1995 almost all new cars have been fitted with electronic immobilisation systems as standard, the number of readers required has reached a completely new order of magnitude. Because the market for powered vehicles is also very price sensitive, cost reduction and miniaturisation by the integration of a small number of functional modules has become worth pursuing. Because of this, it is now possible to integrate the whole analogue section of a reader onto a silicon chip, meaning that only a few external components are required. We will briefly described the *U2270B* as an example of such a reader IC.

The reader IC U2270B by TEMIC serves as a fully integrated HF interface between a transponder and a microcontroller.

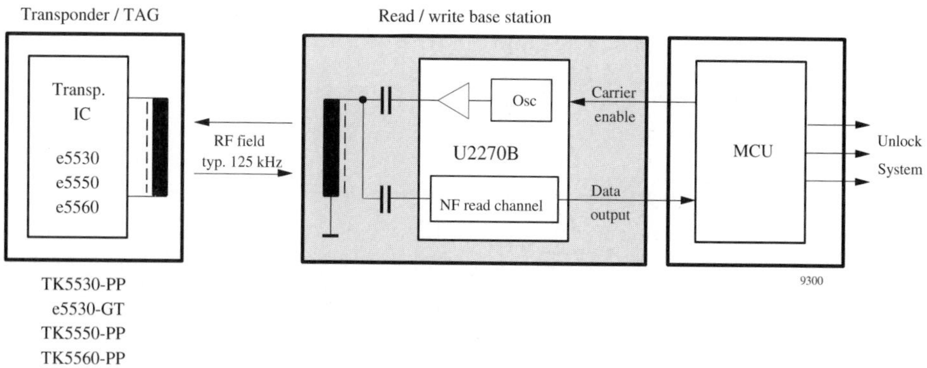

Figure 11.10: The low-cost reader IC U2270B represents a highly integrated HF interface. The control unit is realised in an external microprocessor (MCU), (Reproduced by permission of TEMIC Semiconductor GmbH, Heilbronn)

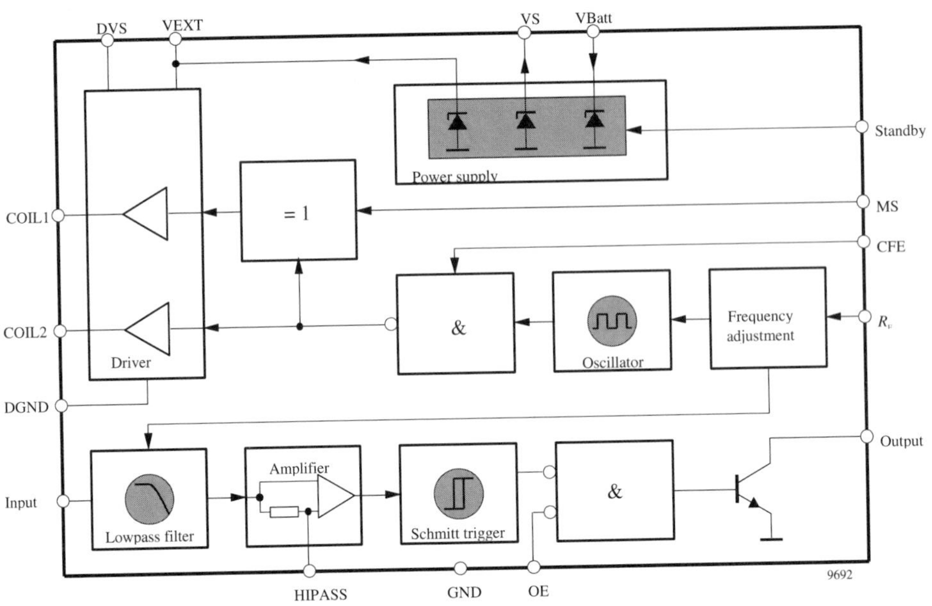

Figure 11.11: Block diagram of the reader IC U2270B. The transmitter arm consists of an oscillator and driver to supply the antenna coil. The receiver arm consists of filter, amplifier and a Schmitt trigger: (Reproduced by permission of TEMIC Semiconductor GmbH, Heilbronn)

The IC contains the following modules: on-chip oscillator, driver, received signal conditioning and an integral power supply.

The on-chip oscillator generates the operating frequency in the range 100 – 150 kHz. The precise frequency is adjusted by an external resistor at pin R_F. The downstream driver generates the power required to control the antenna coil as push-pull output. If necessary, a baseband modulation signal can be fed into pin CFE as a TTL signal and this switches the HF signal on/off, generating an ASK modulation.

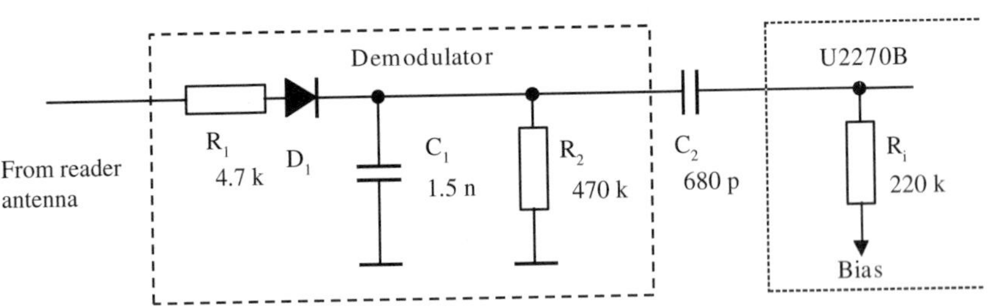

Figure 11.12: Rectification of the amplitude modulated voltage at the antenna coil of the reader. (Reproduced by permission of TEMIC Semiconductor GmbH, Heilbronn)

The *load modulation* procedure in the transponder generates a weak amplitude modulation of the reader's antenna voltage. The modulation in the transponder occurs in the baseband, i.e. without the use of a subcarrier. The transponder modulation signal can be reclaimed simply by demodulating the antenna voltage at the reader using a diode. The signal, which has been rectified by an external diode and smoothed using an RC low-pass filter is fed into the "Input" pin of the U2270B. Using a downstream Butterworth low-pass filter, an amplifier module and a Schmitt trigger, the demodulated signal is converted into a TTL signal, which can be evaluated by the downstream microprocessor. The time constants of the Butterworth filter are designed so that a Manchester or biphase code can be processed up to a data rate of f_{osc} /25 (approximately 4800 bit/s) [temic].

A complete application circuit for the U2270B can be found in the following chapter.

11.4 Connection of Antennas

11.4.1 Antennas for inductive systems

Reader antennas in inductively coupled RFID systems generate magnetic flux Φ, which is used for the power supply of the transponder and for sending messages between the reader and the transponder. This gives rise to two fundamental design requirements for a reader antenna:

- Maximum current i_1 in the *antenna coil*, for maximum magnetic flux Φ;
- Power matching so that the maximum available energy can be used for the generation of the magnetic flux;
- Sufficient bandwidth for the undistorted transmission of a carrier signal modulated with data.

Depending upon the frequency range, different procedures can be used to connect the antenna coil to the transmitter output of the reader, direct connection of the antenna coil to the power output module using power matching or the supply of the antenna coil via coaxial cable.

11.4.1.1 Connection using current matching

In typical low cost readers in the frequency range below 135 kHz, the HF interface and antenna coil are mounted close together (a few centimetres apart), often on a single printed circuit board. Because the geometrical dimensions of the antenna supply line and antenna are smaller than the wavelength of the generated HF current (2200 m) by powers of ten, the signals may be treated as stationary for simplification. This means that the wave characteristics of a high frequency current may be disregarded. The connection of an antenna coil is thus comparable to the connection of a loudspeaker to an NF output module from the point of view of circuitry.

The reader IC U2270B, which was described in the preceding section, can serve as an example of such a low cost reader.

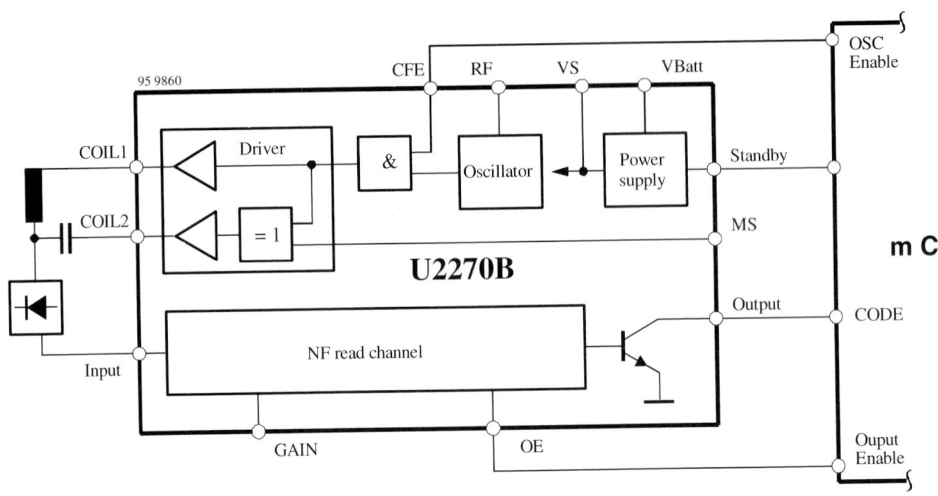

Figure 11.13: Block diagram for the reader IC U2270B with connected antenna coil at the push-pull output. (Reproduced by permission of TEMIC Semiconductor GmbH, Heilbronn)

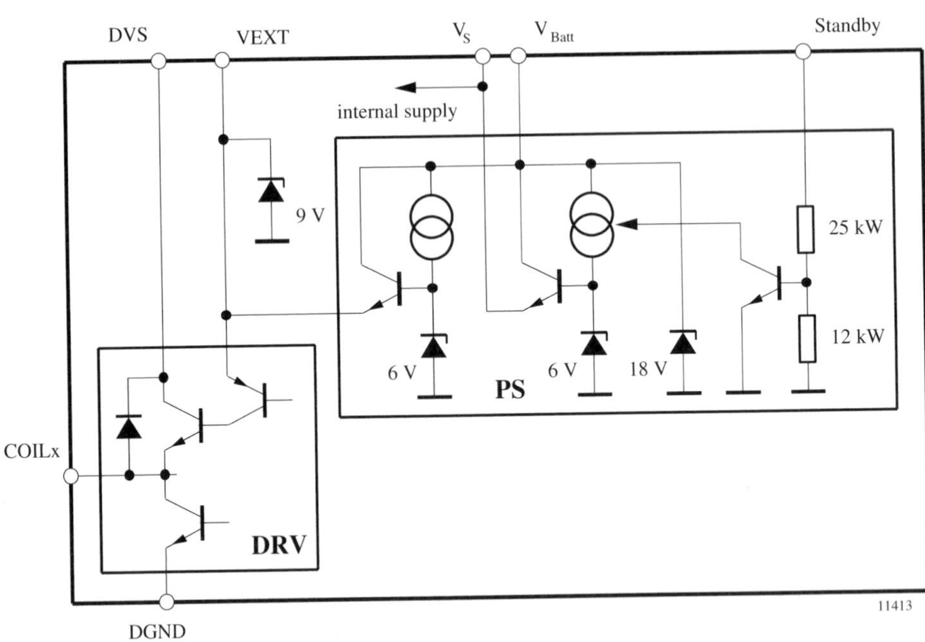

Figure 11.14: Driver circuit in the reader IC UU2270B. (Reproduced by permission of TEMIC Semiconductor GmbH, Heilbronn)

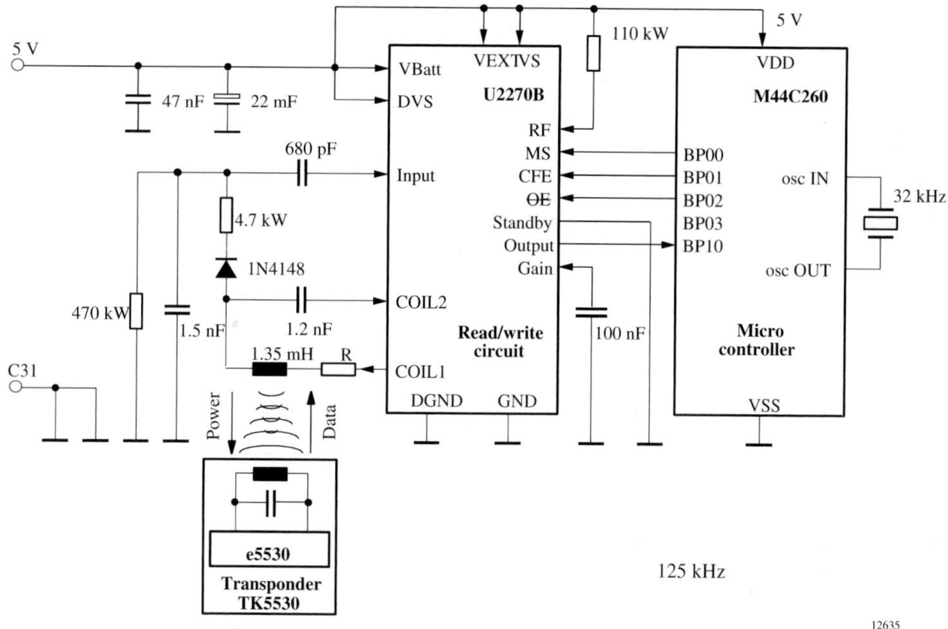

Figure 11.15: Complete example application for the low cost reader IC U2270B. (Reproduced by permission of TEMIC Semiconductor GmbH, Heilbronn)

Figure 11.13 shows an example of an antenna circuit. The antenna is fed by the push-pull bridge output of the reader IC. In order to maximise the current through the antenna coil, a *serial resonant circuit* is created by the serial connection of the antenna coil L_S to a capacitor C_S and a resistor R_S. Coil and capacitor are dimensioned such that the resonant frequency f_0 is as follows at the operating frequency of the reader:

$$f_0 = \frac{1}{2\pi\sqrt{L_S \cdot C_S}} \tag{11.3}$$

The coil current is then determined exclusively by the series resistor R_S.

11.4.1.2 Supply via coaxial cable

At frequencies above 1 MHz, or in the frequency range 135 kHz if longer cables are used, the HF voltage can no longer be considered stationary, but must be treated as an *electromagnetic wave* in the cable. Connecting the antenna coil using a long, unshielded two core wire in the HF range would therefore lead to undesired effects, such as power reflections, impedance transformation and parasitic power emissions, due to the wave nature of a HF voltage. Because these effects are difficult to control when they are not exploited intentionally, shielded cable – so-called *coaxial cable* – is normally used in radio technology. Sockets, plugs and coaxial cable are uniformly designed for a cable impedance of 50 Ω and, being a mass produced product, are correspondingly cheap. RFID systems generally use 50 Ω components.

The block diagram of an inductively coupled RFID system using 50 Ω technology shows the most important HF components:

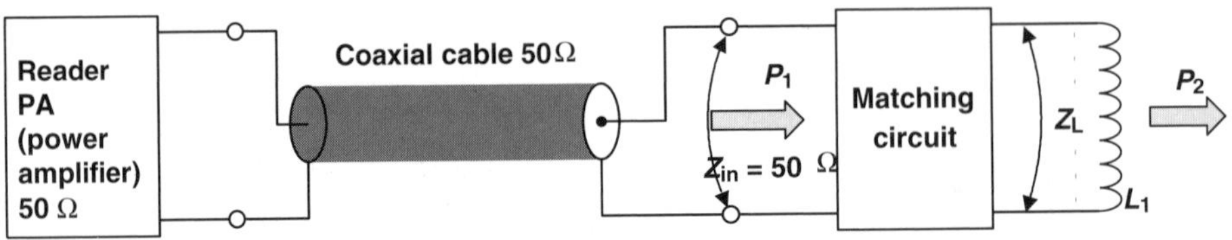

Figure 11.16: Connection of an antenna coil using 50 Ω technology

The antenna coil L_1 represents an impedance Z_L in the operating frequency range of the RFID system. To achieve power matching with the 50 Ω system, this impedance must be transformed to 50 Ω (matched) by a passive *matching circuit*. Power transmission from the reader output module to the matching circuit is achieved (almost) without losses or undesired radiation by means of a coaxial cable.

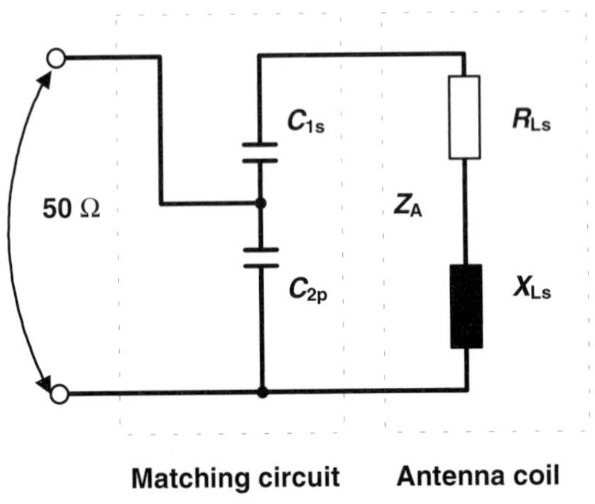

Figure 11.17: Simple matching circuit for an antenna coil

A suitable matching circuit can be realised using just a few components. The circuit illustrated in Figure 11.17, which can be constructed using just two capacitors, is very simple to design [suckrow]. This circuit is used in practice in various 13.56 MHz RFID systems.

Figure 11.18: Reader with integral antenna and matching circuit. (MIFARE®-reader, reproduced by permission of Philips Electronics N.V.)

Figure 11.18 shows a reader with an integral antenna for a 13.56 MHz system. Coaxial cable has not been used here, because a very short supply line can be realised by a suitable layout (stripline). The matching circuit is clearly visible on the inside of the antenna coil (SMD component).

Before we can dimension the circuit, we first need to determine the impedance Z_A of the antenna coil for the operating frequency by measurement. It is clear that the impedance of a real antenna coil is generated by the serial connection of the coil inductance L_S with the ohmic wire resistance RL_S of the wire. The serial connection from XL_S and RL_S can also be represented in the impedance level.

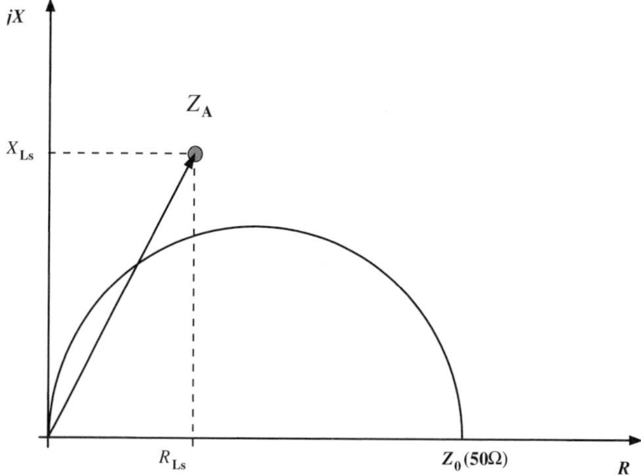

Figure 11.19: Representation of Z_A in the impedance level (Z plane)

The function of the matching circuit is the transformation of the complex coil impedance Z_A to a value of 50 Ω real. A reactance (capacitance, inductance) in series with the coil impedance Z_A shifts the total impedance Z in the direction of the jX axis, while a parallel reactance shifts the total impedance away from the origin in a circular path.

The values of C_{2p} and C_{2s} are dimensioned such that the resulting coil impedance Z_A is transformed to the values desired to achieve 50 Ω.

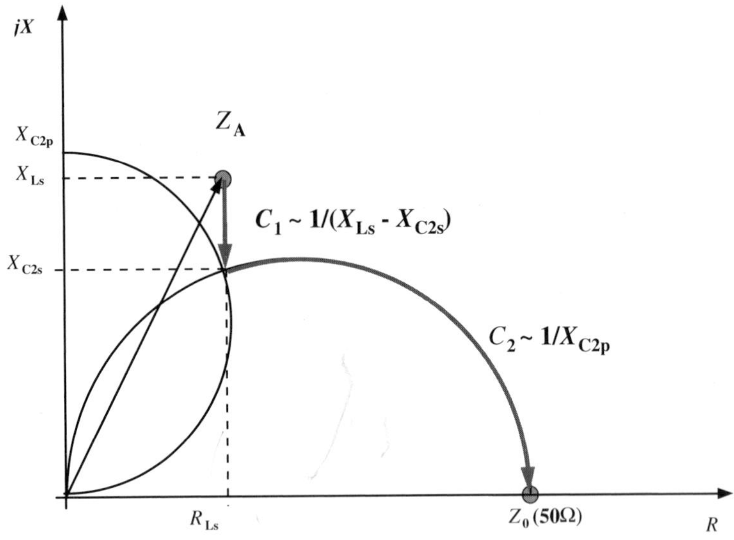

Figure 11.20: Transformation path with C_{ls} and C_{2p}

The matching circuit from Figure 11.17 can be mathematically represented by equation 11.4:

$$Z_0 = 50\Omega = \frac{1}{-j\omega C_{2p} + \left(\frac{1}{\frac{1}{-j\omega C_{1s}} + R_{Ls} + j\omega L_s} \right)} \tag{11.4}$$

From the relationship between resistance and conductance in the complex impedance plane (Z-level), we find the following relationship for C_{2p}:

$$C_{2p} = \sqrt{\frac{Z_0 \cdot R_{Ls} - R_{Ls}^2}{\omega\, Z_0 R_{LS}}} \tag{11.5}$$

As was clear from the impedance plane in Figure 11.20, C_{2p} is determined exclusively by the serial resistance R_{ls} of the antenna coil. For a serial resistance R_{LS} of precisely 50 Ω, C_{2p}

can be dispensed with altogether, however greater values for R_{ls} are not permissible, otherwise a different matching circuit should be selected (see [fricke].

We further find for C_{ls}:

$$C_{1s} = \frac{1}{\omega^2 \cdot \left(L_s - \frac{\sqrt{Z_0 R_{LS} - R_{LS}^2}}{\omega} \right)} \tag{11.6}$$

The antenna current i_{LS} is of interest in this context, because this allows us to calculate the magnetic field strength H that is generated by the antenna coil (see Chapter 4 – "Physical Principles of RFID Systems").

To clarify the relationships, let us now modify the matching circuit from Figure 11.17 slightly:

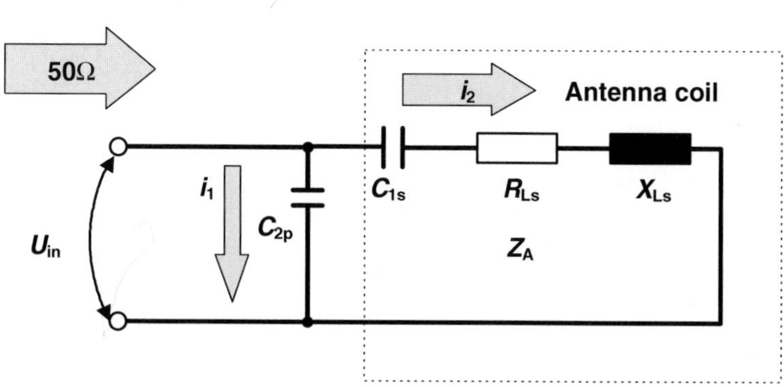

Figure 11.21: The matching circuit represented as a current divider

The input impedance of the circuit at operating frequency is precisely 50 Ω. For this case, and only for this case(!), the voltage at the input of the matching circuit is very simple to calculate. Given a known transmitter output power P and known input impedance Z_0, the following is true: $P=U^2/Z_0$. The voltage calculated from this equation is the voltage at C_{2p} and the serial connection of C_{ls}, R_{ls} and X_{LS}, and is thus known. The antenna current i_2 can be calculated using the following equation:

$$i_2 = \frac{\sqrt{P \cdot Z_0}}{R_{LS} + j\omega L_s - j\frac{1}{\omega C_{1s}}} \tag{11.7}$$

11.4.1.3 The Influence of the Q Factor

A reader antenna for an inductively coupled RFID system is characterised by its resonant frequency and by its *Q factor*. A high Q factor leads to high current in the antenna coil and thus improves the power transmission to the transponder. In contrast, the transmission bandwidth of the antenna is inversely proportional to the Q factor. A low bandwidth, caused by an excessively high Q factor, can therefore significantly reduce the modulation sideband received from the transponder.

The Q factor of an inductive reader antenna can be calculated from the ratio of the inductive coil resistance to the ohmic loss resistance and/or series resistance of the coil:

$$Q = \frac{2\pi \cdot f_0 \cdot L_{\text{coil}}}{R_{\text{total}}} \tag{11.8}$$

The bandwidth of the antenna can be simply calculated from the Q factor:

$$B = \frac{f_0}{Q} \tag{11.9}$$

The required bandwidth is derived from the bandwidth of the modulation sidebands of the reader and the load modulation products (if no other procedure is used). As a rule of thumb, the following can be taken as the bandwidth of an ASK modulated system.

$$B \cdot T = 1 \tag{11.10}$$

where T is the turn-on-time of the carrier signal, where modulation is used.

For many systems, the optimal Q factor is 10 – 30. However, it is impossible to generalise here because, as already mentioned, the Q factor depends upon the required bandwidth and thus upon the modulation procedure used (e.g. coding, modulation, subcarrier frequency).

11.4.2 Antennas for microwave systems

One commonly used antenna design for microwave systems is the so-called *planar antenna* based upon stripline technology. These antennas are characterised primarily by the low constructional height and robust construction. Planar antennas can be cheaply manufactured by printed circuit board etching technology, which achieves high levels of reproducibility. The main component, the transmitter, comprises rectangular *microstrip lines*, of length $l=\lambda/2$. Transmitter elements can be combined in a number of different ways to achieve different directivities and polarisations.

The supply of the transmitter element can be performed using stripline connections to suitable points, by wires through the baseplate (Figure 11.22) or by capacitive coupling via an intermediate layer. An input impedance of 50 Ω (or any other desired value) can be achieved by the selection of the supply point [meinke].

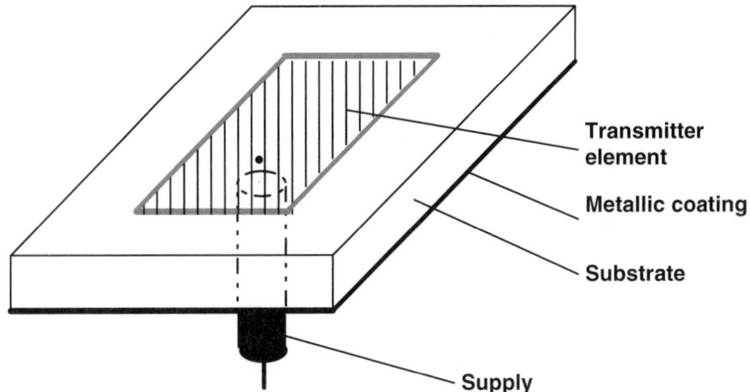

Figure 11.22: Supply of a $\lambda/2$ transmitter unit comprising a planar element with a feed line on the back

11.5 Reader Designs

Different types and designs of readers are available for different applications. Readers can be generally classified into OEM readers, readers for industrial or portable use and numerous special designs.

OEM Readers *OEM readers* are available for integration into customers' own data capture systems, BDE terminals, access control systems, till systems, robots, etc. OEM readers are supplied in a shielded tin housing or as an unhoused board. Electrical connections are in the form of soldered, plug and socket or screw-on terminals.

Figure 11.23: Example of an OEM reader for use in terminals or robots (photo: Long-Range/High-Speed Reader LHR1, reproduced by permission of SCEMTEC Transponder Technology GmbH, Reichshof-Wehnrath)

Typical Technical Data:

Supply voltage:	typically 12 V
Antenna:	external
Antenna connection:	BNC box, terminal screw or soldered connection
Communication interface:	RS232, RS485
Communication protocol:	X-ON/X-OFF, 3964, ASCII
Environmental temperature:	0 – 50°C

12

The Manufacture of Transponders and Contactless Smart Cards

A transponder is made up of two components: the electronic data carrier and the housing. Figure 12.1 gives a simplified representation of the manufacturing process for an inductively coupled transponder.

12.1 Module Manufacture

In accordance with the normal semiconductor manufacturing procedure, the *microchip* is produced on a so-called *wafer*. This is a slice of silicon, which may be 6 inches in diameter, upon which several hundred microchips are produced simultaneously by repeated doping, exposure, etching and washing of the surface.

In the next stage of production, the microchips on the wafer are contacted using metal points and then each of the chips is individually tested for functionality. The chips have additional contact fields for this purpose, which give direct access – i.e. without going through the HF interface – to the chip's memory and security electronics. The chips are placed in so-called *test mode* during this procedure, which permits unlimited direct access to all functional groups upon the chip. The functional test can therefore be performed significantly more intensively and comprehensively than would be possible later on, when communication can only taken place via the contactless technology.

All defective chips are marked with a red ink dot at this stage, so that they can be identified and separated out in the subsequent stages of production. The test mode can also be used to programme a unique *serial number* into the chip, if the chip has an EEPROM. In read only transponders, the serial number is programmed by cutting through predefined connecting lines on the chip using a laser beam.

After the successful completion of the test programme the test mode is deactivated by permanently breaking certain connections (so-called fuses) on the chip by a strong current surge. This stage is important to prevent unauthorised reading of data at a later date by the manipulation of the test contacts on the chip.

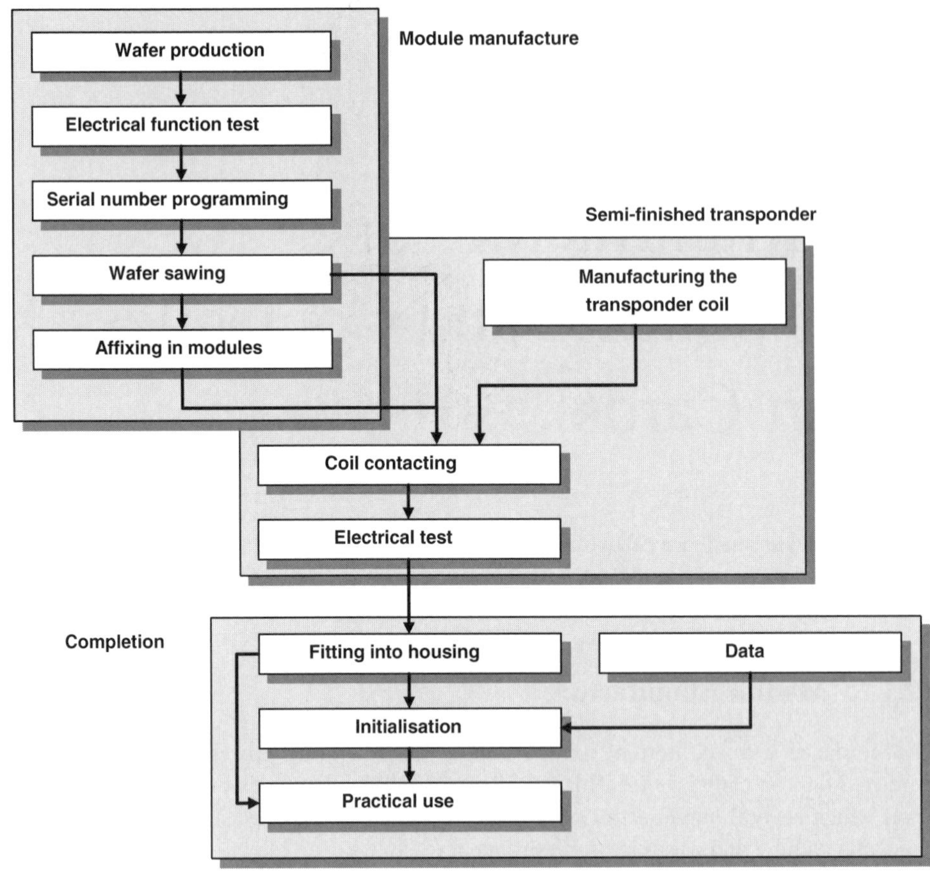

Figure 12.1: Transponder manufacture

After the chips have been tested the wafer is sawn up using a diamond saw to give individual transponder chips. A single chip in this state is known as a *die* (plural: dice). A plastic foil is attached to the reverse of the wafer prior to the sawing operation to prevent the dice from disintegrating (*saw on foil*).

After the sawing operation the dice can be removed from the plastic foil individually and fitted into a module. The connection to the contact surfaces of the module for the transponder coil is by bonding onto the reverse of the connection surfaces. Finally the dice are extrusion coated with a moulding substance. This significantly increases the stability of the brittle and extremely breakable silicon dice. Very small dice, such as those for read only transponders (area of die: 1 – 2 mm^2) are not fitted into a module for reasons of space and cost.

Figure 12.2: Size comparison of a sawn die with a cereal grain. The size of a transponder chip varies between $1 mm^2$ and $15 mm^2$ depending upon its function (photo: HITAG® Multimode-Chip, reproduced by permission of Philips Electronics N.V.)

12.2 Semi-Finished Transponder

In the next stage, the *transponder coil* is produced using an automatic winding machine. The copper wire used is given a coating of low-melting point *baked enamel* in addition to the normal insulating paint. The winding tool is heated to the melting point of the baked enamel during the winding operation. The enamel melts during winding and hardens rapidly when the coil has been removed from the winding tool, causing the individual windings of the transponder coil to stick together. This guarantees the mechanical stability of the transponder coil during the following stages of assembly.

Immediately after the winding of the transponder coil, the coil connections are welded to the contact surface of the transponder module using a spot welding machine. The shape and size of the transponder coil are determined by the format of the finished transponder.

In dice that are not immediately fitted into a module, the copper wire can be bonded directly to the die using a suitable procedure. However, this requires that the wire of the transponder coil is as thin as possible. For this reason, the transponder coil of a glass transponder is wound from wire that is only 30 µm thick.

Once the transponder coil has been contacted, the transponder is electrically functional. Therefore a contactless functional test is carried out at this stage to sort out those transponders that have been damaged during preceding stages. Transponders that have not yet been fitted into housings are called semi-finished transponders, as they can go from this stage into different housing formats.

12.3 Completion

In the next stage, the semi-finished transponder is inserted into a housing. This may take place by injection moulding (e.g. in ABS), casting, pasting up, insertion in a glass cylinder, or other procedures.

After a further functional test, the application data and/or application key can be loaded into the transponder, if required.

12.4 Contactless Smart Cards

Contactless smart cards represent a very common type of transponder. DIN/ISO 7810 specifies the smart card format. The dimensions of a smart card are specified as 85.46 mm × 53.92 mm × 0.76 mm (± tolerances). Particularly significant for the manufacture of *contactless smart cards* is the required thickness of just 0.76 mm, because this places strict limits on the possible dimensions of the transponder coil and chip module.

A contactless smart card may be manufactured from four PVC foils of around 0.2 mm thickness: The two *inlet foils* that are inserted in the inside of the card and two *overlay foils* that form the outside of the cards to be created. Contactless smart cards are produced in sheets of 21, 24 or 48. The foils used thus have an area of around 0.1 to 0.3 m^2.

In the first stage of the operation the overlay foils are printed with the layout of the smart cards. Modern printing machines can produce a high quality colour print, as we know from telephone smart cards. At the same time, the semi-finished transponders (coil contacted to the transponder chip) are placed upon the inlet foils and affixed by cemented joints. There are several possible procedures for the production of the coil – e.g. winding, embedding, etching (like the manufacture of printed circuit boards for electronic circuits) and screen printing (with conductive printing ink).

Winding In the winding technique, the transponder coil is wound by a winding tool in the normal way and affixed using baked enamel. The welding chip module has been welded onto the antenna, the semi-finished transponder is placed on the inlet sheet and mechanically affixed with the aid of cemented joints.

Embedding Inlet manufacture using the embedding technique is a relatively new procedure, that is nevertheless increasing significantly in importance. In this technique, the chip module is first affixed in its intended location on a PVC foil. The wire is embedded into the foil using a sonotrode. The *sonotrode* consists of an ultrasonic emitter with a passage in its head through which the wire is guided onto the foil. The ultrasound emitter is used to locally heat the wire so that this melts into the foil and thus is fixed in form and position. The sonotrode is moved across the inlet foil like in an X-Y plotter, whilst the wire is fed through, so that the transponder can be “drawn” or laid . At the start and the end of the coil a spot welding machine is used to make the electrical connection to the transponder module.

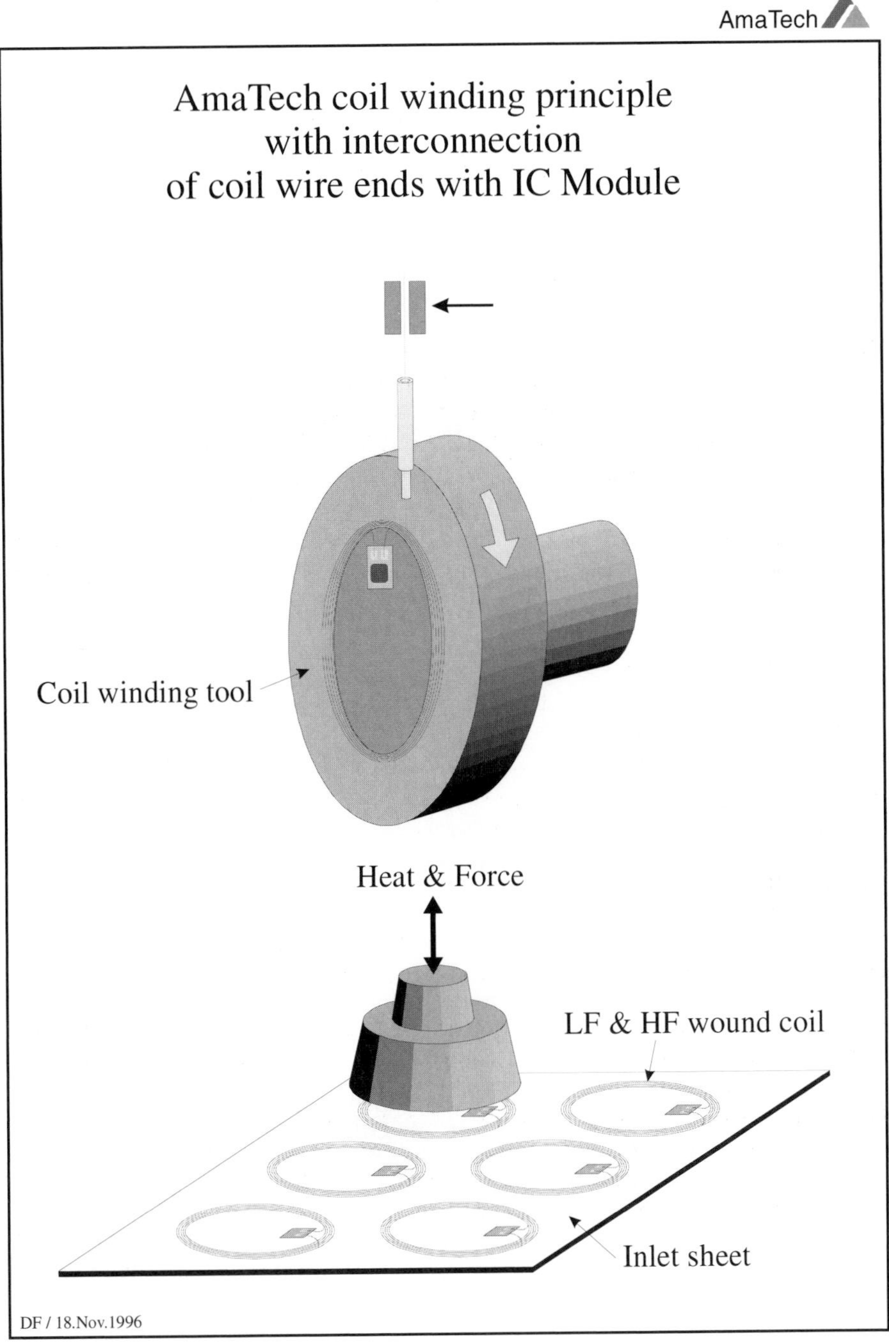

Figure 12.3: Production of a semi-finished transponder by winding and placing the semi-finished transponder on an inlet sheet (Reproduced by permission of AmaTech GmbH & Co. KG, Pfronten)

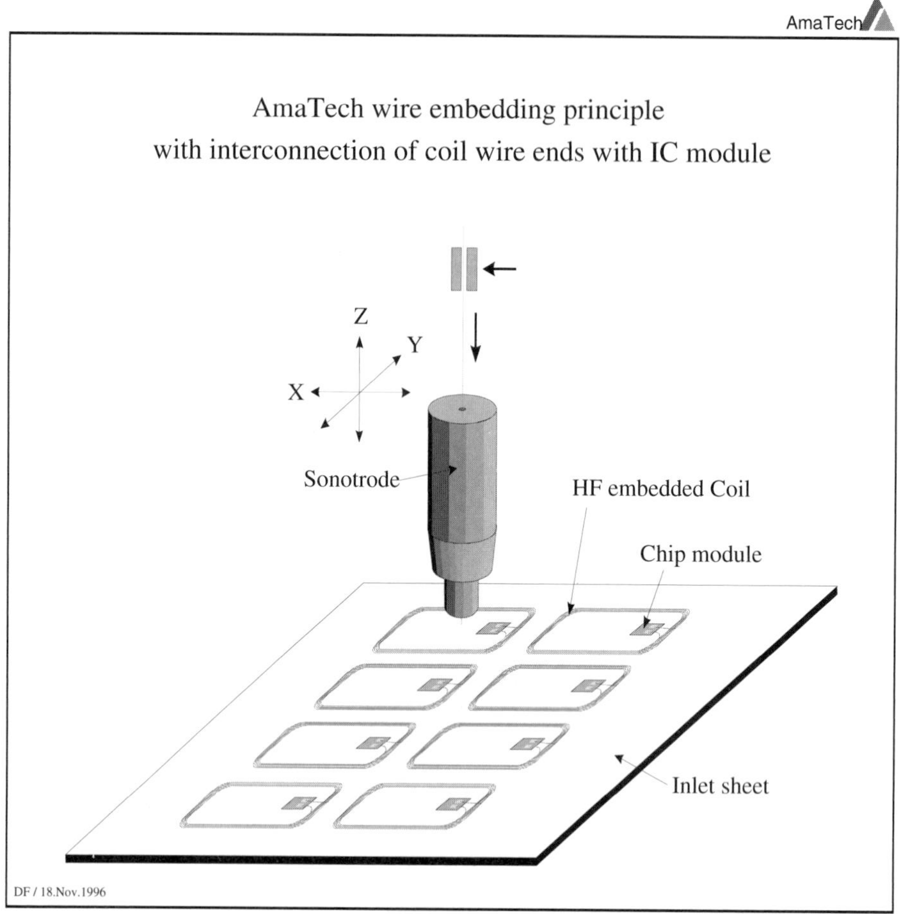

Figure 12.4: Manufacture of an inlet sheet using the embedding principle. (Reproduced by permission of AmaTech GmbH & Co. KG, Pfronten)

Lamination In the next step, the overlay and inlet foils are assembled and joined together with precision. Finally, the foils are placed in a laminating machine. By the conduction of heat, the foils are brought into a soft elastic state at high pressure (approximately 100 – 150 °C). This "bakes" the four sheets to create a permanent bond.

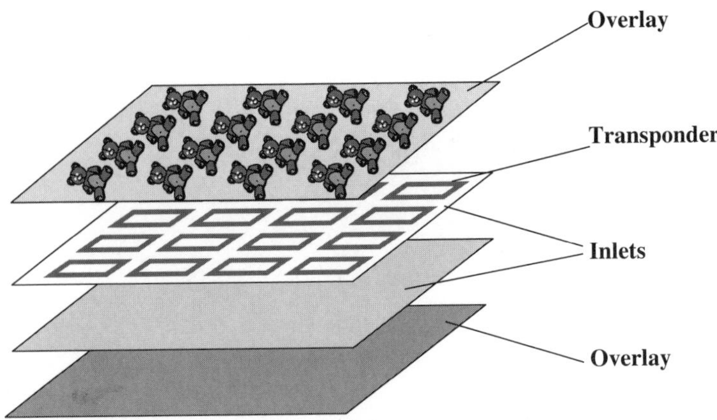

Figure 12.5: Typical construction of a contactless smart card from four PVC sheets, the inlet and overlay foils

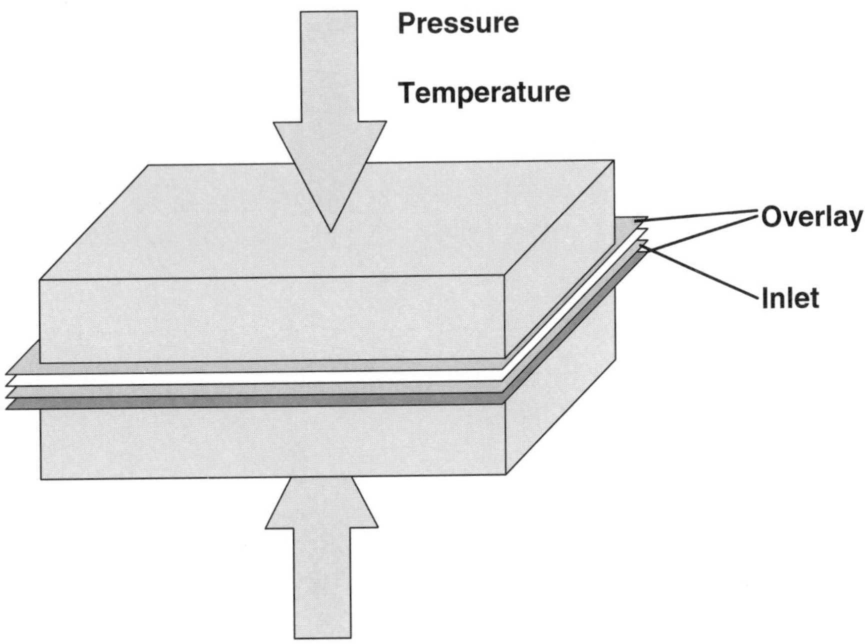

Figure 12.6: During the lamination procedure the PVC sheets are melted at high pressure and temperatures up to 150 °C

After the *lamination* and cooling of the laminated PVC foils, the individual smart cards are stamped out of the multi-purpose sheet. A subsequent functional test ensures the quality of the cards before these can be sent to the customer.

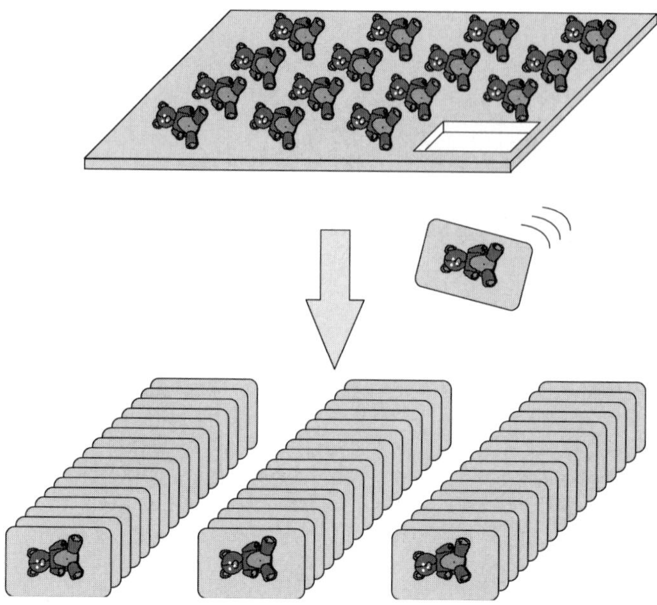

Figure 12.7: After the cooling of the PVC sheets the individual cards are stamped out of the multi-purpose sheets

13

Example Applications

13.1 Contactless Smart cards

The first plastic cards appeared in the USA as early as the beginning of the 1950s, when cheap PVC replaced cardboard. In the years that followed, plastic credit cards became widespread. Incidentally, the first credit card was issued by Diners Club in 1950.

The rapid development of semiconductor technology made it possible to integrate data memory and protective logic onto a single silicon chip in the 1970s. The idea of incorporating such an integrated memory chip into an identification card was patented in 1968 by Jürgen Dethloff and Helmut Grötrupp in Germany. However, it was not until almost 15 years later that the great breakthrough was achieved with the introduction of the telephone smart card by the French company PTT. Several millions of telephone smart cards were in circulation in France in 1986 [rankl]. These first generation smart cards were memory cards with contacts. A significant improvement was achieved when entire microprocessors were successfully integrated into a silicon chip, and these chips incorporated into an identification card. This made it possible to run software in a smart card, thus opening up the possibility of realising high-security applications. Thus, smart cards for mobile telephones and the new bank cards (EC with chip) were realised exclusively using microprocessor cards.

Since the mid 1980s, repeated attempts have been made to launch contactless smart cards onto the market. The operating frequency of 135 kHz that was normal at the time and the high power consumption of the silicon chips on the market necessitated transponder coils with several hundred windings. The resulting large coil cross section, and the additional capacitors that were often required, impeded manufacture in the form of ID-1 format plastic cards, and transponders were usually cast into inconvenient plastic shells. Due to this limitation, contactless smart cards played a minor role in the smart card market for a long time.

In the first half of the 1990s, transponder systems were developed with an operating frequency of 13.56 MHz. The transponders required for these systems required just 5 windings. For the first time it was possible to produce transponder systems in the 0.76 mm thick ID-1 format. The great breakthrough in Germany occurred in 1995, with the introduction of the "Frequent Traveller" contactless customer loyalty card in ID-1 format by the German company Lufthansa AG. It was noteworthy that these cards, manufactured

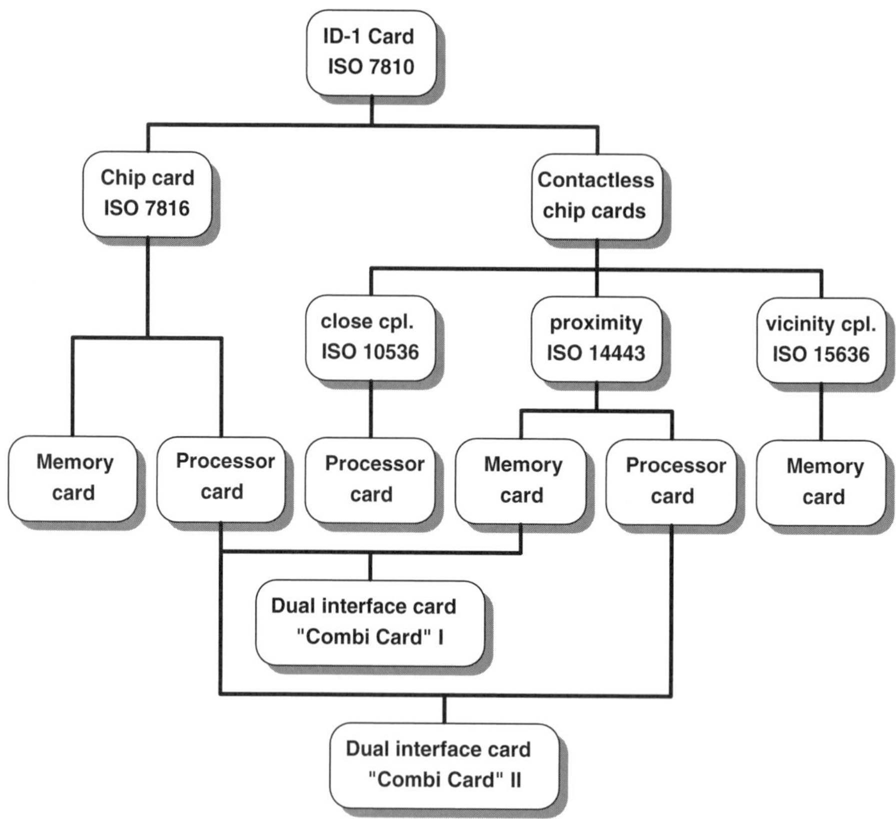

Figure 13.1: The large "family" of smart cards, including the relevant ISO standard

by the Munich company Giesecke & Devrient, still had a magnetic strip, a hologram and were embossed with the customer number and name. A more in-depth description of this project is included in Section 13.3 "Ticketing".

Today, contactless smart cards are divided into three groups based upon the applicable standards: close coupling, remote coupling (inductively coupled) and vicinity coupled (inductively coupled) smart cards. Whilst vicinity coupling cards are only available in the form of memory cards, microprocessor cards have been available in the form of inductively coupled cards in small pilot projects since 1997.

Currently, the main fields of application for contactless smart cards are payment systems (public transport, ticketing) or passes (ID cards, company pass). In the long term we can expect that contactless smart cards will largely replace cards with contacts in their classical fields of application (telephone cards, EC cards). In addition, contactless technology will allow smart cards to be used in completely new fields – fields we may not yet have even thought of.

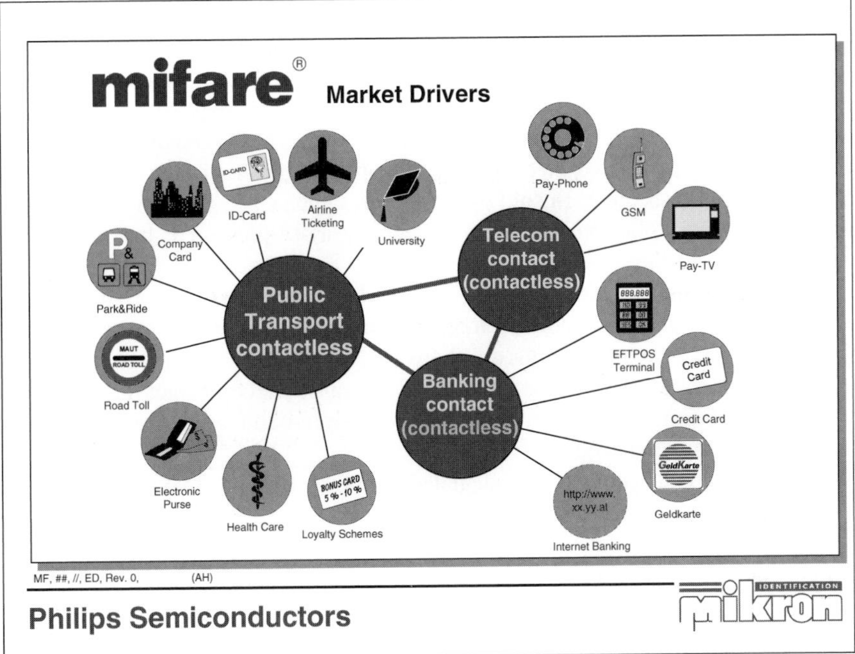

Figure 13.2: The main fields of application for contactless smart cards are public transport and change systems for telephone boxes or consumer goods (groceries, cigarettes). (Reproduced by permission of Philips Electronics N.V.)

13.2 Public Transport

Public transport is one of the applications where the greatest potential exists for the use of RFID systems, particularly contactless smart cards. In Europe and the USA traffic associations are still operating at a huge loss, sometimes as much as 40% of turnover [czako], which must be made up by subsidies from the community and country in question. Due to the increasing shortage of resources, long term solutions must be sought that will cut these losses by reducing costs and increasing income. The use of contactless smart cards as electronic travel passes could make an important contribution to improving the situation (AFC = *automatic fare collection*). In the field of fare management in particular there is a great deal of room for improvement.

13.2.1 The starting point

The unhealthy financial situation of transport companies naturally has many different causes. However, the following factors are worth mentioning in connection with electronic travel passes:

- Transport companies incur high costs through the sale of travel passes by automatic dispensers. For example, the sale of a travel pass through an automatic dispenser in Zürich costs Sfr 0.45, where the average sales price is Sfr. 2.80 [czako]. Sixteen % of the sales price is lost from the outset by the provision of the dispenser, maintenance and repairs alone (filling with notes and coins, repairs, damage by vandalism).
- In vehicles, too, expensive electronic ticket printers or mobile devices are required. Sometimes the tickets are even sold by the driver, which causes long waiting times while passengers board, plus the additional security risk presented by the continuous distraction of the driver.
- Paper tickets are thrown away after use, although the manufacture of fraud-proof tickets for transport companies is becoming more and more expensive.
- In German cities in particular, losses of up to 25% must be taken into account due to fare-dodgers [czako]. This is because German transport companies have very liberal travelling conditions and permit entry to the underground system and buses without travel passes first being checked.
- Association discounts can only be calculated on the basis of costly random counts, which leads to imprecision in the calculation.

13.2.2 Requirements

Electronic fare management systems have to fulfil very high expectations and requirements, particularly with regard to resistance to degradation and wear, write and read speed and ease of use. These expectations can only be satisfactorily fulfilled by RFID systems. The most common format for contactless smart cards is the ID1 format and, recently, wrist watches.

13.2.2.1 Transaction time

The time taken for the purchase or verification of a travel pass is particularly critical in transport systems in which the pass can only be checked inside the vehicle. This is a particular problem in buses and trams. In the underground railway, passes can be checked at a turnstile, or by conductors. A comparison of different methods shows the clear superiority of RFID systems in terms of transaction times.

Table 13.1: Passenger processing times for different technologies. Source: transport companies in Helsinki, taken from [czako]

Technology	Passenger processing time
RFID I (remote coupling)	1.7 sec
Visual verification by driver	2.0 sec
RFID II (close coupling)	2.5 sec
Smart card with contacts	3.5 sec
Cash	>6 sec

13.2.2.2 Resistance to degradation, lifetime, convenience

Contactless smart cards are designed for a lifetime of 10 years. Rain, cold, dirt and dust are a problem for neither the smart card nor the reader.

Contactless smart cards can be kept in a briefcase or handbag and are therefore extremely convenient to use. Transponders can also be fitted into wristwatches.

13.2.3 Benefits of RFID systems

The replacement of conventional paper tickets by a modern electronic fare management system based on contactless smart cards provides a multitude of benefits to all those involved. Although the purchase costs of a contactless smart card system are still higher than those of a conventional system, the investment should repay itself within a short period. The superiority of contactless systems is demonstrated by the following benefits for users and operators of public transport companies.

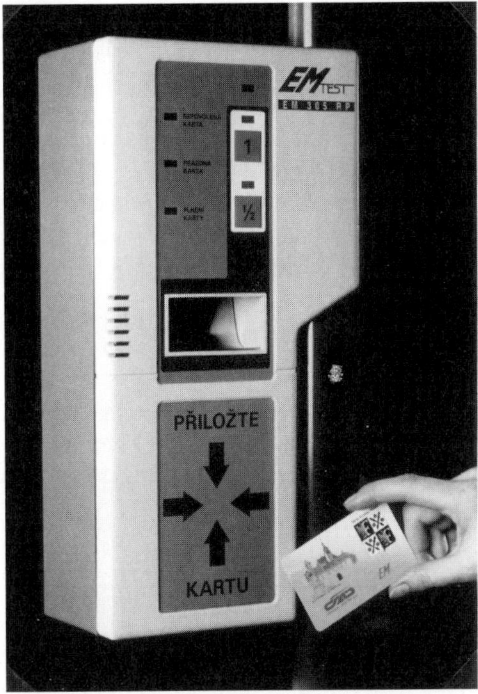

Figure 13.3: Contactless reader in a public transport system. (Photo: Frydek-Mistek project, Czechoslovakia, source: Reproduced by permission of EM Test)

Benefits for passengers

- Cash is no longer necessary, contactless smart cards can be loaded with large amounts of money, passengers no longer need to carry the correct change.
- Prepaid contactless smart cards remain valid even if fares are changed.
- The passenger no longer needs to know the precise fare, the system automatically deducts the correct fare from the card.
- Monthly tickets can begin on any day in the month. The period of validity begins after the first deduction from the contactless card.

Benefits for the driver

- Passes are no longer sold, resulting in less distraction of driving staff.
- No cash in vehicle.
- Elimination of the daily income calculation.

Benefits for the transport company

- Reduction in operating and maintenance costs of sales dispensers and ticket devaluers.
- Very secure against vandalism (chewing gum effect).
- It is easy to change fares, no new tickets need to be printed.
- The introduction of a closed (electronic) system, in which all passengers must produce a valid travel pass, can significantly reduce the number of fare dodgers.

Benefits for the transport association

- It is possible to calculate the performance of individual partners in the association. Because precise data is obtained automatically in electronic fare management systems, the discount for the association can be calculated using precise figures.
- Expressive statistical data is obtained.

Benefits for the treasury

- Reduction of the need for subsidies due to cost reductions.
- Better use of public transport due to the improved service has a positive effect on takings and on the environment.

13.2.4 Fare systems using electronic payment

Transport association regions are often divided into different fare zones and payment zones. There are also different types of travel pass, time zones and numerous possible combinations. The calculation of the fare can therefore be extremely complicated in conventional payment systems and can even be a source of bewilderment to local customers.

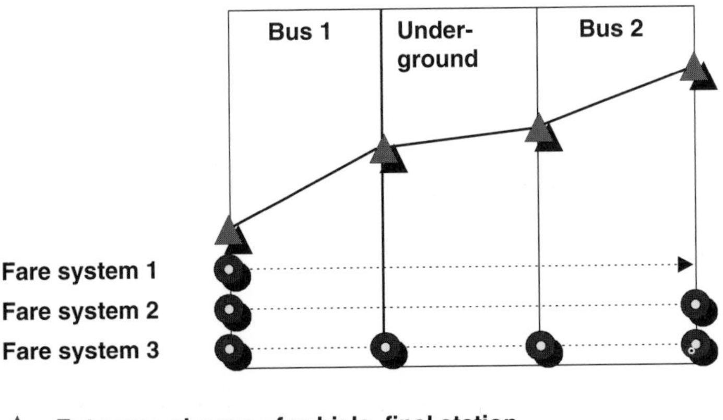

Figure 13.4: Use of the different tariff systems in a journey by public transport. The journey shown involves two changes between the underground and bus network. The number of times the smart card is read depends upon the fare system used

Electronic fare management systems, on the other hand, facilitate the use of completely new procedures for the calculation and payment of fares. There are three basic models for electronic fare calculation:

Fare system 1 Payment takes place at the beginning of the journey. A fixed amount is deducted from the contactless smart card, regardless of the distance travelled.

Fare system 2 At the beginning of the journey the entry point (check-in) is recorded on the contactless card. Upon disembarking at the final station (check-out), the fare for the distance travelled is automatically calculated and deducted from the card. In addition, the card can be checked at each change-over point for the existence of a valid "check-in" entry. To foil attempts at manipulation, the lack of a "check-out" record can be penalised by the deduction of the maximum fare at the beginning of the next journey.

Fare system 3 This model best suited for interlinked networks, in which the same route can be travelled using different transport systems at different fares. Every time the passenger changes vehicles a predetermined amount is deducted from the card, bonus fares for long distance travellers and people who change several times can be automatically taken into account.

Best price calculation In this system all journeys made are recorded on the contactless card for a month. If a certain number of journeys was exceeded on one day or in the month as a whole, then the contactless card can automatically be converted into a cheaper 24 hour or monthly card. This gives the customer maximum flexibility and the best possible fares. Best price calculation improves customer relations and makes a big contribution to customer satisfaction.

13.2.5 Market potential

It is estimated that around 50% of all contactless cards sold are used in the public transport sector [hamann.u]. The biggest areas of use are the large population centres in Asia (Seoul, Hong Kong, Singapore, Shanghai), and European cities (Paris, London, Berlin).

In 1994 and 1995 around 1 million contactless smart cards were produced per year worldwide for public transport applications. In the period 1996 to 1997 the volume rose to over 40 million cards per year [droschl]. The expected volume for 1998 alone is around 100 million contactless smart cards worldwide for public transport applications [hamman.u]. Given annual growth rates of 60% or more, we can expect the annual demand for contactless smart cards to have risen to 250 million by the turn of the century.

The highest growth rates for contactless smart cards in public transport applications will be in the Asiatic-Pacific area, because of the new infrastructures being created here using the latest technologies [droschl].

13.2.6. Example projects

13.2.6.1 Korea – Seoul

The largest electronic travel pass system (AFC) yet to use contactless cards was commissioned at the start of 1996 in the metropolis of *Seoul*, South Korea. The Korean "Bus Card" is a prepaid card, issued with a basic value of 20,000 ₩ (~ 17 euro). Fares are calculated according to fare system 1. A bus journey costs an average of 400 ₩ (~ 0.35 euro), but every time the passenger changes vehicles they must pay again.

Figure 13.5: Use of a contactless smart card in Seoul. A contactless terminal is shown in communication with a contactless smart card in the centre of the picture. (Reproduced by permission of Intec)

Figure 13.6: Contactless smart card for paying for journeys in a scheduled bus in Seoul (Reproduced by permission of Klaus Finkenzeller, Munich)

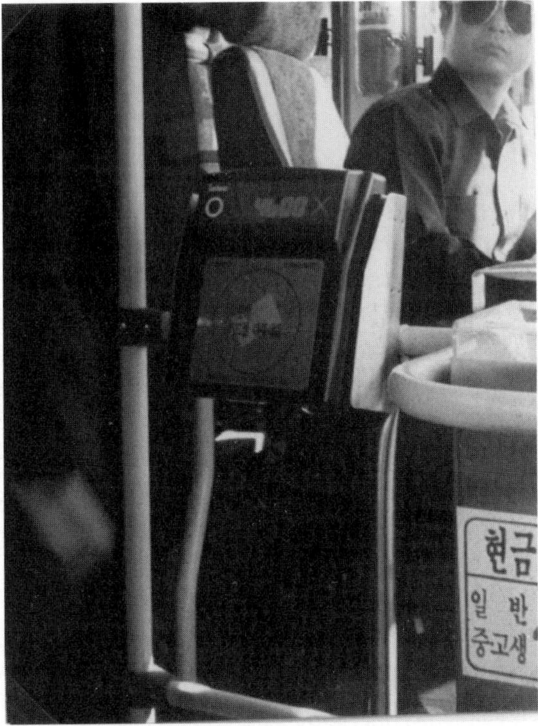

Figure 13.7: Reader for contactless smart cards at the entrance of a scheduled bus in Seoul (Reproduced by permission of Klaus Finkenzeller, Munich)

The card can be used on all 453 lines and recharged at identified kiosks as required. The transport association "Seoul Bus Union" is made up of 89 individual operator companies with a total of over 8700 buses, which were all equipped with contactless terminals by the middle of 1996. When the Kyung-Ki province that surrounds the capital city was included in the scheme, a further 4000 buses and a total of 3500 charging points were fitted with terminals by 1997 [droschl]. The RFID technology used in this project is the MIFARE® system (inductively coupled, 10 cm, 13.56 MHz), which is very popular in public transport applications.

It is predicted that four million Bus Cards will be in circulation by the end of 1997. The huge success of this system has convinced the government of Seoul to introduce a compatible system for the underground railway system.

13.2.6.2 Germany – Lüneburg, Oldenburg

One of the first smart card projects in Germany's public transport system is the *Fahrsmart* project in the KVG Lüneburg – VWG Oldenburg transport association. The subsidised *Fahrsmart* pilot project was launched by the Ministry for Education and Research in this area as early as 1990/91. Around 20 000 smart cards with contacts were issued to customers for this project. However, significant flaws in the installed systems became evident during this pilot project; the biggest problem was that the registration time of over three seconds per passenger was considered to be excessive.

At the beginning of 1996 a new field test was launched, the Fahrsmart II system based upon contactless smart cards. The RFID technology used was the MIFARE® system by Philips/Mikron. System integration, i.e. the commissioning of the entire system was performed by Siemens VT (Berlin).

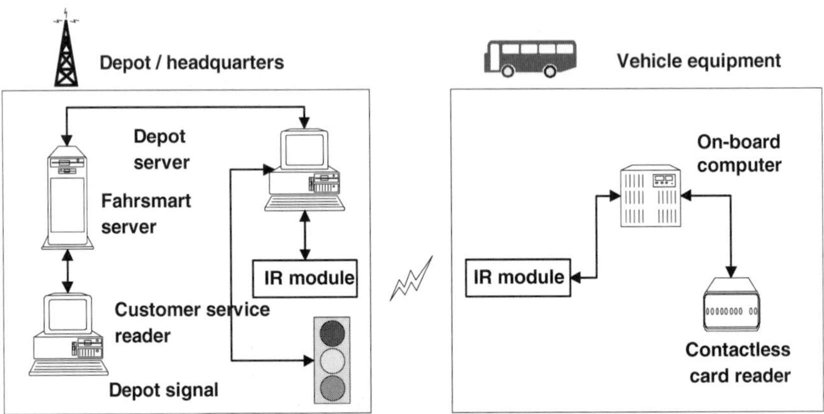

Figure 13.8: System components of the Fahrsmart system. The vehicle equipment consists of a reader for contactless smart cards, which is linked to the on-board computer. Upon entry into the station, the record data is transferred from the on-board computer to a depot server via an infrared link

Figure 13.9: Fahrsmart II contactless smart card, partially cut away. The transponder coil is clearly visible at the lower right-hand edge of the picture (Reproduced by permission of Giesecke & Devrient, Munich)

The Fahrsmart system automatically calculates the cheapest price for the customer (best price guarantee). The passenger must check in at the start of the journey using their personal smart card and check out at the end of the journey. The journey data obtained are collected in the on-board computer and stored on the smart card for verification.

When the vehicle returns to the depot at the end of the day the current day's data is sent from the vehicle computer to the station server via an infrared interface (Figure 13.8). The processed data is then transferred to the central Fahrsmart server via an internal network. To calculate the monthly invoice, the Fahrsmart server analyses the usage profile of each individual passenger and calculates the cheapest ticket for the distance travelled (individual journey – weekly pass – monthly pass etc.)

13.3 Ticketing

13.3.1 Lufthansa Miles & More card

The Lufthansa "Ticketless flying" project is the German showcase project for contactless smart card systems. The pilot test involving 600 regular fliers from May to December 1995 ran so smoothly that by March 1996 the "Miles & More" programme was extended to all *Lufthansa* cards ("HON Club", " SENATOR" and "Frequent Traveller"), and the Lufthansa owned service company AirPlus was established. Since Autumn 1996, all of the approximately 250 000 regular fliers possess the new contactless "ChipCard". This new contactless smart card – in conjunction with the Lufthansa central computer in Munich-Erding – replaces both the old paper ticket and the conventional boarding pass.

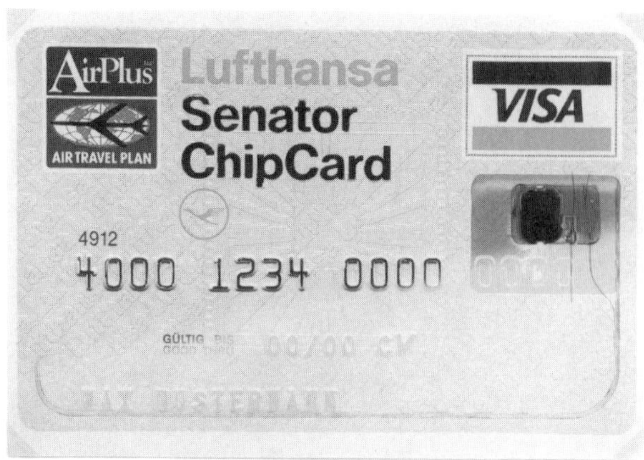

Figure 13.10: Miles & More – Senator ChipCard, partially cut away. The transponder module and antenna are clearly visible at the right-hand edge of the picture underneath the hologram. (Reproduced by permission of Giesecke & Devrient, Munich)

The RFID system selected for this project was the MIFARE® system by Philips/Mikron. The terminals were developed by Siemens-Nixdorf, whilst the contactless smart cards were manufactured by the Munich company Giesecke & Devrient. In addition to a contactless transponder module, the cards also incorporate a magnetic strip, raised lettering, signature strip, hologram and an optional (contact based) telephone chip.

Figure 13.11: Passenger checking in using the contactless Miles & More Frequent Flyer Card. (Reproduced by permission of Lufthansa)

From the point of view of the passenger, the system operates as follows: The card holder books a flight by telephone via a travel agent, quoting his individual card number. With the new cards, the booking may be performed up to one hour before departure. An electronic ticket is created that combines personal data and the flight data, and this is saved in the Lufthansa computer. To check in at the airport (at the last minute) the smart card owner only has to present his contactless smart card briefly at the Chip-In-Terminal. Normally, he can do this without even removing the card from his briefcase. The system verifies the booking, and the flight data appear on the screen. At this point the passenger is given the option of either confirming the suggested flight including seat reservation or selecting an alternative flight on the touch screen monitor. The passenger receives a printed receipt specifying his seat number and boarding gate, plus other details. This procedure allows the passenger to check in with his hand luggage in less than 10 seconds, which helps to prevent queues at the terminal. When he arrives at the boarding gate he merely presents his ChipCard once again and can then board.

The new ChipCard solution benefits both passengers and Lufthansa and Airplus. The passengers are overjoyed about the time saving on the ground, the ability to book at short notice and the convenient and simple operation. Clear rationalisation successes can be seen at Lufthansa and AirPlus, above all due to the self service of customers, which significantly reduces labour intensive handling and verification tasks. Furthermore, the ability to accept short notice bookings is increased by the time saving on the ground, thus increasing competitiveness compared to alternative travelling options (according to [g&d1]).

13.3.2 Ski tickets

Anyone entering a *ski lift* has to be in possession of a valid daily or weekly pass. These tickets were originally made of cardboard and validated by a date stamp. Checking paper tickets is very labour intensive because each ticket must be checked visually for validity. Furthermore, it is inconvenient for individual skiers to have to fish around in their anoraks for a sodden paper ticket with cold fingers before every journey on the lift.

Figure 13.12: Contactless reader as access control and till device at a ski lift. (Reproduced by permission of Legic Identsystems, CH-Wetzikon)

RFID technology offers an ideal alternative by replacing paper tickets with contactless smart cards or disk transponders. When the transponder is sold a deposit of 5 – 10 Euro is usually retained. After use, the transponder can be returned and the deposit refunded. The lift operator can revalue the transponder using special readers and it can thus be reused.

The read range of the system is designed to be great enough that the transponder tickets no longer need to be held in the hand, but can remain in an anorak pocket.

All ski lift entrances are protected by a turnstile, which is released by the read electronics upon detection of a valid transponder. In order to read the skier's transponder however it is carried, every entrance is monitored by two antenna opposite each other.

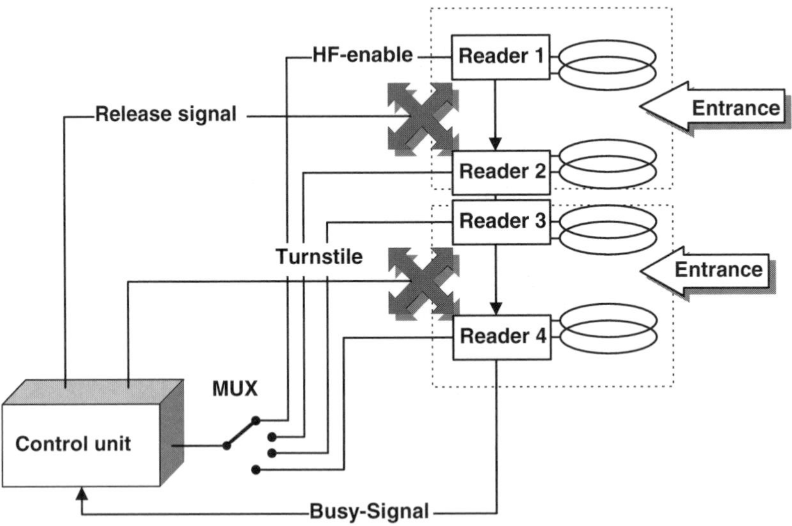

Figure 13.13: To achieve mutual decoupling the readers are switched alternately in time-division multiplex operation

The size of the magnetic antenna is a problem, because it must be very large due to the desired read range. The resulting magnetic coupling between the two reader antennas is so great at a distance of several meters that the resulting mutual interference makes it impossible to read a transponder. To circumvent this problem, ski ticket systems activate only one reader at a time, while the HF field of all other readers is switched off altogether. To achieve this, a multiplexer sends a start signal to one reader after the other in a cyclical sequence, which causes the addressed reader to switch on its HF field and check for the presence of a transponder. If the reader discovers a transponder, then it activates a *busy signal*. The busy signal tells the control unit to suppress the cyclical start signal for the duration of the busy signal. The active reader is now free to perform the data exchange it has begun with the transponder. After the end of this transaction, the active reader stops transmitting the busy signal, whereupon the multiplexer can continue its cyclical interrogation.

13.4 Access Control

Access control systems using PVC cards have been around for a long time. The first such systems used punched PVC cards, which were subsequently replaced by infra-red passes (IR barcode), magnetic strip passes, Wiegand passes (magnetic metal strips) and finally smart cards incorporating microchips [virnich] [schmidhäusler].

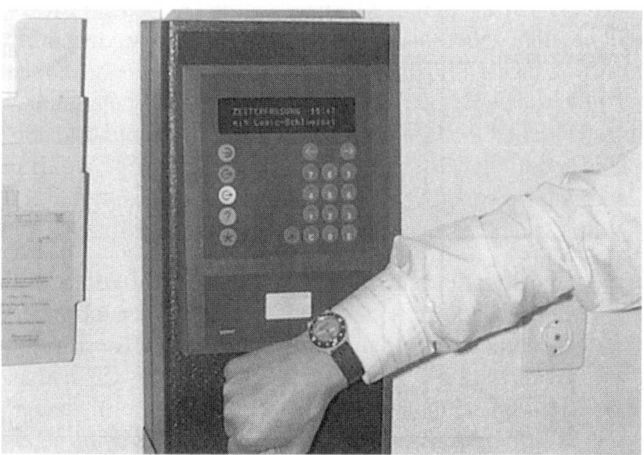

Figure 13.14: Access control and time keeping are combined in a single terminal. The clock with an integral transponder performs the function of a contactless data carrier. (Reproduced by permission of Legic Identsystems, CH-Wetzikon)

The main disadvantage of the procedures described is the inconvenient operating procedure, because the cards must always be inserted into a reader the right way round. Access control using contactless systems permits much greater flexibility because the transponder only needs to pass a short distance from the reader antenna. Passes can be made in the form of contactless smart cards, key rings and even wristwatches.

A great advantage of contactless access control systems is that the reader is maintenance free and is not influenced by dust, dirt or moisture. The antenna can be mounted "under plaster", where it is completely invisible and protected against vandalism. Hands-free readers are also available for mounting in turnstiles or to increase convenience. In these designs, the transponders do not even need to be removed from the pocket or jacket clip.

Other applications in the field of access control are cat flaps operated by a transponder in the cat's collar, and the use of read-only transponders as anti-theft sensors for opening or closing doors and windows [miehling].

13.5 Transport Systems

13.5.1 Eurobalise S21

Although Europe is moving closer together cross-border transport still presents an obstacle to Europe's railways. Different signals and train security systems force trains to incur the cost of carrying multiple sets of equipment on locomotives and tractive units. It is often necessary to expend precious time by changing tractive vehicles at the border, burdening trains with a competitive disadvantage compared to flying or travelling by road [lehmann].

For this reason, the European Union is backing the purchase of a unified European train security and control system, the *ETCS* (European Train Control System). The ETCS will facilitate interoperable cross-border traffic and improve the competitiveness of railways by implementing the latest train control technology.

The ETCS comprises four main systems:

EURO-Cab A vehicle device, in which all connected elements are linked to the secure vehicle computer EVC (European Vital Computer) by a special ETCS bus system.

EURO-Radio A GSM radio link between the vehicles and a radio centre by the track, the RBC (Radio Block Center).

EURO-Loop A system for linear data transfer over distances up to several hundred meters. The system is based upon so-called leakage cable, i.e. coaxial cables for which the sheathing is designed to be partially permeable to the electromagnetic field. The frequency ranges of this application lie between around 80 MHz and 1 GHz [ernst]. EURO-Loop is primarily used to transfer information for the evaluation of discretely transmitted data.

EURO-Balise A system for the discrete transmission of data. Depending upon design, local data (location marking, gradient profiles, speed limits) or signal related data for the route are transmitted to the vehicle [lehmann].

The *Eurobalise* subsystem is particularly important, because it is a crucial prerequisite for the full introduction of the ETCS. In January 1995, after lengthy experiments, the technical framework data for the EURO-Balise were determined. It is an inductively coupled RFID system with anharmonic feedback frequency.

The power supply to the system is taken from a passing a tractive unit by inductive coupling at the ISM frequency 27.115 MHz. Data is transferred to the tractive unit at 4.24 MHz, and the system is designed to reliably read the data telegram at train speeds of up to 500 km/h.

Figure 13.15: EURO-Balise in practical operation. (Reproduced by permission of Siemens Verkehrstechnik, Braunschweig)

Basic data for the EURO-Balise:

Coupling:	inductive
Power transmission frequency, vehicle → balise:	27.115 MHz
Data transmission frequency, balise → vehicle:	4.24 MHz
Modulation type:	FSK
Modulation index:	1
Data rate:	565 kbit/s
Telegram length:	1023 or 341 bit
Useful data size:	863 or 216 bit
Read distance:	230 to 450 mm
Maximum sideways offset:	180 mm
Coverage with snow, water, ore:	noncritical

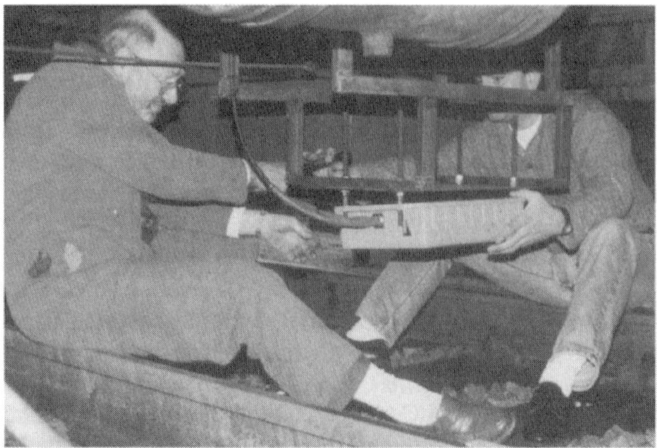

Figure 13.16: Fitting a read antenna for the EURO-Balise onto a tractive unit. (Reproduced by permission of Siemens Verkehrstechnik, Braunschweig)

Four different balise types have been developed by Siemens:

- Type 1 transmits a permanently programmed telegram.
- Type 2 transmits a telegram that can be programmed by the user via the contactless interface. For example, this may be line data such as gradient and speed profiles.
- Type 3 transmits a telegram generated by a line device (transparent balise). Type three is primarily used in connection with signals.
- Type 4 makes it possible to download data as vehicles drive past.

13.5.2 International container transport

International freight transport containers have been identified using the alphanumeric identification procedure specified in the international standard ISO 6346 since the end of the 1960s. This identification mark consists of four letters, the owner's code, a six digit numeric serial number and a test digit and is painted onto the outside of the container at a specified position.

ABZU 001 234 3

Figure 13.17: Container identification mark, consisting of owner's code, serial number and a test digit

Almost all of the 7 million containers in use worldwide employ the identification procedures specified in this standard and thus have their own, unmistakable identification number. The process of manually recording the container identification number and entering it into the computer of a transhipment plant is extremely susceptible to errors. Up to 30% of identifications have been falsely recorded at some point. Automatic data transmission can help to solve this problem by the reading of a transponder attached to the container. In 1991 the international standard ISO 10374 was drawn up to provide a basis for the worldwide use of this technology.

The bands 888 to 889 MHz and 902 to 928 MHz (North America) and 2.4 to 2.5 GHz (Europe) are used as the operating frequencies for the transponders. The transponders must respond on all three of the frequency ranges used. Backscatter modulation (modulated reflection cross section) with an FSK modulated subcarrier is the procedure used for the data transfer from the container to the reader. The subcarrier frequencies are 20 kHz and 40 kHz. A total of 128 bits (16 bytes) are transmitted within just 2 ms.

The reader's signal is not modulated (read only transponder). The specified maximum reader distance is 13 m.

ISO 10374 specifies the following information that can be stored in the transponder:

- Owner's code, serial number and test digit;
- Container length, height and width;
- Container type, i.e. suitcase container, tank container, open top container and others;
- Laden and tare weight.

A battery provides the power supply to the electronic data carrier in the transponder (active transponder). The lifetime of the battery corresponds with the lifetime of the container itself, i.e. around 10 to 15 years.

The same technology is used in the identification of goods wagons in North American and European railway transport. A European standard is in preparation for the automatic identification of European interchangeable containers [siedelmann].

13.6 Animal Identification

13.6.1 Stock keeping

Electronic identification systems have been used in stock keeping for almost 20 years [kern-97] and are now state of the art in Europe. In addition to internal applications for automatic feeding and calculating productivity, these systems can also be used in inter-company identification, for the control of epidemics and quality assurance and for tracing the origin of animals. The required unified data transmission and coding procedures are provided by the 1996 ISO Standards 11784 and 11785 (see chapter 9 "Standardisation – Animal Identification"). The specified frequency is 134.2 kHz, and FDX and SEQ transponders can both be used.

There are four basic procedures for attaching the transponder to the animal: collar transponders, ear tag transponders, injectible transponders and the so-called bolus.

Collar transponders can be easily transferred from one animal to another. This permits the use of this system within a company. Possible applications are automatic feeding in a feeding stall and measuring milk output.

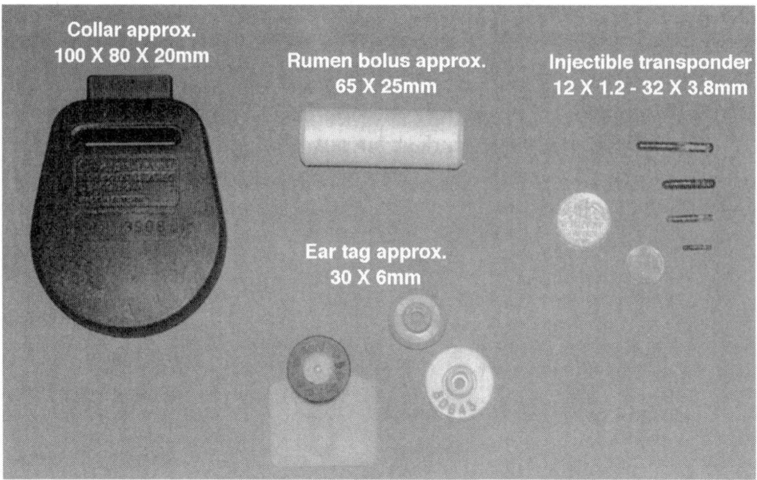

Figure 13.18: Size comparison of different variants of electronic animal identification transponders: collar transponder, rumen bolus, ear tags with transponder, injectable transponder. (Reproduced by permission of Dr. Michael Klindtworth, Bayrische Landesanstalt für Landtechnik, Freising)

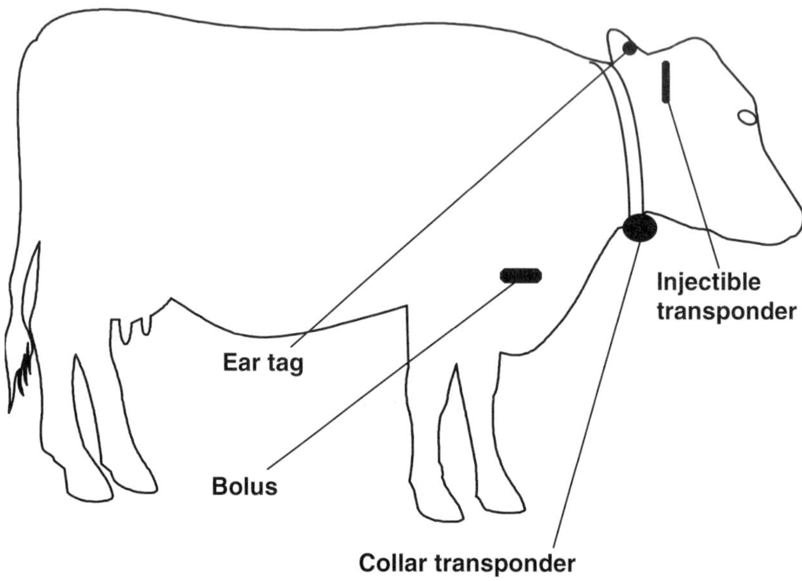

Figure 13.19: The options for attaching the transponder to a cow

Ear tags incorporating an RFID transponder, compete with the much cheaper barcode ear tags. However, the latter are not suitable for total automation, because barcode ear tags must be passed a few cm from a hand reader to identify the animal. RFID ear tags, on the other hand, can be read at a distance of up to 1 m.

Injectible transponders were first used around 10 years ago. In this system, the transponder is placed under the animal's skin using a special tool. A fixed connection is thereby made between the animal's body and the transponder, which can only be removed by an operation. This allows the use of implants in inter-company applications, such as the verification of origin and the control of epidemics.

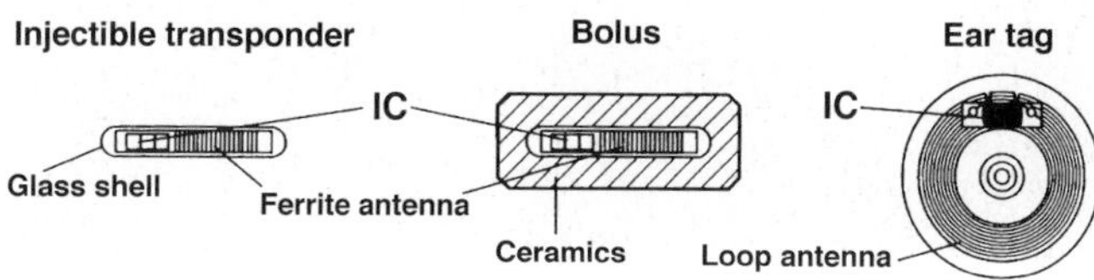

Figure 13.20: Cross sections of various transponder designs for animal identification. (Reproduced by permission of Dr. Georg Wendl., Landtechnischer Verein in Bayern e.V., Freising)

The implant is in the form of a glass transponder of 10, 20 or 30 mm in length. The transponder is supplied in a sterile package or with a dose of disinfectant. The dimensions of the glass transponder are amazingly small, considering that they contain the chip and a coil wound around a ferrite rod. A typical format is 23.1 mm x 3.85 mm [ti-96].

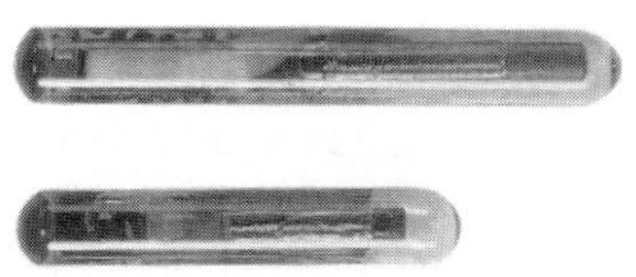

Figure 13.21: Enlargement of different types of glass transponder. (Reproduced by permission of Texas Instruments)

Various instruments and *injection needles* are available for performing the injection:

- "Single-shot" devices use closed hollow needles ("O" shape), which are loaded individually. Single use needles containing transponders in a sterile package are also available. The hollow needles are sharpened at the tip, so that the skin of the animal is ripped open when the needle is inserted. The blunt upper part of the needle tip presses the cut flap of skin to one side so that the insertion point is covered up again when the needle has been removed, allowing the wound to heal quickly [kern-94].

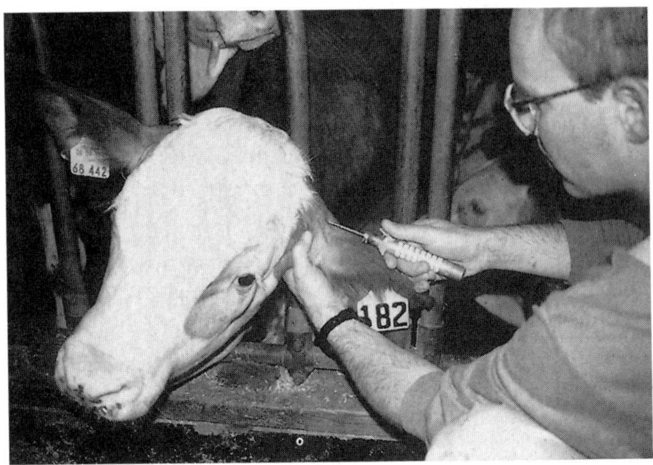

Figure 13.22: Injection of a transponder under the scutulum of a cow. (Reproduced by permission of Dr. Georg Wendl, Landtechnischer Verein in Bayern e.V., Freising)

- The "Multi-shot" device has a magazine for several transponders, thus dispensing with the need to load the device. Open-ended hollow needles ("U" shaped) are used, as these are easier to clean, disinfect and check than closed hollow needles and can therefore be used several times.
  The injection does not hurt the animal and can be carried out by practised laymen. However, attention should be given to hygiene to ensure that the wound heals safely.

However, an injected transponder represents a foreign body in the animal's tissues. This can lead to problems in the locational stability of the transponder within the animal's body, and may therefore cause problems when reading the transponder. From our experience of war injuries we know that shrapnel can often wander several decimetres through the body during a person's lifetime. An injected transponder can also "wander" around. To solve this problem, the Bayerischen Landesanstalt für Landtechnik in Weihenstephan, a branch of the Technical University in Munich, has been investigating various injection sites since 1989 [kern-94]. As a result of these studies, injection under the *scutulum* is currently favoured over the use of the right ear, with the injection being directed towards the occipital bone. According to findings of the Landanstalt, this position is also suitable for measuring the animal's body temperature.

The so-called *bolus* is a very useful method of fitting the transponder. The bolus is a transponder mounted in an acid resistant, cylindrical housing, which may be made of a ceramic material. The bolus is deposited in rumen, the omasum that is present in all ruminants, via the gullet using a sensor. Under normal circumstances the bolus remains in the stomach for the animal's entire lifetime. A particular advantage of this method is the simple introduction of the transponder into the animal's body, and in particular the fact that it does not cause any injury to the animal. The removal of the bolus in the slaughter house is also simpler than the location and removal of an injected transponder [kern-97].

Figure 13.23: Automatic identification and calculation of milk production in the milking booth. (Reproduced by permission of Dr. Georg Wendl, Landtechnischer Verein in Bayern e.V., Freising)

Figure 13.24: Output related dosing of concentrated feed at an automatic feed booth for milk cows. In the illustration the cow is identified by the transponder at its neck. (Reproduced by permission of Dr. Georg Wendl, Landtechnischer Verein in Bayern e.V., Freising)

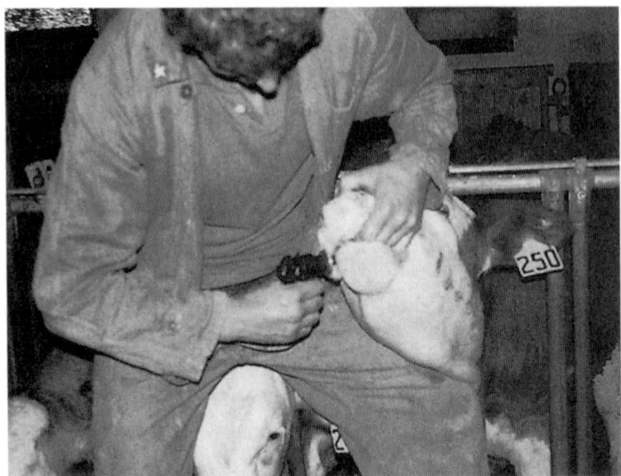

Figure 13.25: Oral application of a bolus transponder. (Reproduced by permission of Dr. Michael Klindtworth, Bayerische Landesanstalt für Landtechnik, Freising)

Figure 13.26: Example of automated animal recognition in practice: Grouping calves properly for feeding often requires much time and effort. Here a machine takes on this task: The animals can receive an individually adjustable amount of milk in several small portions. (Reproduced by permission of Dr. Michael Klindtworth, Bayerische Landesanstalt für Landtechnik, Freising)

It is clear that the injected transponder and the bolus are the only foolproof identification systems available to stock keepers. A more detailed comparison of the two systems [kern-97] shows that the bolus is particularly suited for use in the extensive type of stock keeping that is prevalent in Australia or South America. In intensive stock keeping methods, commonly used in central Europe, both systems appear to be suitable. The degree to which bolus, injection or even RFID ear tags will become the industry standard means of identification remains to be seen.

13.6.2. Carrier pigeon races

Participating in races is a significant part of carrier pigeon breeding. In these races, hundreds of pigeons are released at the same place and time, at a location a long distance from their home. Pigeons are judged by the time they take to return home from the point where they were released. One problem is the reliable recording (confirmation) of arrival times, because in the past the breeders themselves recorded the times using a mechanical confirmation clock.

To solve the problem of timing, the pigeons are fitted with rings that incorporate a read only transponder based upon a glass transponder. As the pigeons are loaded onto the transporter for transport to the release site, the serial numbers of the transponders are read to register the animals for participation in the race. Upon the pigeon's arrival at its home pigeonry a reader installed in the pigeonhole records the serial number and stores it, together with the precise arrival time, in a portable control unit. Judging takes place by the reading of the devices at the operating point.

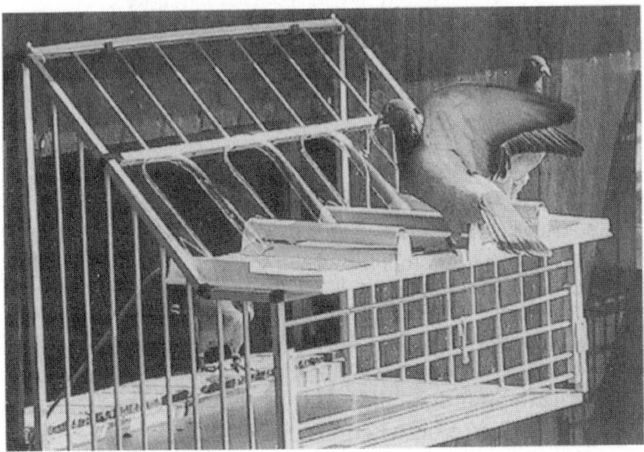

Figure 13.27: Pigeon upon arrival at its own pigeonry. Upon the pigeon's entry, the transponder in the ring is read. (Reproduced by permission of Legic Identsystems, CH-Wetzikon)

However, the ingenuity of some of the breeders was greatly underestimated when this system was first introduced. It was not long before some breeders were not only able to read the transponder codes from the pigeon ring, but could also fool the reader using a simulation device in the home pigeonry. The technology involved was fairly simple, all that was required was an extremely simple read only transponder, whose "serial number" could be altered using external DIP switches. Thus, some breeders were able to significantly accelerate the "flight speeds" of their champions.

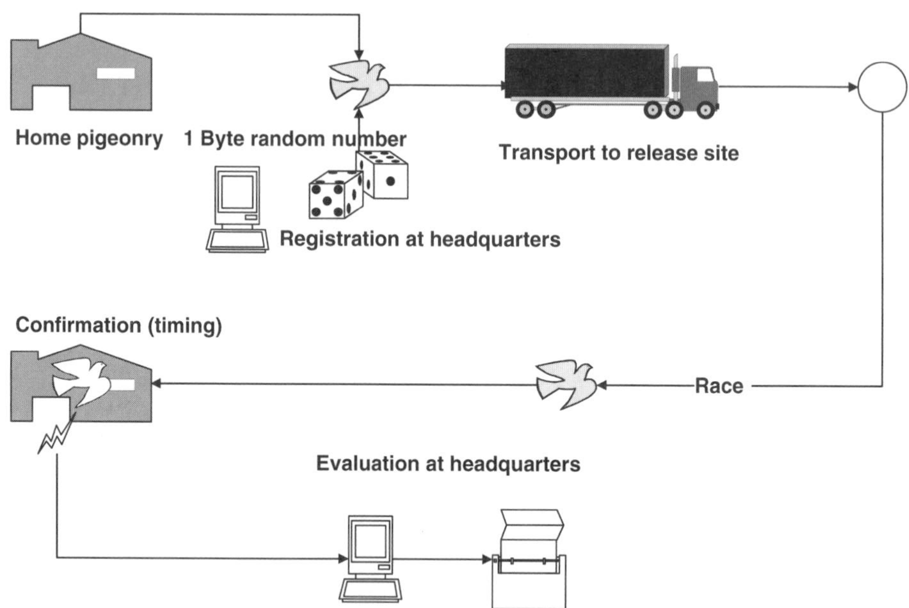

Figure 13.28: The generation of a random number which is written to the transponder before the start protects against attempted fraud

An effective measure to protect against such attempts at fraud is the incorporation of an additional writeable EEPROM memory into the transponder. The memory size is just 1 byte to keep the chip size and cost of circuitry low. Before the start, a previously determined random number, for which there are 2^8=256 possibilities, is written to this byte in the transponder at the headquarters. It is crucial that the breeder does not have access to his bird while it is being transported to the release site after the transponder has been programmed. This prevents the random number from being read. When the pigeon reaches its home pigeonry, its arrival is confirmed electronically. The time, together with the transponder code and the secret *random number* are stored. When the records are evaluated at the headquarters, the random number read upon arrival is compared with the number programmed at the start. The measured times are only validated if the two figures are identical, otherwise it is assumed that an attempted fraud has taken place.

The procedure described is clearly adequate to successfully prevent attempted fraud. With 256 possibilities for the random number the probability that this will be guessed correctly in a single attempt is only 0.4%.

In order to keep the weight and dimensions of the pigeon transponder low, glass transponders are used in this application, which are cast into a plastic ring. These plastic rings can be fastened to the pigeon's leg without hindering the animal or causing it any discomfort.

13.7 Electronic Immobilisation

The sharp rise in *vehicle theft* at the beginning of the 1990s – particularly in Germany – boosted the demand for effective anti-theft systems. Battery operated remote control devices with a range of 5 to 20 metres had already been available on the market for years. These are small infrared or RF transmitters operating on the UHF frequency 433.92 MHz, which are primarily used to control the central locking system and an integral alarm. An (electronic) immobiliser may also be coupled to the remote control function. In this type of anti-theft device, however, the mechanical lock can still be used to gain access to the vehicle – in case the remote control device fails to work due to the failure of the battery in the transmitter. This is the greatest weakness of this type of system, as the system cannot check whether the mechanical key is genuine. Vehicles secured in this manner can therefore be opened with a suitable tool (e.g. picklock) and started up by an unauthorised person.

Since the middle of the 1990s, transponder technology has provided a solution that can be used to check the authenticity, i.e. the genuineness, of the key. This solution has proved ideal for the realisation of the electronic immobilisation function via the ignition lock. Today, transponder technology is usually combined with the above-mentioned remote control system: the remote control operates the vehicle's central locking and alarm system, whilst transponder technology performs the immobilisation function.

13.7.1 The functionality of an immobilisation system

In an *electronic immobilisation* system a mechanical ignition key is combined with a transponder. The miniature transponder with a ferrite antenna is incorporated directly into the top of the key (see Figure 13.29).

The reader antenna is integrated into the *ignition lock* in such a manner that when the ignition key is inserted, the (inductive) coupling between reader antenna and transponder coil is optimised. The transponder is supplied with energy via the inductive coupling and is therefore totally maintenance free. Electronic immobilisers typically operate at a transmission frequency in the LF range 100 – 135 kHz. ASK modulation is the preferred modulation procedure for the data transfer to the transponder, because it allows reader and transponder to be manufactured very cheaply [doerfler]. Load modulation is the only procedure used for data transmission from the transponder to the reader.

When the ignition key is turned in the ignition lock to start the vehicle, the reader is activated and data is exchanged with the transponder in the ignition key. Three procedures are employed to check the authenticity of the key:

Figure 13.29: Ignition key with integral transponder. (Reproduced by permission of Philips Electronics N.V.)

- Checking of an individual serial number:
  In almost all transponder systems the transponder has a simple individual *serial number* (unique number). If the normal number of binary positions is used, significantly more different codes are available than worldwide car production (2^{32} = 4.3 billion, $2^{48} = 2.8 \times 10^{14}$). Very simple systems (1st generation immobilisation) read the transponder's serial number and compare this with a reference number stored in the reader. If the two numbers are identical the motor electronics are released. The problem here is the fact that the transponder serial number is not protected against unauthorised reading and, in theory, this serial number could be read by an attacker and copied to a special transponder with a writeable serial number.
- Rolling code procedure:
  Every time the key is operated a new number is written to the key transponder's memory. This number is generated by a pseudo-random generator in the vehicle reader. It is therefore impossible to duplicate the transponder if this system is used. If several keys are used with one vehicle then each key runs through its own pseudo-random sequence.
- Cryptographic procedures (authentication) with fixed keys:
  The use of cryptographic procedures offers much greater security (2nd generation immobilisation). In the *authentication* sequence (challenge response) knowledge of a secret (binary) key is checked, without this key being transmitted (see Chapter 8 "Data

Security"). In vehicle applications, however, unilateral authentication of the key transponder by the reader in the ignition lock is sufficient.

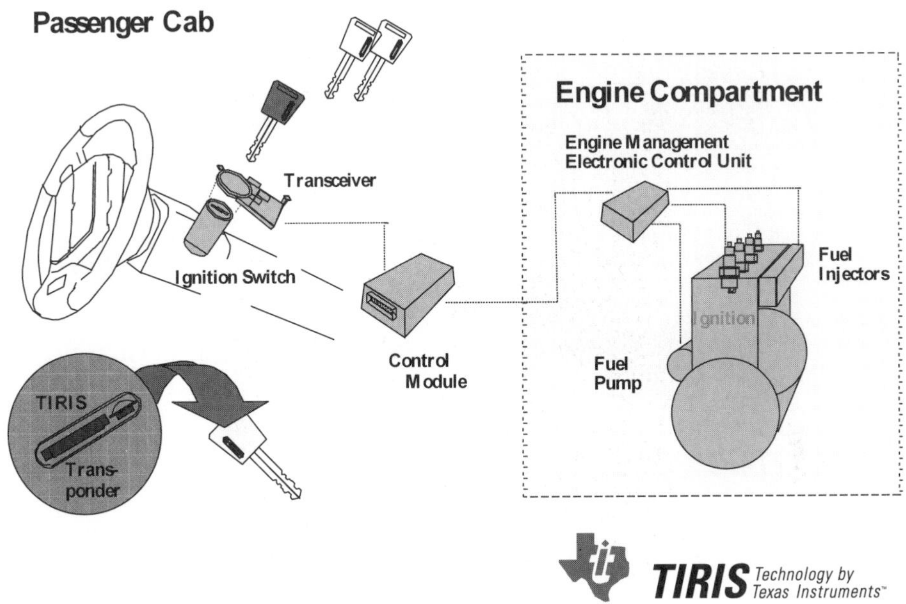

Figure 13.30: Functional group of an electronic immobilisation system. The RFID reader authenticates itself with regard to the motor electronics to prevent the manipulation of the reader. The motor electronics control the ignition, fuel and starter and thus can block all the crucial functions of the vehicle (Reproduced by permission of Texas Instruments)

The RFID reader now communicates with the vehicle's *motor electronics*, although this communication is protected by cryptographic procedures. The motor electronics control all important vehicle functions, in particular the ignition system and fuel system. Simply short circuiting or disconnecting certain cables and wires is no longer sufficient to circumvent an electronic immobilisation system. Even attempting to fool the motor electronics by inserting another ignition key of the same type into the ignition lock is bound to fail because of the authentication procedure between reader and motor electronics. Only the vehicle's own key has the correct (binary) key to successfully complete the authentication sequence with the motor electronics.

The installation of such an electronic immobiliser to the engine management system can only be performed at the factory by the vehicle manufacturer, thus guaranteeing optimal interaction between engine control system and security device. The individual key data is programmed in the factory by laser programmable fuses on the chip or by writing to an OTP-EEPROM. The vehicle manufacturer is also responsible for implementing appropriate security measures to prevent criminals from unlawfully procuring replacement parts [wolff]. With few exceptions, electronic immobilisation systems have been fitted to all new cars as standard since the beginning of 1995 [anselm96].

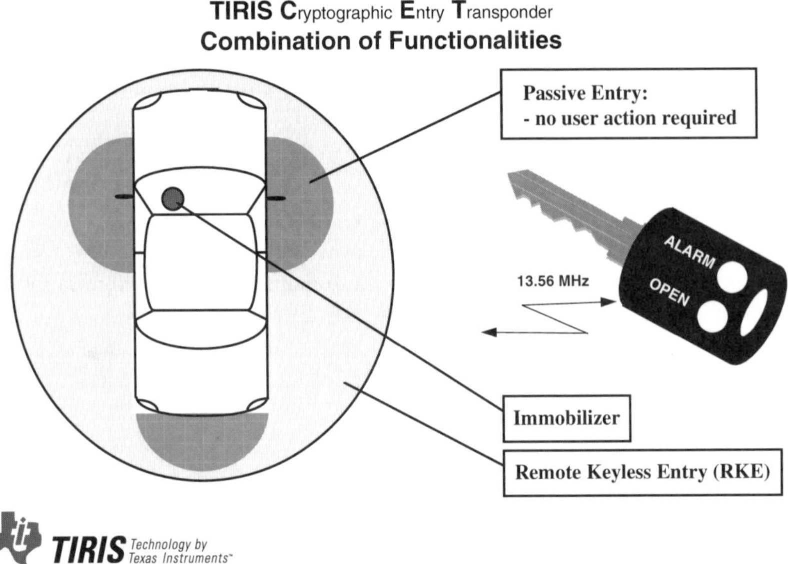

Figure 13.31: Electronic immobiliser and door locking system are integrated into a transponder in the ignition key. In the ignition lock and in the vicinity of the doors (passive entry) the transponder is supplied with power by inductive coupling. At greater distances (remote keyless entry) the transponder is supplied with power from a battery (round cell in the top of the key) at the push of a button ("OPEN"). (Reproduced by permission of Texas Instruments)

13.7.2 Brief success story

In 1989 the Berlin wall and the border to Eastern Europe were opened and the years following 1989 were characterised by dramatic increases in vehicle thefts in Germany. From 48 514 thefts in 1988, the figure had risen to 144 057 thefts just five years later in 1993 – almost a threefold increase. This prompted the German Federal Supervisory Office for Insurance to declare a change to the General Insurance Conditions for Motor Vehicle Insurance (AKB) at the beginning of 1993.

According to the old conditions, vehicle owners with fully comprehensive insurance could, under certain conditions, claim the full price for a new car if his vehicle was stolen, although the resale value of the stolen vehicle and thus the damage suffered was significantly less than this [Wolff]. The value of a vehicle after just a few months falls a long way short of the price of a new car.

Under the new conditions, only the cost of replacing the vehicle, i.e. its actual market value, is refunded in the case of loss (accident, theft, ...). Furthermore, if the loss is due to theft an excess is deducted from the payment, which may be waived if the vehicle is fitted with an approved anti-theft device [Wolff]. The vehicle owner's own interest in having an effective anti-theft device was significantly increased by the new insurance conditions.

The effectiveness of electronic immobilisation has been clearly demonstrated by the decreasing trend in vehicle thefts in Germany. In 1994 there had already been a slight fall of about 2000 to 142 113, compared to the record figure from 1993. Two years later – 1996 – 110 764 thefts were reported. This represents a fall of 22% in just 2 years.

Another factor is that since 1995 electronic immobilisers have been fitted to all new cars – with a few exceptions – in the factory as standard. If we consider vehicles secured in this manner alone, then we can expect a reduction in the theft rate by a factor of 40(!)

In this connection it is interesting to examine investigations by insurance companies into vehicle thefts where electronic immobilisers were fitted [anselm95] [anselm96] [caspers]:

Of 147 stolen vehicles in 1996, 70% of thefts were performed using the original key, which the thief had obtained by breaking into homes, garages and workshops, or by stealing from offices, bags and changing rooms or by the fraudulent renting and misappropriation of rental or demonstration cars. In the remaining 30 % of cases, the vehicles either disappeared under circumstances that indicated the cooperation of the owner (without this being proved in individual cases), or vehicles were loaded onto lorries and transported away by professionals.

There has not been one case since 1995 where the electronic immobiliser has been "cracked" or beaten by a thief.

13.7.3 Predictions

The next generation of immobilisers will also incorporate a passive, cryptologically secured access system. In this system, a reader will be fitted in each of the vehicle's doors. Sequential systems (TIRIS®) will be able to achieve a remote range, in which the transponder is supplied by a battery, so that the vehicle's central locking system can be operated from a greater distance away. This is similar in its function to the combination of an immobiliser and central locking remote control on a single transponder.

13.8 Container Identification

13.8.1 Gas bottles and chemical containers

Gas and chemicals are transported in high quality rented containers. Selecting the wrong bottle during refilling or use could have fatal consequences. In addition to product specific sealing systems, a clear identification system can help to prevent such errors. A machine readable identification system gives additional protection [braunkohle]. A large proportion of containers supplied today are identified by barcodes. However, in industrial use the popular barcode system is not reliable enough, and its short lifetime means that maintenance is expensive.

Transponders also have a much higher storage capacity than conventional barcodes. Therefore additional information can be attached to the containers such as owner details, contents, volumes, maximum filling pressure and analysis data. The transponder data can also be changed at will, and security mechanisms (authentication) can be used to prevent unauthorised writing or reading of the stored data.

Inductively coupled transponders operating in the frequency range < 135 kHz are used. The transponder coil is housed in a ferrite shell to shield it from the *metal surface* (see also Chapter 4, Section 4.1.11.3 "Ferrite Shielding in a Metallic Environment").

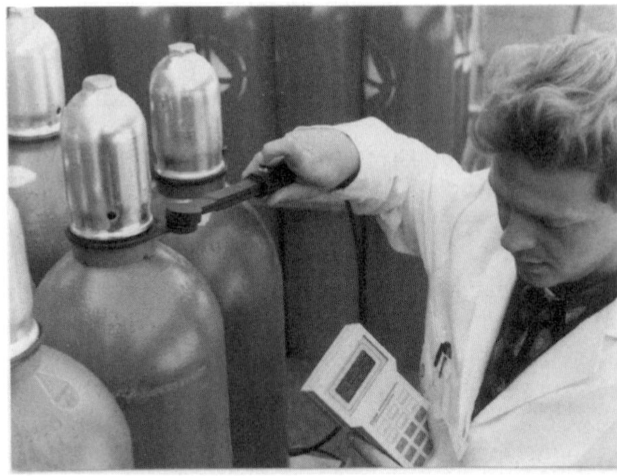

Figure 13.32: Identification of gas bottles using a portable reader. The reader (scemtec SIH3) is designed to function with transponders from different manufacturers. (Reproduced by permission of Messer Griesheim)

The manufacturing process for the transponders is subject to exacting standards: the transponders are designed for an extended temperature range from minus 40°C to plus 120° C, their height is just 3 mm. These transponders must also be resistant to damp, impact, vibrations, dirt, radiation and acids [bührlen].

Because the transmission procedure for transponders used in *container identification* has not been standardised, various systems are available. Because a device has been developed that can process all the transponder types used, the user can choose between the different transponder systems – or may even use a combination of different systems.

Mobile and stationary readers are available. Stationary readers can be incorporated into a production system which automatically recognises and rejects wrong containers. After filling, the current product data is automatically stored on the transponder. When this system is used in combination with database management, the number of containers used by a customer for a given gas consumption can be drastically reduced, because excessive standing times or storage periods can be easily recognised and corrected. In addition, all the stations that the container passes through on its way to the customer and back can be automatically recorded by the use of additional readers. So, for example, it is possible to trace customers who return the containers dirty [braunkohle]. For gas, where there is not much potential for product differentiation between manufacturers, the associated cost savings can convey an important competitive advantage [bührlen].

In total, over eight million gas bottles in Germany alone are waiting to be fitted with transponders. For Europe, this figure is approximately 30 million. In addition to gas bottles, transponders are also used for rental containers, beer kegs and boxes and transportation containers for the delivery industry.

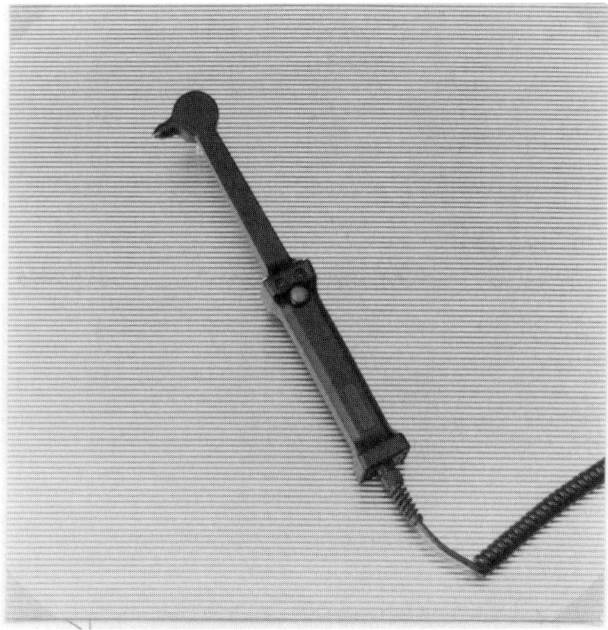

Figure 13.33: Portable antenna for reading inductively coupled transponders mounted on gas bottles or other containers. (Reproduced by permission of SCEMTEC Transponder Technology GmbH, Reichshof-Wehnrath)

13.8.2 Waste disposal

Because of increasingly rigorous environmental legislation, the cost of waste disposal is increasing all the time. Costs associated with creating new waste disposal sites and maintaining existing sites are being passed on to individual households and industrial companies. Automatic measurement of the amount of waste produced helps to distribute the costs fairly. For this reason, more and more cities are using RFID systems to optimise communal *waste disposal*, and are thus putting the conditions in place for replacing the flat rate charge for waste disposal with a charge based upon the quantity of waste produced. The waste disposal companies will only charge for the amount that has actually been removed.

To achieve this goal, a transponder is fitted to the dustbin and automatic reader systems are installed in rubbish collection vehicles. As soon as the dustbin is placed on the vehicle's emptying device its transponder is read. In addition, either the weight or the volume of rubbish is calculated, depending upon the preference of the community. A counter, to show how often the bin has been emptied in the year, is also feasible [euro-id].

The identifier read by the transponder is stored in a smart card in the vehicle's on-board computer together with the data collected. At the end of a round the driver passes the card to the operations centre so that the collected data can be processed. Individual households no longer pay a monthly flat rate, but each receive an individual bill [prawitz].

Figure 13.34: Dustbin transponder for fitting onto metal surfaces. (Reproduced by permission of Deister Electronic, Barsinghausen)

Figure 13.35: Reader antenna for installation in the dustcart. A plastic dustbin fitted with a transponder is shown in the background. (Reproduced by permission of Deister Electronic, Barsinghausen)

In Germany RFID systems are already in use in various cities including Bremen, Cologne and Dresden and in numerous communities.

13.9 Sporting Events

In large scale sporting events such as major marathons, the runners who start at the back of the field are always at a disadvantage, because their times are calculated from the moment the race is started. For many runners it takes several minutes before they actually cross the starting line. In very large events with 10 000 participants or more, it might be 5 minutes before the last runners have crossed the starting line. Without individual timing, the runners in the back rows are therefore at a severe disadvantage.

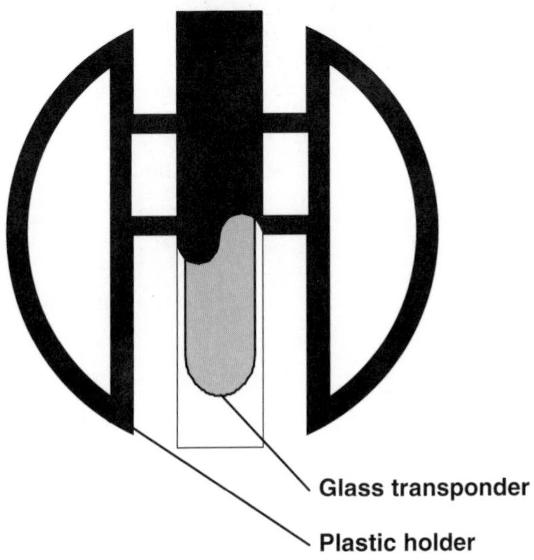

Figure 13.36: The transponder consists of a glass transponder, which is injected into a plastic housing that is shaped according to its function. The diagram shows the partially cut away plastic housing

To rectify this injustice, each runner carries a transponder with them. The system is based upon the idea that each runner places his feet repeatedly on the ground and thus comes very close to a *ground antenna*. In experimental events it was found that using a ingenious arrangement of multiple antennas in an array and a chip in the shoe over 1000 runners can be registered up to eight times in a minute with a start width of just 4 m [champion-chip].

The transponder is based upon a glass transponder operating in the frequency range 135 kHz, embedded into a specially shaped (ABS) injection moulded housing. To get the transponder as close as possible to the ground – and thus to the antenna of the time measurement device – this is attached to the runner's shoe using the shoelaces.

Figure 13.37: The ChampionChip transponder is fastened to the runner's shoe with the shoelace. (Reproduced by permission of ChampionChip BV, NL-Nijmegen)

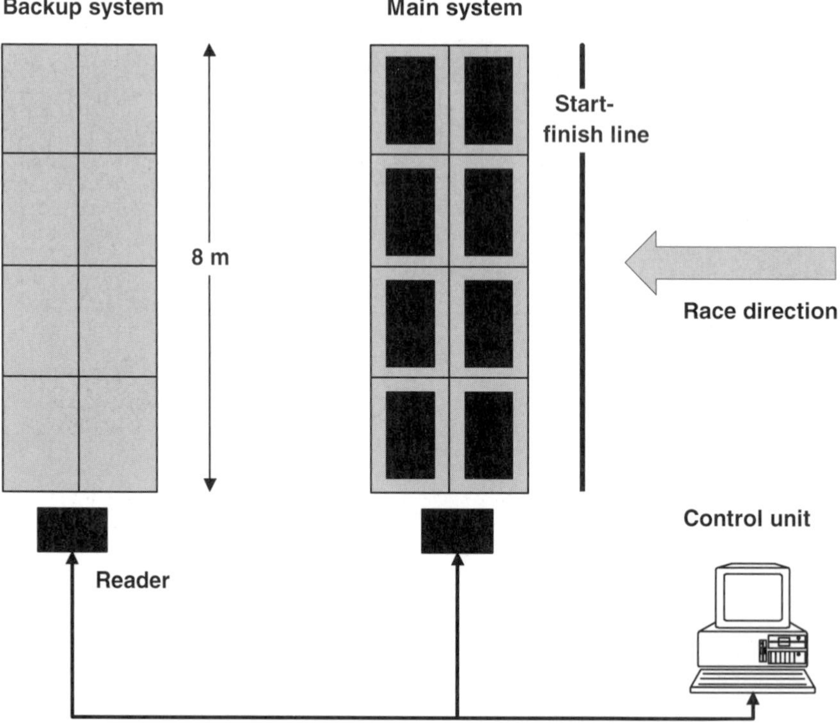

Figure 13.38: A control station consists of a main system and a reserve system. The systems are made up of arrays of antennas in mats

The reader antennas are cast into thin *mats* and can thus be placed on the ground and still be protected from all environmental influences. The dimensions of a single mat are 2.10 × 1.00 m. At a normal running speed a net time resolution of ± 1 second is possible, derived from the time the runner remains within the read range of a mat. The accuracy for cyclists improves to ± 0.2 seconds. The measured time is immediately displayed on a screen, so that the reader can read his current intermediate time or final time as he passes a control station.

The runner can make a one off purchase of the transponder for 38 DM and then use it wherever compatible timing systems are used.

The performance of a transponder based timing system has been demonstrated at the following events: Rotterdam Marathon – 10 000 participants, Shell Hanseatic Marathon, Hamburg – 11 500 participants and the Berlin Marathon with 13 500 participants [champion-chip].

Figure 13.39: Runners passing the control station at the end of the 101st Boston Marathon. In the foreground we can see the mats containing the readers. The times can be displayed on a screen immediately. (Reproduced by permission of ChampionChip, NL-Nijmegen)

13.10 Industrial Automation

13.10.1 Tool identification

As well as its metal cutting tool industry, Germany's woodworking industry also plays a dominant role in the world market. The modern woodworking and furniture manufacturing industry is dominated by *CNC technology* because this enables manufacturers to manufacture at a low cost and remain competitive.

CNC machines equipped with tool holders and automatic tool changers fulfil tasks that are increasingly associated with small batch production. This increases the proportion of manufacturing costs incurred by retooling and tool-change times.

Another consideration is the fact that a CNC woodworking machine differs from a metalworking machine because of its higher rotation and path speeds. Rotation speeds from 1000 min^{-1} to more than 20 000 min^{-1} (!) are attained in wood and plastic processing. The risk of accidents for man and machine is therefore very high during the tool-change operation, for example hazards may be caused by the wrong fitting of the CNC machine's chain magazine [leitz], [töppel].

Figure 13.40: CNC milling tool with transponder in the retention bolts. (Reproduced by permission of Leitz GmbH & Co., Oberkochen)

This potential hazard can be eliminated by fitting a transponder in the *taper shaft* or in the *retention bolts* of the toolholder. All relevant tool data are preprogrammed into the transponder by the tool manufacturer. The machine operator fits the transponder tools into the CNC machine's toolholder in any order. Then the CNC machine initiates an automatic read sequence of all tools in the toolholder, during which the tools are first ordered into toolholder positions and then all geometric and technical data for the tools is transmitted correctly to the tool management system of the CNC control unit. There is no manual data entry, which eliminates the possibility of human error [leitz]. The danger of accidents due to excessive speeds, the selection of the wrong rotation direction or the incorrect positioning of the tool in relation to the workpiece is thus eliminated.

Inductively coupled transponders operating in the frequency range < 135 kHz are used. The transponder coil is mounted on a ferrite core to shield it from the *metal surface* (see also Chapter 4, Section 4.1.8.3 "Ferrite shielding in a metallic environment" and Chapter 2,

Figure 13.41: Various woodworking tools with transponder data carrier in the taper shaft. (Reproduced by permission of EUCHNER + Co., Leinfelden-Echterdingen)

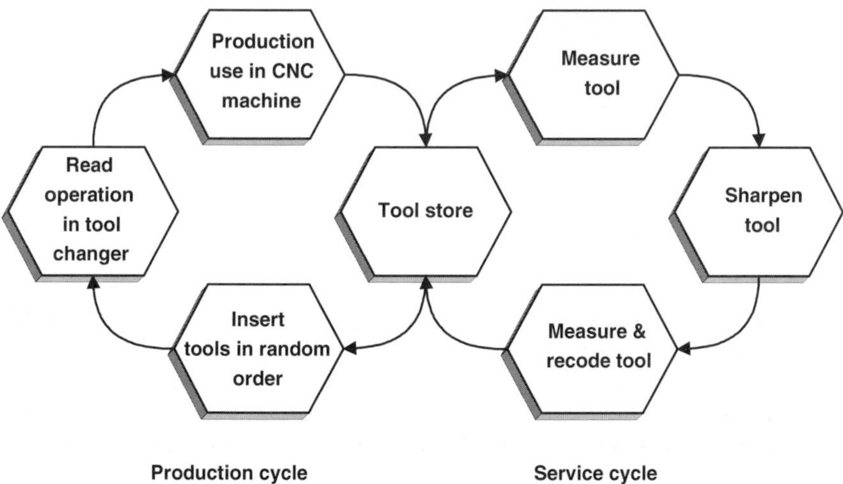

Figure 13.42: Representation of the tool cycle when using transponder coded CNC tools

Section 2.2 "Transponder Types"). The transponder must have a minimum of 256 bytes of memory, which is written with an ASCII string containing the required tool data. An example of a data record is illustrated in Table 13.2 (from [leitz]).

Table 13.2: Example of a data record for a tool transponder

Customer	Möbelwerk XY
LEITZ ID no.	130004711 D25x60
Manufacturing ref.	Y21
Place of manufacture	UHE
Rotation direction	3
Max. rotation speed	24 000
Min. rotation speed	18 000
Ideal rotation speed	20 000
Radius correction	25 011
Longitudinal correction	145 893
Greatest radius	25 500
Greatest length	145 893
Maximum travel	3000
Current travel	875
Tool number	14
Tool type	1
Number of sharpenings	2
Angle of clearance	20
Cutting rake	15
Free text	Finishing cutter HM Z=3

Modern transponder coded CNC tools can be incorporated into a cost saving production and service cycle. The service cycle is incorporated, smoothly and simply, into the production cycle as follows:

The worn tool is first examined and measured in detail to determine its condition. The tool is then serviced, sharpened and balanced on the basis of this data. After every maintenance sequence the tool length and radius is updated and written to the transponder, so that correctly dimensioned workpieces are produced by both new and sharpened tools without intervention by the operator.

13.10.2 Industrial Production

Production processes underwent a process of continuous rationalisation during the development of industrial mass production. This soon led to production line assembly, with the same stage of production being performed at a certain position on the assembly line time after time. A production process of this type can only produce objects that are identical in function and appearance. However, machines that produce large quantities of a single product with no variants can achieve huge time savings.

If different variants of a product are to be produced on an assembly line in an automated procedure, the object must be identified and its state clearly recognised at every work station so that the correct processes can be performed. Originally, the objects were provided with process cards for this purpose, which gave the operating personnel all the information required at a particular work station, for example the desired paint colour. In electronic form this is only possible using coding pegs affixed to the revolving palettes so that palette numbers can be read for use by the electronic control system. The position of these coding pegs could be sensed by inductive proximity switches [weisshaupt]. This procedure has recently been improved by the use of bar code labels, which are merely stuck onto the individual objects.

RFID technology now provides an additional option, and can not only be read, but also written. Now, in addition to recording the identity of an object, it is also possible to track its current state (e.g. processing level, quality data) and the past and future (desired end state) of the object.

Using modern identification techniques, production systems can now be realised which can produce variants of a product, or even different products, down to a batch size of one [weisshaupt]. The automotive industry is a good example: Because vehicles are predominantly produced to order and it is rare for two vehicles to be identical automatic material flow tracking is crucial to smooth operation. A vehicle must be clearly identified in the individual manufacturing stages to avoid an unwanted air conditioning system from being fitted or the wrong paint colour being applied during painting, or similar [homburg].

There are two possible methods of controlling a system based upon object data; central and decentral control.

Central control In this method material flow and object state are continuously monitored during the process and stored in a database in a central computer. This builds up an image of the current process data and system state in the process control system of the central computer. It is irrelevant here whether the state of objects in the process is determined using barcodes, radio, optical character recognition, RFID or any other type of information coding and transmission.

The monitoring of the process must be completely infallible, otherwise there is a danger that an object will become out of control. The restarting of the system after a fault or the crashing of the control software can be particularly critical moments.

Central control systems based upon a powerful central database are usually used when it is necessary to access the information from different points simultaneously, when a transparent image of process data must be available continuously for other purposes or if important data needs to be stored permanently [homburg-p&f]. Apart from production, typical applications include stores technology, logistics or the collection of operating data.

Decentral control The use of readable and writeable data carriers opens up the possibility of controlling a system locally, i.e. completely independently of the central process computer. Each object carries a complete data record with it which contains information about its identity, its current state, its history and future – material and data flow are interlinked. For this to be successful, it must be possible not only to read the relevant data from the object at each processing station, but also to change and update this information. Of all known identification technologies this can only be achieved with the necessary reliability using writeable RFID transponders.

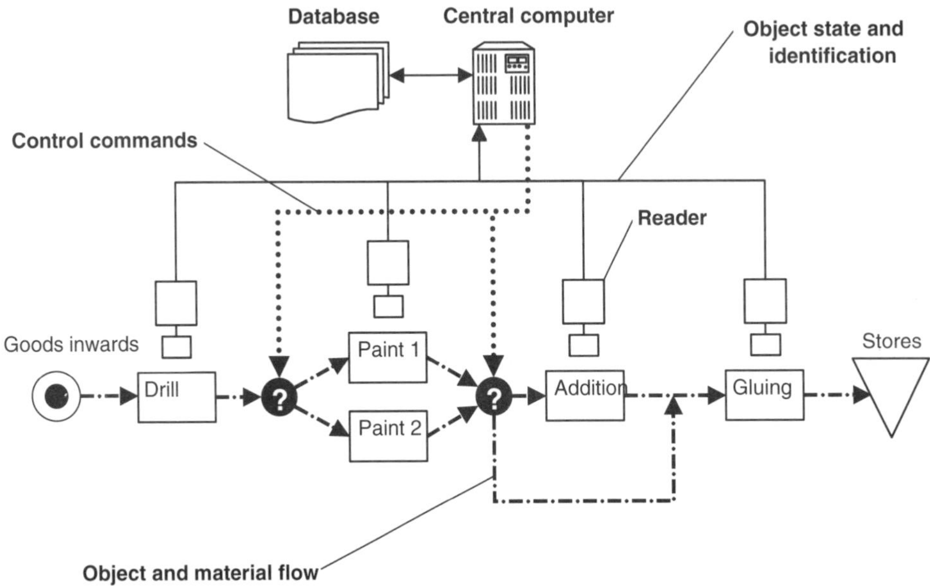

Figure 13.43: The object and data flow in a central control system is performed using completely separate routes. The central computer has a powerful database in which all process data is stored

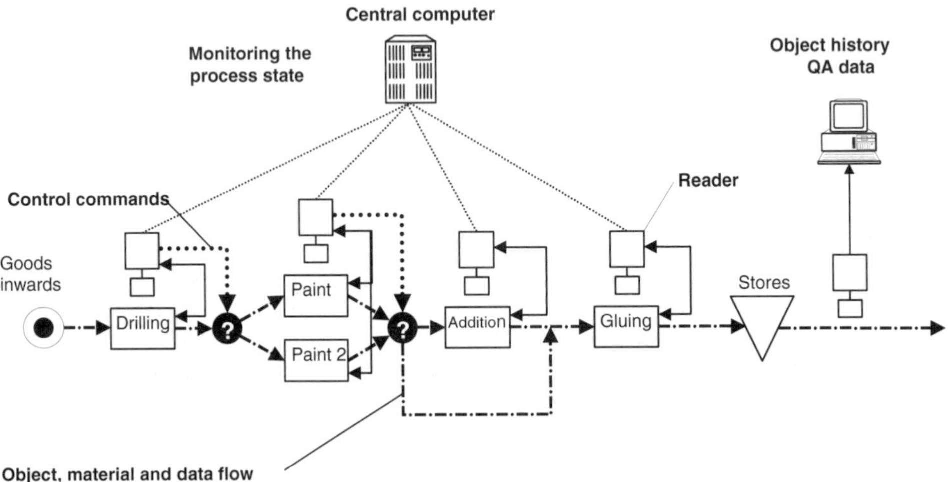

Figure 13.44: In a decentralised system the data is carried along with the objects

The option of changing the object data in the transponder at each processing station, means that it is possible to create an information flow between the individual processing stations, which takes the pressure off the control system. Production and processing operations are becoming faster all the time, so the fact that information can be carried with the object and is available in the right place can become a decisive factor in speeding processes up. Due to possible overheads associated with accessing a remote database, readers in systems that operate in seconds are becoming increasingly unable to keep pace, perhaps to set points or to initiate the correct processing operation [homburg-p&f]

13.10.2.1 Benefits from the use of RFID systems

- Coordination of data and material flow (spatial separation, easing the burden on the control system)
- Quality control
  In modern production lines the quality of products is tested at test points located at a number of stations. When the product is inspected at the end of the production process, it must be possible to unambiguously attribute the quality data gathered earlier to the correct object. With writeable transponders that travel with the product this is easy to achieve because all the quality data obtained during the production process is carried with the object.
- System security
  Shifting object data from the central computer to the object significantly increases system security. Even after software crashes or failures of the central computer the relationship between an object and its current data can be established anywhere at any time. If necessary, objects can also be withdrawn from the production process, without losing the data. If the object is later put back into the process, work can continue without problems or faults occuring.
- Data security
  Protecting the data stored in the transponder using the checksum procedure (e.g. CRC, see Chapter 7.1 "Checksum Procedure") ensures the complete security of the data. Read errors are recognised as such and the data ignored.
- Flexibility
  The use of writeable transponders facilitates a much more flexible control of the manufacturing process. For example, the setup data for universally programmable robots and production machines can be written to the transponder carried with the object during the preparatory stage and is available immediately where it is needed. Using this technique, products can be manufactured right down to a batch size of 1, without having to set up a complex communication with the central computer for each object.
- Harsh environmental conditions
  RFID systems are completely insensitive to dust, moisture, oils, coolants, cuttings, gases, high temperatures and similar problems that can occur in a manufacturing environment. Glass and plastic transponders usually comply with protection type IP67, that is they are totally dustproof and waterproof.
  Even particularly dusty or dirty environmental conditions which would make the use of barcode readers impossible due to the rapid blocking of scanner optics, pose no problems for RFID systems.

13.10.2.2 The selection of a suitable RFID system

When selecting a suitable RFID system for use in the production process the characteristics of the different memory technologies should be considered (see also Chapter 10.3 "Memory Technology"):

EEPROM Data stored in an EEPROM is retained for several years without a power supply. The energy required for writing to or reading from a transponder using EEPROM technology is transmitted by inductive coupling. Because the transponder does not require a battery it may be very small. The guaranteed number of write access operations to a memory address is around 100 000 cycles, which is greater than the lifetime of the transponder. However, a new type of EEPROM technology is now available on the market which can be reprogrammed more than 10^6 times. This can be increased still further by the use of FRAM memory technology. Over 10×10^9 write cycles have been achieved using this technology.

SRAM Unlike EEPROMS, SRAM memory cells require a constant power supply to retain stored data. Therefore, transponders using this memory technology always have their own battery. Data transmission between reader and transponder employs either inductive coupling or the backscatter procedure (microwave). SRAM memory can be reprogrammed any number of times with high write times. However, the integral battery limits the temperature range of this transponder to the range 0°C to 60°C.

Table 13.3: Comparison between the two memory technologies for transponders

	EEPROM / FRAM	SRAM
Memory size	16 Byte – 32 kByte	1 kByte – 512 kByte
Data transmission	inductive coupling	inductive coupling, backscatter
Power supply	inductive coupling	battery
Typical number of write cycles	EEPROM: 100 000 – 1 000 000 FRAM: 10·109	unlimited
Typical temperature range	-20°C – 120°C	0°C – 60°C
Applications	• Applications with a limited number of reprogramming operations (EEPROM) • Applications with expanded temperature range	• Applications with any number of reprogramming operations, e.g. in assembly systems • Use in the 'normal' industrial temperature range • Need for higher memory size in short transaction time • Large transponder range required (or low positional accuracy)

13.10.2.3 Example projects

Let us now consider a few examples of the use of RFID systems in manufacturing. It is no coincidence that most of the examples described here are taken from the automotive industry. Companies in this industry are continuously striving to optimise the production process.

In the Dingolfing factory (Southern Germany) of BMW the bodies of the 7th and 5th series were originally identified manually at the identification points using barcode readers. To cut costs, a microwave identification system was installed (2.45 GHz, transmission power of reader: 10 mW) at the end of 1996. A transponder is now fitted to the bonnet of each painted body as it enters the production process and data for the model (e.g. chassis number) is written to this transponder. A total of around 3000 transponders are in circulation. Around 70 readers are installed in the assembly area at individual identification points in the various assembly stages. As soon as the body enters the interrogation zone of a reader the read process is initiated by an inductive proximity switch. The required data is read from, or if necessary written to, the transponder. The transponders are equipped with a battery designed for a lifetime of 8 years. They have a memory size of 32 kBytes, and the range of up to 4 m is sufficient in all stages of assembly [pepperl-bmw], [pepperl-k&f].

Figure: 13.45: After successful identification of the vehicle its specific data is interrogated and displayed (Reproduced by permission of Pepperl & Fuchs, Mannheim)

In Mercedes Benz's Tuscaloosa factory in the USA, inductively coupled transponders (125 kHz, read only system) are used for the identification of skids for vehicle bodies. After the skid has travelled through the painting line several times it must be cleaned. This selection process can be performed without unnecessary costs by data capture using transponders.

In the production of its 1E generation, vehicle manufacturer General Motors produces 26 different engine models under one roof in its Flint factory (Michigan, USA). The product range incorporates a multitude of engine types, which includes 1997 to 1998 models, 5.0 to 5.7 litre engines, engines for automatic or shift transmission, engines for export, engines designed to run on environmental fuel and petrol, engines for cars and lorries. By equipping the product carriers with transponders (13.56 MHz, 8 kByte memory) it is possible to track and identify all engine models throughout the production process. Using RFID it is possible to trace any engine in the factory within seconds. Around 50 readers are installed in the production area for this purpose [escort-gm].

The John Deere Company in Waterloo (Iowa, USA), the worldwide market leader in the production of agricultural machinery, employs an inductively coupled RFID system in the manufacture of tractors. The tractor bodies are equipped with special transponders which can resist temperatures up to 225°C and can even communicate with a reader so that they can be used in the painting ovens [escort-deere]. The data carriers (13.56 MHz, 8 kByte memory) can be easily fitted to the rear axle of the tractor chassis. Because most tractors are manufactured to order, the use of modern identification technology allows tractors to be adapted to individual customer requirements.

The use of RFID systems in the meat processing industry is a very interesting example. Barcode systems cannot be used due to the high temperatures of above 100°C during canning and the long cooling periods. The company J. M. Schneider Meats, one of the largest meat processing companies in Canada with 15 factories, therefore uses an inductively coupled RFID system in the processing sequence for product identification and tracking. At the beginning of the process the meat is stacked onto mobile shelves. The meat is conveyed into the chill room via the smoking chambers on these shelves and in the last stage of processing it is heated to above 100°C for conservation. It is vital that the company always knows precisely where the individual shelves are and which process they are currently undergoing. Transponders are therefore attached to the individual shelves (13.56 MHz) and the important data, such as for example the location of the shelves, meat type and weight data, is written to the transponder. The delivery time of the meat, which depends upon its use-by date, can thus be tracked via the RFID system [escort-schneider].

A further application that should clarify how RFID systems can help to increase the quality of a product by the selection of tolerances [pepperl-fab] is illustrated in Figure 13.46. This application involves the assembly of a precision clutch release stop. The system consists of a pallet rotary system, two robots and a manual work station. All the pallets are equipped with transponders. The individual components that make up the stop are measured by one of the robots and these measurements are used to select components to minimise the play in the finished stop. Information about the matching of individual components to each other is written as data to the transponder and thus carried along with the individual components. The second robot is an assembly robot which assembles the individual components into a stop. At this point the data is read from the transponder, so that the robot always assembles the correct individual components.

Figure 13.46: Individual components of a clutch release stop on a revolving pallet. Transponder and reader antenna are visible in the lower part of the photo (Reproduced by permission of Pepperl & Fuchs, Mannheim)

The use of RFID systems can also bring benefits in storekeeping and order processing. One of the leading pharmaceutical companies, Sanacorp in Munich, has an electronically controlled stockkeeping and order processing system which allows products to be automatically collected in accordance with the delivery note. More than 6000 consignment containers (plastic containers) pass through the stores every day and need to be identified at individual loading points. In the old system using barcode labels or reflective code labels, up to 100 errors occured daily, meaning that the falsely identified consignment containers passed through all of the loading points on the way to the goods outwards, thus delaying the entire consignment. To guarantee the infallible recognition of the consignment containers these were fitted with transponders (134 kHz, SEQ), which were welded to the base of the plastic trough. The reader antennas are located under the conveyor belts at the relevant stations. As soon as a consignment container enters the interrogation zone of a reader the transponder is read and the stored data is transferred to the stock-control computer. The central computer is informed of where each consignment container is located, whether delays are occuring during loading and how busy the indivdual loading points are. The rapid collection of goods from the stores is important because customers, who are mainly pharmacists, expect the consignment to arrive on time and be complete. This can only be guaranteed by a technically infallible order picking procedure [sander].

14

Market Overview

14.1 Selection Criteria

There has been an enormous upturn in the demand for *RFID systems* in recent years. This is illustrated by the increasing number of contactless smart cards in use as electronic tickets in *public transport* systems. Just five years ago it was unthinkable that tens of millions of contactless tickets would now be in use worldwide. The number of fields in which contactless identification systems can be used has also multiplied in recent years.

The developers of RFID systems have taken this change into account and countless systems are now available on the market the technical parameters of which are optimised for very different fields of application – *ticketing*, *animal identification*, *industrial automation* or *access control*. These fields of application often overlap, making it difficult to classify suitable systems. To make matters more difficult, with a few exceptions (animal identification, close coupling smart cards), no binding standards have yet been formulated for RFID systems.

It is almost impossible, even for an expert, to maintain an overview of the product range of RFID systems on offer today. This makes it difficult for the user to select the most suitable system.

There follow a few comments regarding the points that should be considered in the selection of an RFID system.

14.1.1 Operating frequency

RFID systems in the frequency range between 100 kHz and around 30 Mhz operate using inductive coupling. Microwave systems in the frequency ranges of 2.45 or 5.8 GHz, on the other hand, use electromagnetic fields for coupling.

The specific *absorption rate* (damping) for water or nonconductive substances at 100 kHz is lower by a factor of 100 000 than it is at 1 GHz. Thus almost no absorption or damping takes place. Lower frequency HF systems are therefore much more common due to the fact that they offer better penetration through objects [schürmann-94]. One example of this is the bolus, a transponder that is placed in the first stomach (rumen) of cows and can be read externally at an operating frequency of < 135 kHz.

Microwave systems have a significantly greater *range* than inductive systems, typically 2 – 15 m. However, in contrast to inductive systems, microwave systems require an additional back-up battery. The transmission power of the reader is generally insufficient to provide enough power to operate the transponder.

Another important factor is sensitivity to *electromagnetic interference fields*, such as those generated by welding robots or powerful electric motors. Inductive transponders are at a clear disadvantage here. For this reason, microwave systems have become dominant in production lines and painting systems in the automobile industry. In addition, microwave systems have a high storage capacity (up to 32 kBytes) and high temperature resistance (up to 250°C).

14.1.2 Range

The required range of an application depends upon several factors:

- Positional accuracy of the transponder;
- Minimal distance between several transponders in practical operation;
- Speed of the transponder in the interrogation zone of the reader.

So, for example, in contactless counting applications – tickets for public transport – the positioning speed is very low because the transponder is placed into the reader by hand. The minimum distance between several transponders is the distance between two passengers entering a vehicle. The optimal range is 5 – 10 cm for these systems. A greater range could lead to problems, because the reader would sometimes detect the tickets of several passengers simultaneously. It would therefore no longer be possible to reliably match the ticket to the passenger.

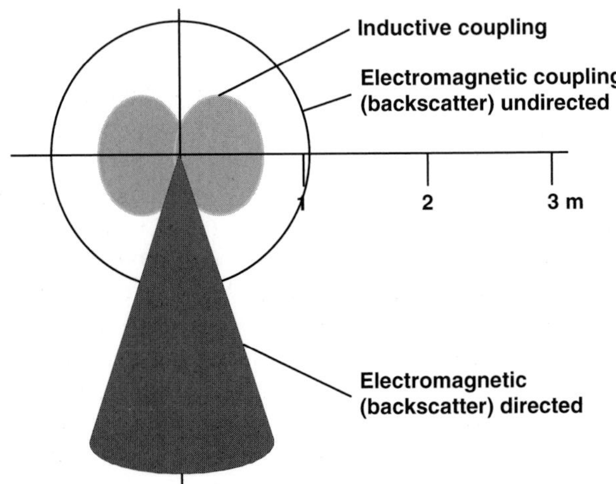

Figure 14.1: Comparison of the interrogation zones of different systems

Different vehicle models with different dimensions are often constructed simultaneously on the production lines of the automotive industry. This gives rise to great variations in the distance between the transponder on the vehicle and the reader [Bachthaler]. The write/read distance of the RFID system used must therefore be designed for the maximum required range. The distance between the transponders must be such that only a single transponder is within the interrogation zone of the reader. Microwave systems with a *directed beam* offer clear advantages over the broad, undirected fields of inductively coupled systems.

The speed of the transponder in relation to the reader, together with the maximum write/read distance, determines the length of time the transponder spends in the interrogation zone of the reader. In vehicle identification systems, the necessary range of the RFID system is such that at the maximum speed of the vehicle, the time spent in the interrogation zone is sufficient for the transmission of the required data.

14.1.3 Security requirements

A great deal of thought should be given to the sequrity requirements – i.e. with regard to encryption and authentication – for a planned RFID application, so that the possibility of nasty surprises in the operational phase can be excluded from the start. For this reason, consideration should be given to the question of how attractive a system is to a potential attacker, i.e. what potential exists to obtain financial or other benefits by manipulating the system. To estimate this incentive, we divide applications into two groups:

- Industrial or closed applications;
- Public applications related to money or assets.

Let us now look at two very different example applications:

A typical example of an industrial or closed application would be an assembly line in the automotive industry. Only authorised persons have access to the RFID system, which means that the circle of potential attackers can be monitored. Although a malicious *attack* on this system could disrupt the operating sequence by changing or falsifying data, there would be no personal benefit to the attacker. The probability of an attack can therefore be assumed to be zero, so a cheap low end system without security logic can be used.

Our second example is a ticketing system for use in the public transport network. Such a system, especially the data carrier in the form of a contactless smart card, is freely accessible to all. The circle of potential attackers therefore cannot be monitored. A successful attack on such a system could cause severe financial losses to the public transport company in question, perhaps due to the organised sale of falsified travel passes, even before we consider the loss of image to the company. For such applications a high end transponder with authentication and encryption procedures is vital. For applications with maximum security requirements, for example banking applications with cash dispensers, only transponders with microprocessors should be used.

14.1.4 *Memory size*

The chip size of the data carrier – and thus its price category – is primarily determined by its *memory size*. Price sensitive applications with a low local information requirement use fixed coded read only data carriers. Only the identity of an object can be defined in these data carriers. Further data is stored on the central database of a master computer. Systems that write current data back to the transponder require a transponder with EEPROM or RAM memory technology.

EEPROM memory is usually found in inductively coupled systems. Memory sizes ranging from 16 bytes to 8 kBytes are available.

Battery backed up SRAM memories, on the other hand, are mainly found in microwave systems. The available memory sizes range from 256 bytes to 64 kbytes.

14.2 System Overview

The following overview is a small section of the broad spectrum of systems available on the market and should give an insight into the complexity of the RFID market.

The selection in no way represents an evaluation or preference for the systems represented, but was derived from the coincidental availability of the technical data sheets to the author at the time this book was created.

Table 14.1: Overview of RFID systems on the market

System manufacturer:	Coupling, operating method, energy, distance:	Memory: gross / net:	Security logic	Downlink reader → transponder:	Uplink transponder → reader:
3M	EAS transponder,	1 bit electro-magnetic	n.i.	n.i.	n.i.
AEG, trovan	128 kHz, ind., 0 – 70 cm	ROM, 64/39 bit, read-only	n.i.	(read only)	subharmonic, 64 kHz, BPSK, Manchester, 8375 bit/s
Avonwood EUREKA	132 kHz, ind., 0 – 1 m, Li battery	5 – 128 Byte	n.i.	n.i.	subharmonic, 66 kHz, BPSK, Manchester
Baumer-Ident, OIS-P/M	2.45 or 5.6 GHz battery, 0 – 5 m	RAM, 8 kByte	n.i.	2.45 or 5.6 GHz 8 kbit/s	n.i.
Baumer-Ident, OIS-R	433 MHz, battery, 0 – 5 m	RAM, 128 kByte	n.i.	433 MHz	anharmonic, 455 kHz, 2.5 kbit/s

Table 14.1 continued: Overview of RFID systems on the market

System manufacturer:	Coupling, operating method, energy, distance:	Memory: gross / net:	Security logic	Downlink reader → transponder:	Uplink transponder → reader:
Baumer-Ident, OIS-I	4 MHz, ind., 0 – 12 cm	EEPROM, 8 kByte	n.i.	4 MHz	anharmonic, 455 kHz, approx. 660 bit/s
Buscom	20 kHz, ind.,	n.i.	n.i.	20 kHz, PSK	anharmonic FSK – 250/300 kHz
Cd. Micron comm MicroStamp Engine™	2.45 GHz, 500 mW, em., 0 – 15 cm, battery	4 kByte ROM, 512 Byte RAM	8 bit CPU, ASK, spread spectrum signal, 308 kBit/s	backscatter, subcarrier 596 kHz 149 kBit/s	microchip
CFG, 188 000	125 kHz, ind., 0 – 70 cm	EEPROM, 128 / 96 Byte	56 bit password, (R/W)	125 kHz, ASK, PP code, (pulse position)	load modulation
deister-electronic, IDC, ICC, IFC	125 kHz, ind.,	EEPROM, 8 – 128 Byte	Internal clock generator	125 kHz, ASK, PL code (pulse length)	Load modulation
DASA, BagTag	132 kHz, ind.,	FRAM, 64 Byte	can read 1 – 6 transponders at the same time!	132 kHz, 3300 bit/s	anharmonic 2.7 – 4.2 MHz, f-multiplex, 16 500 bit/s
Diehl Ident, Telides	5.8 GHz, 25 mW, batt., 0 – 10 m,	30 bit	n.i.	(read only)	n.i.
Daisy	134 kHz, ind., 0 – 8 cm	n.i.	n.i.	FDX-B as per ISO 11784/5	FDX-B as per ISO 11784/5
Endrich, SID	125 kHz, ind., 0 – 50 cm	EEPROM, 128 / 116 byte	password for each 4 Byte, check word, 8 byte serial number, encryption	125 kHz, ASK, Manchester code	load modulation, 2000 kbit/s
EM microelectronic-Marin, H4005	100 – 200 kHz	ROM, 16 Byte	ISO 11784, ISO 11785 compatible	(read only)	load modulation
V4070	115 – 135 kHz	30 bit OTP	96 bit key, 32 bit serial no. authentication	ASK	load modulation, Manch. code

Table 14.1 continued: Overview of RFID systems on the market

System manufacturer:	Coupling, operating method, energy, distance:	Memory: gross / net:	Security logic	Downlink reader → transponder:	Uplink transponder → reader:
Esselte Meto	16 Hz, ind., EAS	1 bit, electro-magnetic	n.i.	(EAS)	harmonic, 5 kHz, 7.5 kHz
Euchner, CIS2	1 MHz, ind., 0 – 1.5 cm	EEPROM, 0 – 1 kByte	n.i.	ASK	n.i.
CIS3	125 kHz, ind., 0 – 2.5 cm	EEPROM, 16 Byte	n.i.	ASK, biphase code	n.i.
CIS3-IBS	125 kHz, ind., 0 – 2.0 cm	FRAM, 2 – 32 kByte	n.i.	ASK, biphase code	n.i.
EURO I. D. ID 250	125 kHz, ind., 0 – 40 cm	17 Byte	n.i.	n.i.	subharmonic, 62.5 kHz
ID 101 – 300	124 kHz, ind., 0 – 40 cm	8 Byte	n.i.	(read only)	subharmonic, 62.0 kHz
Idesco OY, microlog	24 kHz, ind.	EEPROM, 128 Byte	4 byte serial number, 2 byte password	n.i.	n.i.
Inside Technologie, 100L	125 kHz, 0 – 1.5 m	EEPROM, 1 kByte	4 byte key authentication	n.i.	n.i. 3.9 kbit/s
100H	13.56 MHz, 0 – 15 cm	EEPROM, 1 kByte	anticollision, encryption	n.i.	n.i. 106 kbit/s
ISD Ltd., M 1000	915 MHz (US) em., 0 – 5 m	n.i.	serial number	n.i.	backscatter
M 4000	27.125 MHz, ind.	8 Byte	n.i.	n.i.	load modulation
M 5000	13.560 MHz, ind.	8 Byte	n.i.	n.i.	load modulation
M 6000	433 / 458 Mhz, 0 – 2 m passive, 0 – 7 m with battery	n.i.	serial number	n.i.	backscatter
Kaba, Legic MIM 256	13.56 MHz, ind., 0 – 50 cm	EEPROM, 22 – 256 Byte	Serial number authentication	ASK	load modulation with subcarrier

Table 14.1 continued: Overview of RFID systems on the market

System manufacturer:	Coupling, operating method, energy, distance:	Memory: gross / net:	Security logic	Downlink reader → transponder:	Uplink transponder → reader:
Kapsch, Multicard	close coupling, ind., 4.91 MHz	EEPROM	8 bit RISC-CPU	as per ISO 10536 ind., 154 kbit/s	ind., cap., 154k
Knitter-electronic, CIS	125 kHz, ind., 0 – 15 cm	FRAM, 32 byte	write protection	125 kHz, FSK	subharmonic 64 kHz, PSK, 1953 bit/s
MCRF 200	125 kHz, ind.	16 Byte, OTP	n.i.	(read only)	load modulation, FSK, PSK
MCRF 250	125 kHz, ind.	16 Byte, OTP	anticollision	(read only)	load modulation, FSK, PSK
metget 4015	125 kHz, ind.	ROM, 8 byte read-only	n.i.	(read only)	load modulation, Manchester code
Micro-Synsys mic3® Card	4 MHz, ind. “coil on chip” 0 – 1.5 mm	EEPROM 256 Byte	n.i.	n.i.	n.i.
Mitsubishi Melcard MF 5102	409.6 kHz, ind., Li-Batt. 3V, 55 mAh, SEQ	Mask-ROM 8 k Byte, SRAM 320 Byte	8 bit μC	25.6 kbit/s, ASK, ISO 7816-Interface	25.6 kbit/s, ASK (SEQ)
OTI, EYECON	13.56 MHz, ind., 0 – 15 cm	16k Mask-ROM, 8k EEPROM, 128 – 512 Byte RAM	8 bit μC DES, RSA	13.56 MHz, 106 kbit/s	load modulation 106 kbit/s
Panasonic ID256C	125 kHz, ind., 0 – 10 cm	FRAM, 36 / 32 Byte	Memory Locking Options	125 kHz, PSK 7812 bit/s	subharmonic 62.5 kHz, PSK, 7812 bit/s
Pepperl & Fuchs System V/T (MVC-60-250k)	2.45 GHz, (80 mW) em., 0 – 4m Li battery	SRAM, 7678 / 7552 Byte	n.i.	76.8 kbit/s	backscatter, 76.8 kbit/s
System V	250 kHz, ind., 0 – 10 cm	n.i.	n.i.	7.5 kBaud	n.i.

Table 14.1 continued: Overview of RFID systems on the market

System manufacturer:	Coupling, operating method, energy, distance:	Memory: gross / net:	Security logic	Downlink reader → transponder:	Uplink transponder → reader:
Philips Gratkorn (was Mikron) MIFARE® light	13.56 MHz, ind., 0 – 10 cm	EEPROM, 48 / 24 Byte	32 bit serial number authentication encryption, anticollision	13.56 MHz, ASK, 106 bit/s	load modulation 106 bit/s
MIFARE® 1	-“-	1024/ 748 Byte	-“-	-“-	-“-
MIFARE® plus	-“-	15 kByte Mask-ROM, 8k EEPROM, contacts	-“- + 8 bitμC,	-“-	-“-
MIFARE® pro	-„-	20 kByte Mask-ROM, 8k EEPROM, 256 Byte RAM, contacts	8 bit μC, triple-DES, coprocessor	-„-	-„-
I•CODE	13.56 MHz, ind., 0 – 1.5 m	348 bit, 1 bit EAS function	64 bit serial number, anticollision	n.i.	n.i.
HITAG®	125 kHz, ind., 0 – 1 m	EEPROM, 256 Byte	32 bit serial number encryption	125 kHz, ASK 4 kbit/s	load modulation, Manchester
Racom®, LF	125 kHz, ind., 0 – 10cm	FRAM, 512 Byte	authentication password encryption	125 kHz, FSK 7812 kbit/s	subharmonic 62.5 kHz, PSK, Miller-Code
SEPT, La Poste	6.78 MHz, ind., battery	n.i.	μP 68 HC 05	FSK, 19.2 kbit/s	n.i.
SGS Thomson ST16RF42	13.56 MHz, ind.	16 k Byte mask ROM 2 k EEPROM 384 byte RAM 7816 contacts	n.i.	ASK, 106 kbit/s	Load modulation
Siemens, SLE 44R35 (MIFARE®)	13.56 MHz, ind., 0 – 10 cm	1024/748 byte	32 bit serial number authentication encryption anticollision	13.56 MHz, ASK, 106 bit/s	Load modulation 106 bit/s

Table 14.1 continued: Overview of RFID systems on the market

System manufacturer:	Coupling, operating method, energy, distance:	Memory: gross / net:	Security logic	Downlink reader → transponder:	Uplink transponder → reader:
SLE44R42S (MIFARE® plus)	-"-	14 k Mask ROM 4 k EEPROM 7816 contacts	-"- + 8 bit,	-"-	-"-
MOBY-F	125 kHz, ind., 0 – 7 cm	240 Byte EEPROM	n.i.	n.i.	n.i.
MOBY-L	4 MHz, ind., 0 – 5 cm	512 Byte EEPROM	n.i.	0.1 kbyte/s	n.i.
MOBY-I	1.81 MHz, ind., 0 – 1 m	128 Byte EEPROM, 32 k RAM	n.i.	n.i.	n.i.
MOBY-E	13.56 MHz, ind., 0 – 10 cm	752 Byte EEPROM	n.i.	n.i.	n.i.
MOBY-V	433 MHz, em., battery, 0 – 80 cm	32 k RAM	n.i.	>1 kbyte/s	n.i.
SOFIS	2.45 GHz, em., 0 – 1.3 m	20 bit fix. SAW	n.i.	(read only)	surface wave
Sokymat Titan 4000	125 kHz, ind.	128 Byte EEPROM	password	ASK	load modulation
Unique 1200	125 kHz, ind., 0 – 20 cm	8 Byte, OTP Laser-ROM	n.i.	ASK	load modulation
Sony, FeliCa	13.56 MHz, ind., 0 – 10 cm	1 kByte	authentication encryption	modified ASK 250 kbit/s	n.i.
TagMaster AB, Confident S1251, S1255	2.45 Ghz, em., 0–4 m (read) 0–0.5m (write), Li battery	8 – 75 Byte	anticollision	(random interval mode) 4 kBit/s	backscatter, 16 kBit/s
Temic, TK 5530	125 kHz, ind.	16 Byte PROM (Laser cutting)	n.i.	read-only	load modulation FSK, PSK, / Manchester, bi-phase, max. 15 kbit/s

Table 14.1 continued: Overview of RFID systems on the market

System manufacturer:	Coupling, operating method, energy, distance:	Memory: gross / net:	Security logic	Downlink reader → transponder:	Uplink transponder → reader:
TK 5550	-„-	33 Byte EEPROM	write-lock-bits	ASK, Pulse-Pause coding	-„-
TK 5560	-„-	40 Byte EEPROM	unilateral authentication	-„-	-„-
Texas Instruments TIRIS®	134 kHz, ind., SEQ	170 Byte EEPROM	n.i.	n.i.	FSK 134.2 124.2 kHz

Abbreviations:

em. Electromagnetic coupling (backscatter)

ind. Inductive coupling

n.i. No information available

The listed data is derived from the following data sheets and company documentation:

3M Deutschland Gmbh:
Product description: *3M Sicherungssysteme für Bibliotheken*, 3M Deutschland GmbH, D-Neuss (undated).
AEG:
Data sheet: *Elektronische Identifikation System trovan*, AEG, D-Ulm (undated).
Avonwood Developments LTD:
Various data sheets: *Eureka 111 – passive tag systems, Introducing the next generation!, The 311 tag – a low frequency active transponder, The 411 tag – a low frequency active transponder, The 511 tag – a low frequency active transponder*, avonwood developments LTD, GB Wimborne Dorset, 1998.
Baumer Ident:
Data sheet: *OIS® Identsysteme*, Baumer Ident, D-Weinheim (undated).
Buscom:
Product information: *Buscom Fahrgeld-Management Systeme auf Basis der Proximity Technologie*, BUSCOM Deutschland, D-Bornheim (Cologne) 1996.
Data sheet: *BUSCOM*, Oulu (undated).
CFG:
Data sheet: *Contactless identification products – RF identification tags, badges and reader/writers, Series 188 000*, CFG, CH-Morges (undated).
Company documentation: *Mikroelektronik: Unsere Kompetenz sichert Ihren Erfolg*, CFG Microelectronic SA, CH-Morges (undated).
Deutsche Aerospace (DASA):
Data sheet: *Baggage Handling – BagTag Radio Frequency Identification System*, Deutsche Aerospace, D-Ulm (undated)

Diehl Ident:
Data sheet: *Telides – die neue wirtschaftliche Art der automatischen Identifikation,*
Data sheet: *DAISY – eröffnet neue Möglichkeiten in der Tieridentifikation*, Diehl Ident GmbH, D-Röthenbach an der Pegnitz, 1995.
endrich Identifikationssysteme:
Data sheet: *SID-TAG, contactless read only chipcard*, endrich Bauelemente Vertriebs GmbH, D-Nagold (undated).
EM Microelectronic-Marin SA:
Data sheet: *H4005 – ISO 11784/11785 compliant read only contactless identification device*, EM Microelectronic-Marin SA, CH-Marin 1996.
Company documentation: *Contactless Identification Devices*, EM Microelectronic-Marin SA, CH-Marin (undated).
Data sheet: *V4070 - Crypto Contactless Identification Device*, EM Microelectronic-Marin SA, CH-Marin 1997.
Euchner:
Anfragen bei Vertrieb/Technik und verschiedene Datenblätter, Fa. Euchner, D-Leinfelden Echterdingen, 1997.
EURO I. D.:
Company documentation: *Identifikation ist unser Geschäft*, Fa. EURO I. D. Identifikationssysteme GmbH & Co. KG, D-Weilerswist, Mai 1997.
Idesco OY:
Data sheet: *Idesco Microlog® 1k Memory*, Idesco, Finland-Oulu (undated).
Inside Technologies:
Data sheet: *InCrypr® 100 Product Range – contact / contactless smart card ICs*, Inside Technologies, F-Aix-En-Provence, October 1996.
ISD:
Various data sheets: *model 1000 / 4200 / 500 / 6000 – remote identification system*, ISD, GB-Twickenham (undated).
Kaba AG:
Data sheet: *Legic® – Das intelligente Identifikationsmittel*, Bauer Kaba AG, CH-Wetzikon, (undated).
Kapsch:
Data sheet: *KAPSCH-Multicard, Produkt-Kurzcharakteristik*, Kapsch, Industrielle Elektronik / Systemtechnik, A-Wien, 24.04.1995.
Knitter-electronic:
Data sheet: *CIS-Serie kontaktlos*, Knitter-switch, D-Baldham, 06/1994
metget AB:
Data sheet: *contactless ISO-card MT4016*, metget AB, S-Ronneby, 01.11.1995.
Micro Sensys:
Product information: *mic3® Card - kontaktlose Chipkarte mit RF-ID Transponder*, Micro Sensys GmbH, D-Erfurt, (undated)
Esselte Meto
Data sheet: *EAS-Datenblatt* Esselte Meto, D- Hirschhorn Neckar (undated).
Mitsubishi:

Data sheet: *Technische Information für MELCARD Contactless Cards, CSM Transceiver und Developers Kit*, Mitsubishi Europe GmbH, Juli 1994.
OTI:
Data sheet: *OTI Read/Write RF smart-card*, OTI, Israel-Rosh Pina, (undated).
Panasonic:
Data sheet: *RF-ID Card ID256C*, Panasonic / Matsushita Electronics Corporation, Japan-Kyoto, 1/1996.
Pepperl + Fuchs:
Handbuch: *Mikrowellen-Identifikationssystem*, Pepperl & Fuchs GmbH, Geschäftsbereich Fabrikautomation, D-Mannheim, 1995.
Kurzübersicht: *Identifikationssysteme*, Pepperl+Fuchs GmbH, Geschäftsbereich Fabrikautomation, D-Mannheim, (undated).
Philips Mikron:
Data sheet: *MIFARE® – System Overview*, Philips Semiconductors Gratkorn (Mikron), A-Gratkorn (Graz), 10/1996.
Data sheet: *hitag – System Overview*; Philips Semiconductors Gratkorn (Mikron), A-Gratkorn (Graz), 10/1996.
Data sheet: *I•Code – ICs for Smart Labels*, Philips Electronics N. V., NL-Einhoven, 11/1997.
Microchip:
Data sheet: *microID™ 125 kHz RFID System Design Guide*, microchip technology inc., USA Chandler AZ, September 1998
Micron communications, INC:
Data sheet: *MicroStamp® 4000 Interrogator*, micron communications INC., USA-Boise, ID., 1997
Racom:
Data sheet: *Racom LF contactless Smart Cards*, Racom Systems Inc., 6080 Greenwood Plaza Boulevard, Englewood, Colorado 80111, 1996.
SEPT Caen:
Data sheet: *No-contact card*, Service d'études communes de La Poste et de France Télécom, 42 rue des Coutures, BP 6243, F-14066 Caen Cedex, Sept. 1991.
SGS-Thomson:
Data sheet: *CMOS based smartcard IC with 2048 Byte EEPROM for contactless/contact applications*, SGS-Thomson, F-Rouset, 1997.
Siemens:
Data sheet: *ICs for Chip Cards – SLE 44R35/MIFARE®*, Siemens Semiconductor Group, 1994.
Data sheet: *ICs for Chip Cards – SLE 44R42S Combi Card-IC*, Siemens Semiconductor Group, 09/1996.
Overview: *MOBY-F – Das Identifikationssystem nicht nur für die Logistik*, Siemens AG, Bereich Antriebs- und Automatisierungstechnik, D-Fürth 1997.
Product catalogue: *Kombinationstechnik – Identifikationssysteme MOBY*, Siemens AG, Bereich Automatisierungstechnik, D-Fürth, April 1997.
Data sheet: *SOFIS – das sichere Ortungs- und Auto-ID-System für Verkehrsunternehmen*, Siemens AG, Bereich Verkehrstechnik, D-Berlin, (undated).

Sokymat:
Data sheets: *Unique 1200 – contactless read only laminated disc tag* und *Titan 4000 - contactless read-write chipcard*, Sokymat IDent Component GmbH, D-Bergisch Gladbach, 08/1995.
Sony:
Data sheet: *FeliCa – remote card system*, Sony Dorporation, Japan-Tokyo, (undated).
TagMaster AB:
Data sheets: *Script Tag™ S1251 – General purpose R/W Card, Mark Tag™ S1255 – Multiple access R/O Card*, TagMaster AB, S-Kista, 01/1998
Temic:
Various data sheets: *Read-only-Transponder TK5530, read/write transponder TK 5550, Read/write-crypto-Ttransponder*, TEMIC Telefunken microelectronic, D-Heilbronn, September 1996.
Texas Instruments:
Data sheet: *TIRIS® – Multipage Transponder*, Texas Instruments, D-Freising, 04/1996.

14.3 Contact Addresses, Technical Periodicals

The author himself can only be contacted by post at the address of the publishing company:

Klaus Finkenzeller
c/o Carl Hanser GmbH & Co
Kolbergstr. 22
D-81679 München

and via the Internet:

Homepage: http://rfid.notrix.de
e-mail: bm242372@fbmev.de
or: dl5mcc@qsl.net

14.3.1 Industrial associations

Contactless smart cards can be used as electronic tickets in public transport products, helping to improve speed and convenience and aiding the implementation of flexible strategies. The KONTIKI working group aims to analyse technological and application-related developments, to develop practical application options for public transport and to use these to make recommendations to transport companies and associations. The idea is that users, manufacturers, consultants, associations and organisations work together to create interdisciplinary solutions. The working group is now active across Europe.

Work is carried out in subgroups, and the results are presented centrally. The working group also provides contacts and consultants to potential users for future smart card projects. Contact address:

Arbeitskreis kontaklose Tickets i.G.
KONTIKI
c/o TCAC GmbH
Buchenstraße 16b
D-01097 Dresden
Telephone: 03 41 / 8 02 59 11
Fax: 03 51 / 8 02 59 13
e-mail: TCAC-GmbH@t-online.de
Homepage: http://kontiki.org

Many manufacturers of RFID systems are members of AIM International, an industrial association for automatic identification and data capture systems.

AIM International Inc.
Mr. Brian P Wynne
634 Alpha Drive
Pittsburgh, PA 15238-2802
USA

Tel: +1 412 963 8009
Fax: +1 412 963 8753
e-mail: adc@aimi.org
Homepage: http://aimi.org

The following local associations are available:

AIM AFRICA
Ms. Veronica Schnippenkotter
P O Box 1952
Jukskei Park 2153
South Africa
Tel: +27 11 462 7437
Fax: +27 11 462 7437
Homepage: http://www.aimafr.iafrica.com

AIMR ARGENTINA
c/o Compudata S.A.
Mr. Roberto Marinez Taylor
Av. 53, No.762
(1900) La Plata
Argentina
Tel: +54 21 22 5922
Fax: +54 21 25 3287

AIM AUSTRALIA
Ms. Jenny Bathurst
Automatic Data Capture Association (ADCA)
Locked Bag 2555
Chatswood Delivery Centre
Chatswood NSW 2067 Australia
Tel: +61 2 9422 2566
Fax: +61 2 9422 2555
e-mail: adca@ozemail.com.au
Homepage: http://www.adca.com.au

AIM BRAZIL
c/o AbeCD
Mr. Mauricio Boaventura
Rua Hungria, 664-C 11. Andar-cj. 111
CEP 01455-000 Sao Paulo
Brazil
Tel: +55 11 815 7028
Fax: +55 11 212 9541
Homepage: http://www.xpnet.com.br/aimbrasil

AIM-Deutschland
Erwin Kretz
Akazienweg 26
D-68623 Lampertheim-Neuschloß

Hotline: +49 / 62 06 / 1 31 77 Faxline: +49 / 62 06 / 1 31 73

AIM EUROPE
Mr. Ian G Smith
The Old Vicarage
Haley Hill
Halifax HX3 6DR
West Yorkshire, England
United Kingdom
Tel: +44 1422 368368
Fax: +44 1422 355604
Homepage: http://www.aim-europe.org

AIM JAPAN
Mr. Osamu Manabe
Alos Gotanda Bldg 6F
1-10-7 Higashigotanda
Shinagawa-ku Tokyo 141
Japan
Tel: +81 3 3440 9085

Fax: +81 3 3440 8086
e-mail: aimjapan@mars.dtinet.or.jp
Homepage: http://www.aimjapan.or.jp

AIM KOREA
Mr. Song Jae O
489-22 Kalhyun Dong
Eun Pyung Ku
Seoul, 122-050
Korea
Tel: +82 2 359 7364
Fax: +82 2 359 7365

AIM MEXICO
Mr. Oscar Marquez
Asturias No 31
Col. Alamos
03400 Mexico D F
Mexico
Tel: +52 5 519 1553
Fax: +52 5 530 5482
Homepage: http://www.aim-mexico.com

AIM SINGAPORE
Mr. Dominic Tan
c/o Monarch Marking
Shun Li Industrial Complex
705 Sims Drive #02-10113
Singapore
Tel: +65 744 8300
Fax: +65 744 0835

AIM USA
Mr. Larry W Roberts
634 Alpha Drive
Pittsburgh, PA 15238-2802
USA
Tel: +1 412 963 8588
Fax: +1 412 963 8753

14.3.2 Technical journals and events

The following technical journals deal with the subject of RFID (amongst other subjects):

German language:
ident, Das Forum für Automatische Datenerfassung
Umschau Zeitschriftenverlag – Breidenstein GmbH
Stuttgarter Str. 18 – 24
D-60329 Frankfurt am Main
Telefone: +49 / 69 / 26 00-0 Fax: +49 / 69 / 26 00-6 09

English language technical journals in the U.S.A.:
idSYSTEMS
174 Concord St.
Peterborough, NH 03458
U.S.A.
Telephone: (603) 924-9631
Fax: (603) 924-7408
e-mail: editors@idsystems.com
Homepage: http://www.idsystems.com

And in Europe:
Automatic I.D. News
Tower House
Sovereign Park
Lathkill Street
Market Harborough
Leicestershire LE16 9EF
United Kingdom
Homepage: http://www.idnewseurope.com

Furthermore, every Spring in Dortmund the SMAID (Spring Meeting on Auto-ID) takes place, which is an important meeting place for manufacturers, researchers and users of Auto ID technologies. This event is organised by the University of Dortmund:

Universität Dortmund
Lehrstuhl für Förder- und Lagerwesen
a-ID-a / M. Wölker
Emil-Figge-Str. 73
D-44227 Dortmund

Telefone: +49 / 231 / 7 55-44 93 Fax: +49 / 231 / 7 55-30 92

The Scantech-EXPO Europe has established itself as the trade fair for barcode and RFID systems, and takes place annually in November at various sites (Cologne, Paris). For information, please contact:

SCANTECH EXPO Europe
Advanstar Exhibitions
Advanstar House

Park West
Sealand Road
Chester CH1 4RN, UK
Fax: +44 / 1244 / 37 00 11

14.3.3 RFID on the Internet

A good overview of current standardisation, the work of AIM and a comprehensive company overview can be found on the RFID homepage of AIM:
http://www.rfid.org

Current information about new developments and products in the field of RFID and EAS, an overview of the most important RFID patents and an introduction into the technical principles of the most important RFID procedures can be found in Transponder News

http://rapidttp.co.za/transponder/
http://rapidttp.com/transponder

The RFID Bulletin Board provides a discussion forum on the Internet. The RFID Bulletin Board aims to provide a neutral forum for the free exchange of information between RFID users, developers and anyone who is interested in any aspect of RFID. Questions, contributions and discussion regarding technical or commercial aspects of RFID, comments regarding events, questions about applications, standardisation for RFID systems and similar are welcomed:

http://www.us.parsimony.net/forum/forum490/

The RFID-Webring is a voluntary amalgamation of very different Internet pages relating to the subject of RFID from either a technical or financial point of view. For further information, see the Webring's homepage:

Homepage: http://www.webring.org/cgi-bin/webring?ring=rfid;home

15

Appendices

15.1 Relevant Standards and Regulations

CEPT/ERC 70-03: *CEPT/ERC Recommendation 70-03 Relating to the use of Short Range Devices (SRD)*. Joint regulation for SRDs in member states of CEPT.
CEPT T/R 60-01 *Low-power radiolocation equipment for detecting movement and for alert* (EAS). Technical Recommendation.
CEPT T/R 22-04: *Harmonisation of frequency bands for Road Transport Information Systems (RTI)* (toll systems, freight identification). Technical Recommendation.
DIN/ISO 9798: *Information technology – Security techniques – Entity authentication*. Fundamentals and description of authentication procedures.
ISO 10373: *Identification Cards – Test methods*. Test methods for (contactless) chip cards for testing the card body and the microchip.
ISO 10374: *Container – Automatische Identifizierung* (*Freight containers – Automatic identification*). Automatic identification of freight containers by a 2.45 GHz transponder system.
ISO 10536: *Identification cards – Contactless integrated circuit(s) cards*. Contactless chip cards using close coupling technology.
ISO 11784: *Radio-frequency identification of animals - code structure*. Identification of animals using RFID systems. Description of the data structure.
ISO 11785: *Radio-frequency identification of animals – technical concept*. Identification of animals using RFID systems. Description of RF transmission procedures.
ISO 14443: Future standard for contactless chip cards (proximity integrated chipcards) with an operating frequency of 13.56 MHz.
ISO 15693: Future standard for contactless chip cards (vicinity integrated chipcards) with a read range up to 1 m.
ISO 69873: *Werkzeuge und Spannzeuge mit Datenträgern – Maße für Datenträger und deren Einbauraum.*
DIN VDE 0848: *Sicherheit in elektromagnetischen Feldern (Teil 2 – Schutz von Personen im Frequenzbereich 30 kHz bis 300 GHz, Teil 4A2 – Schutz von Personen im Frequenzbereich 0 Hz bis 30 kHz).*
EN 300 220: *Radio Equipment and Systems (RES); Short range devices, Technical characteristics and test methods for radio equipment to be used in the 25 MHz to 1000 MHz frequency range with power levels ranging up to 500 mW.*

EN 300 330: *Radio Equipment and Systems (RES); Short range devices (SRDs), Technical characteristics and test methods for radio equipment in the frequency range 9 kHz to 25 MHz and inductive loop systems in the frequency range 9 kHz to 30 MHz*; Technical figures and test procedures for inductive loop systems in the frequency range 9kHz bis 30 MHz.
EN 300 440: *Radio Equipment and Systems (RES); Short range devices, Technical characteristics and test methods for radio equipment to be used in the 1 GHz to 25 GHz frequency range with power levels ranging up to 500 mW.*
ETS 300 683: *EMC standard for Short Range Devices*, EMV product family standard for LPD applications.
EN 50061: *Sicherheit implantierbarer Herzschrittmacher*. Regulations for the protection against malfunction due to electromagnetic influences (corresponds with VDE 0750).
VDE 0750: See EN 50061.
VDI 4470 - Part 1: *Anti-theft Systems for Goods – Detection Gates – Inspection Guidelines for Customers.*
VDI 4470 - Part 2: *Anti-theft Systems for Goods – Deactivation Devices – Inspection Guidelines for Customers.*

15.1.1 Sources of supply for standards and regulations

In Germany, DIN, ISO, VDE, VDI, ETS and other standards can be obtained from:

Beuth Verlag GmbH,
Burggrafenstr. 3
10772 Berlin

Telecommunication standards (EN, I-ETS) can be obtained from the following address:

European Telecommunications Standards Institute (ETSI)
ETSI Publication Office
Mrs Marina Lystoe
Route des Lucioles
F-06921 Sophia Antipolis
CEDEX-France
Tel: +33 92 94 42 41, Fax: +33 93 95 81 33

15.2 References

[anselm95] Dr.-Ing. Anselm, Dieter, *Diebstahl von Kraftfahrzeugen mit Wegfahrsperren*, Allianz-Zentrum für Technik, Munich

[anselm96] Dr.-Ing. Anselm, Dieter, *Voller Erfolg der elektronischen Wegfahrsperre*, Allianz-Zentrum für Technik, Munich 21.03.1996

[atmel] Atmel, RFID-ASIC Fact Sheet, March 1994

[bachthaler] Bachthaler, Reiner, *Auswahlkriterien für elektronische Datenspeicher*, appeared in: ident, Vol. 3/1997

[baur] Baur, Erwin, *Einführung in die Radartechnik*, Teubner Studienskripten, Stuttgart 1985, ISBN 3-519-00106-3

[bosse] Bosse, Georg, *Grundlagen der Elektrotechnik – Das elektrostatische Feld und der Gleichstrom*, B.I.-Hochschultaschenbücher Band 182, Mannheim 1969

[braunkohle] Journals: *Braunkohle, Elektronische Kennzeichnung von Gefahrstoffen*, Edition 2/1997 (March, April)

[bruhnke] Bruhnke, Michael, *Kontaktlose Chipkartentechnologie in der Automobilindustrie (Immobilizer)*, Lecture scrip for ChipCard 96, Eching 1996

[bührlen] Bührlen, Martin, *Mikron-Chip macht Gasflaschen intelligent*, appeared in: Card-Forum, Vol. 11/1995

[caspers] Dr. Caspers, Friedrich, *Aktuelle Themen der Kfz-Versicherung*, Allianz Versicherungs-AG, Munich, 25.03.1997

[champion-chip] Advertising text: *Real-Time ChampionChip – Das Zeitmeß- und Identifikationssystem für den aktiven Sport*, Sport Team, Drebber

[couch] Couch II, Leon W., *Digital and analog communication systems*, Prentice-Hall Inc, London 1997, ISBN 0-13-599028-9

[czako] Czako, Josef, *Neue Innovationsplattform für Verkehrsunternehmen*, Tagungsband – OMNICARD 1997, in Time – Berlin 1997.

[dobrinski] Dobrinski, Krakau, Vogel, *Physik für Ingenieure*, B. G. Teubner, Stuttgart 1984, ISBN 3-519-06501-0

[doerfler] Dr. Doerfler, *Mikroelektronische Authentifizierungssysteme für die Serienausstattung von Kfz*, appeared in: GME-Fachbericht 13, *Identifikationssysteme und kontaktlose Chipkarten*, vde-Verlag, Berlin 1994

[droschl] Droschl, Georg, *Der Markt für kontaktlose Chipkarten: Von der Vision zur Realität*, Tagungsband – OMNICARD 1997, in Time – Berlin 1997.

[dziggel] Dziggel, Klaus Peter, *The SOFIS Auti ID Identification System*, Lecture manuscript for SMAID 97, Uni Dortmund, 1997

[ernst] Ernst, Horst, *EURO-Balise S21 – Meilenstein für das ETCS*, ETR – Eisenbahntechnische Rundschau 45, October 1996

[ero] E.R.O., *Report of project team SE24 on the sharing between the inductive systems and radiocommunication systems in the band 9 ... 135 kHz*, 6 October 1995

[euro-id] EURO I.D., Datenblatt: *Anwendungsbeispiele für das trovan® RF-Identifikationssystem – Dienstleistungen – Abfall – Logistik*, EURO I.D. Identifikationssysteme GmbH & Co. KG, Weilerswist

[fleckner] Fleckner, Harald, *Dioden und ihre Anwendung in Frequenzvervielfachern für den Mikrowellenbereich*, appeared in: UKW-Berichte 1/1987, Verlag UKW-Berichte, Baiersdorf

[fricke] Dr.-Ing. Fricke, Hans, Dr.-Ing. habil. Lamberts, Kurt, Dipl.-Ing. Patzelt, Ernst, *Grundlagen der elektrischen Nachrichtenübertragung*, B. G. Teubner Verlag, Stuttgart 1979, ISBN 3-519-06416-2

[fumy] Fumy, Walter, *Kryptographie*, R. Oldenburg Verlag München Wien 1994

[g&d1] Datenblatt Giesecke & Devrient, *Referenzprojekte - kontaktlose Chipkarte RM8k-MIFARE®*, Munich 1997

[gillert] Gillert, Frank, *Quellensicherung auf Basis von EAS-Technologien*, appeared in: ident, Vol. 3/1997

[glesner] Dr. Glesner, Dieter, *Erst simulieren – dann bauen, Rechnerische Behandlung von Magnetantennen*, appeared in: CQ DL, Vol. 1/1997, DARC-Verlag Baunatal

[glogau] Glogau, Ralf, *Geheimsache*, appeared in: DOS, Vol. 12/94, DMV Verlag

[golomb] Golomb, W. Solomon, *Shift Register Sequences*, Aegean Park Press, Laguna Hills – California, 1982, ISBN 0-89412-048-4

[haber] Haberland, Marc, *Gedächtnis ohne Ladungsträger, Ferroelektrische RAMs – die Speicher der Zukunft*, appeared in: Elektronik 25/1996

[hamann.p] Hamann, Peter, *Der Chip als Fahrkarte*, appeared in: Verkehrstechnischer Express 2/96

[hamann.u] Hamann, Ulrich, *Optimierte Halbleiter-Chips für kontaktlose Chipkarten-Applikationen*, Tagungsband – OMNICARD 1997, in Time – Berlin 1997

[hawkes-97] Hawkes, Peter, *Singing in Concert – Some of the possible methods of orchestrating the operation of multiple RFID-Tags enabling fast, efficient reading without singulation*, Amsterdam, 19.02.1997

[herter] Herter, E., Lörcher, W. *Nachrichtentechnik – Übertragung, Vermittlung und Verarbeitung*, 4th Edition, Carl Hanser Verlag, Munich 1987, ISBN 3-446-14593-1

[ident1] ident, Edition 1/96, appeared in UMSCHAU Zeitschriftenverlag, 60037 Frankfurt am Main

[idesco] IDESCO Technical Information, *IDESCO MICROLOG® 1k Memory*, Fa. Idesco, Oulu-Finland

[isd] Integrated Silicon Design PTY LTD (ISD), *Training Manual*, Adelaide – Australia, 1996

[itt75] Intermetall Semiconductors ITT, *Kapazitätsdioden, Schalterdioden, PIN-Dioden - Grundlagen und Anwendungen*, Freiburg 1975

[jörn] Jörn, Fritz, *WIE – Elektronische Diebstahlsicherung*, manuscript (published 1994 in FAZ)

[jurisch] Jurisch, Reinhard, *Coil on Chip – monolithisch integrierte Spulen für Identifikationssysteme*, appeared in: GME technical report *Identifikationssysteme und kontaktlose Chipkarten*, vde-Verlag, Berlin 1994

[kern-94] Kern, Christian, *Injektate zur elektronischen Tieridentifizierung*, working paper 205, published by the Kuratorium für Technik und Bauwesen in der Landwirtschaft e. V. (KTBL), Darmstadt, März 1994, (KTBL-Schriften-Vertrieb im Landwirtschaftsverlag GmbH, Münster-Hiltrup)

[kern-97] Kern, Ch., Wendl G., *Tierkennzeichnung – Einsatz elektronischer Kennzeichnungssysteme in der intensiven und extensiven Rinderhaltung am Beispiel von Deutschland und Australien*, appeared in: Landtechnik Vol. 3/1997

[knott] Knott, Eugene F., *Radar Cross Section*, Artech House, London

[kuchling] Kuchling, Horst, *Taschenbuch der Physik*, Verlag Hari Deutsch, Thun und Frankfurt/Main 1985, ISBN 3-87144-097-3

[lehmann] Lehmann, Ulrich, *Aktivitäten von Siemens zur Einführung der EURO-Balise S21*, SIGNAL + DRAHLT (88) 12/96

[leitz] Fa. Leitz, Firmenschrift: *Intelligente Werkzeuge für mehr Sicherheit und Komfort*, Fa. Leitz, Oberkochen

[lemme] Lemme, Helmuth, *Der Mikrorechner in der Brieftasche*, appeared in: Elektronik 20/1993, 22/1993, 26/1993, Franzis-Verlag, Munchen

[link] Dr.-Ing. Link, Walter, *Identifikation mit induktiven Systemen*, appeared in: ident, Vol. 2/1996 and 1/1997

[longo] Longo, G., *Secure digital communications*, Springer Verlag New-York 1993

[mäusl] Mäusl, Rudolf, *Digitale Modulationsverfahren*, Hüthig Verlag, Heidelberg 1985, ISBN 3-7785-0913-6

[mansukhani] Mansukhani, Arun, *Wireless Digital Modulation*, Applied Microwave & Wireless, Nov./Dec. 1996

[meinke] Meinke, H., Gundlach, F. W., *Taschenbuch der Hochfrequenztechnik*, 5th Edition, Springer Verlag, Berlin/Heidelberg 1992, ISBN 3-540-54717-7

[miehling] Miehling, Martin, *Die Transpondertechnik in der Praxis – Hightech für die Sicherheit*, appeared in: W&S, Vol. 10/1996, Hüthig GmbH, Heidelberg

[morel] Morel, J.-P., Dr. Vilaseca, A., *Doppler-Radar im 10 GHz- Amateurband*, appeared in: UKW-Berichte 4/1991, UKW-Verlag, Baiersdorf 1991

[nührmann] Nührmann, Dieter, *Professionelle Schaltungstechnik*, Franzis Verlag, Munich 1994, ISBN 3-7723-6715-1

[pana] Panasonic, Technical Data Sheet – *Features of ferroelectric nonvolatile memory*

[paul] Dr.-Ing. Reinhold, Paul, *Elektrotechnik 1 – Felder und einfache Stromkreise*, 3rd Edition, Springer Verlag Berlin, Heidelberg 1993, ISBN 3-540-55753-9

[pein] Rüdiger, Pein, *Hilfe bei Prüfungsfragen – Prüfsummenverfahren*, DOS, Vol. 2, 1996

[philmag] Philips Components, Datenblatt, *Ferrite rof antennas for RF-identification transponders*, August 1994

[plotzke] Plotzke, O., Stenzel, E., Frohn, O., *Elektromagnetische Exposition an elektronischen Artikelsicherungsanlagen*, Forschungsgemeinschaft für Energie und Umwelttechnologie – FGEU mbH commissioned by the Bundesanstalt für Arbeitsmedizin, Berlin 1994

[prawitz] Prawitz, Ursula, *Ident-Systeme in der Müllentsorgung: Kostensenkung für Bürger und Kommunen*, appeared in: Ident, Vol. 1/1996

[rankl] Rankl, W., Effing, W., *Handbuch der Chipkarten*, 2nd Edition, Carl Hanser Verlag, Munich 1996, ISBN 3-446-18893-2

[reichel] Dr.-Ing. Reichel, Karl, *Praktikum der Magnettechnik*, Franzis Verlag, Munich 1980, ISBN 3-7723-6661-9

[rothammel] Rothammel, Karl, *Antennenbuch*, 7th Edition 1981, Franckh'sche Verlagsbuchhandlung, W. Keller & Co., Stuttgart 1981, ISBN 3-440-04791-1

[rueppel] Dr. Rueppel, Rainer A., *Analysis and Design of Stream Ciphers*, Springer Verlag, Heidelberg 1986, ISBN 3-540-16870-2

[ruppert] Ruppert, Helmut, *Identifizierungssysteme mit zusätzlichen Sensorfunktionen*, appeared in: *GME-Fachbericht Nr. 13 – Identifikationssysteme und kontaktlose Chipkarten*, vde-Verlag, Berlin 1994

[schmidhäusler] Schmidhäusler, Fritz, *Zutrittskontrolle richtig planen – Techniken, Verfahren, Organisation*, Hüthig Verlag, Heidelberg 1995, ISBN 3-7785-2415-1

[schürmann-93] Schürmann, Josef, *TIRIS – Leader in Radio Frequency Identification Technology*, Texas Instruments Technical Journal, November - December 1993 [schürmann-94] Schürmann, Josef, *Einführung in die Hochfrequenz Identifikations Technologie*, appeared in: GME Technical Report No. 13, *Identifikationssysteme und kontaktlose Chipkarte*, vde-verlag, Berlin, 1994

[seidelmann] Dr. Seidelmann, Christoph, *Funkwellen für Container – Automatische Identifizierung im kombinierten Verkehr*, appeared in: ident 4/1997, Umschau Zeitschriftenverlag, Frankfurt

[sickert] Sickert, Klaus, *Kontaktlose Identifikation – eine Übersicht*, appeared in: GME-Fachbericht Nr. 13, *Identifikationssysteme und kontaktlose Chipkarte*, vde-verlag, Berlin, 1994.

[sickert-90] Sickert, Klaus, *Von der kontaktbehafteten zur kontaktlosen Chipkarte*, appeared in: Weinerth, Hans (Herausgeber), *Schlüsseltechnologie Mikroelektronik – Investitionen in die Zukunft*, Franzis-Verlag, Munich 1990, ISBN 3-7723-7202-3

[siebel] Siebel, Wolf, *KW-Spezial-Frequenzliste*, Siebel Verlag Wachtberg-Pech, 1983

[sofis] Siemens AG., data sheet *SOFIS - das sichere Ortungs- und Auto-ID System für Verkehrsunternehmen*, Siemens AG., Bereich Verkehrstechnik, Berlin (ohne Datumsangabe)

[suckrow] Suckrow, Stefan, *Das Smith-Diagramm*, Funkschau-Arbeitsblätter, appeared in: Funkschau, Heft 10/97, Franzis Verlag, Munich

[temic] TEMIC – Telefunken microelektronic GmbH, *Remote Control and Identification Systems*, Design Guide, D-Heilbronn, August 1977

[ti-96] Texas Instruments Deutschland GmbH, *Standard Transponder Specifications*, 06/1996

[tietze] Tietze, U., Schenk, Ch., *Halbleiter Schaltungstechnik*, 7th Edition, Springer-Verlag, Berlin 1985, ISBN 3-540-12488-6

[töppel] Töppel, Matthias, *Zehn Milliarden Zugriffszyklen – Prozeßgesteuerte Identifikationssysteme*, appeared in: elektro AUTOMATION, Vol. 4/1996, Konradin Verlag, Leinfelden-Echterdingen

[virnich] Dr. Virnich, M., Dr. Posten, K., *Handbuch der codierten Datenträger*, Verlag TÜV Rheinland GmbH, Cologne 1992, ISBN 3-8249-0044-0

[vogt] Vogt Elektronik, Components Handbook – 1990, Passau 1990

[wolff] Wolff, Hartmuth, *Optimaler Kfz-Diebstahlschutz durch elektronische Wegfahrsperren*, appeared in: GME Technical Report No. 13, *Identifikationssysteme und kontaktlose Chipkarten*, vde-Verlag, Berlin 1994

16

Index

LIVERPOOL
UNIVERSITY
LIBRARY